Dielectric Properties of Polymers

Proceedings of a Symposium held on March 29-30, 1971,
in connection with the 161st National Meeting of the
American Chemical Society in Los Angeles, California,
March 28 - April 2, 1971

Edited by Frank E. Karasz

Polymer Science and Engineering
Goessmann Laboratory
University of Massachusetts
Amherst, Massachusetts

ℚℙ PLENUM PRESS • NEW YORK - LONDON • 1972

Library of Congress Catalog Card Number 74-185258

ISBN-13: 978-1-4615-8671-5   e-ISBN-13: 978-1-4615-8669-2
DOI: 10.1007/978-1-4615-8669-2

© 1972 Plenum Press, New York
Softcover reprint of the hardcover 1st edition 1972
A Division of Plenum Publishing Corporation
227 West 17th Street, New York, N.Y. 10011

United Kingdom edition published by Plenum Press, London
A Division of Plenum Publishing Company, Ltd.
Davis House (4th Floor), 8 Scrubs Lane, Harlesden, NW10 6SE, London, England

PREFACE

The study of the electrical properties of polymers constitutes a substantial fraction of the total research interest in macro-molecular behavior. Because of the increasing attention now being paid to theoretical and experimental aspects of the phenomena encountered, a symposium focusing particularly on dielectric behavior was held at the 161st National Meeting of the American Chemical Society in Los Angeles, California, March, 1971. The authoritative papers in this volume were all presented at this symposium, which was sponsored by the Polymer Division of the American Chemical Society. An obvious advantage of such a collection is that it keeps together closely related material which would otherwise tend to be dispersed amongst different journals; a further factor of which most authors represented here have taken advantage is that it gives an opportunity to present research in less attenuated form than is now normally possible in journals.

I wish to thank all the authors for their uniformly excellent contributions and for their cooperation in the publication of this volume. Thanks are also due to Dr. S. Matsuoka for chairing one of the sessions at the symposium.

September, 1971   Frank E. Karasz

v

LIST OF CONTRIBUTORS

B. Baysal, Department of Chemistry, Dartmouth College, Hanover,
    New Hampshire

Richard H. Boyd, Department of Chemical Engineering and Division of
    Materials Science and Engineering, University of Utah,
    Salt Lake City, Utah

M. G. Broadhurst, Institute for Materials Research, National Bureau
    of Standards, Washington, D.C.

Mary Barkley Clark, Salk Institute for Biological Studies, La Jolla,
    California

H. F. Cole, Bell Telephone Laboratories, Denver, Colorado

L. C. Corrado, University of Wisconsin-Green Bay, Manitowoc Campus,
    Manitowoc, Wisconsin

R. A. Creswell, Bell Canada-Northern Electric Research Limited,
    Ottawa, Canada

Richard G. Crystal, Research Laboratories, Xerox Corporation, Xerox
    Square, Rochester, New York

S. B. Dev, Department of Pure and Applied Chemistry, Strathclyde
    University, Glasgow, United Kingdom

F. A. Emerson, Naval Weapons Center, China Lake, California

R. L. Jernigan, Physical Sciences Laboratory, Division of Computer
    Research and Technology, National Institutes of Health,
    Bethesda, Maryland, and Department of Chemistry, University of
    California at San Diego, La Jolla, California

M. A. Kabayama, Bell Canada-Northern Electric Research Limited,
    Ottawa, Canada

B. A. Lowry, Department of Chemistry, Dartmouth College, Hanover,
    New Hampshire

W. J. MacKnight, Chemistry Department and Polymer Science and
    Engineering Program, University of Massachusetts, Amherst

S. Matsuoka, Bell Telephone Laboratories, Murray Hill, New Jersey

A. M. North, Department of Pure and Applied Chemistry, Strathclyde
    University, Glasgow, United Kingdom

M. M. Perlman, Department of Physics, College militaire royal
    de Saint-Jean, Saint-Jean, Quebec, Canada

John M. Pochan, Xerox Corporation, Xerox Square, Rochester,
    New York

Christopher H. Porter, Department of Chemical Engineering and
    Division of Materials Science and Engineering, University
    of Utah, Salt Lake City, Utah

C. W. Reed, General Electric Company, Research and Development
    Center, Schenectady, New York

J. C. Reid, Department of Pure and Applied Chemistry, Strathclyde
    University, Glasgow, United Kingdom

R. J. Roe, Bell Telephone Laboratories, Murray Hill, New Jersey

F. H. Smith, Benger Laboratory, E. I. duPont deNemours and Company,
    Waynesboro, Virginia

W. H. Stockmayer, Department of Chemistry, Dartmouth College,
    Hanover, New Hampshire

David C. Watts, Edward Davies Chemical Laboratory, University
    College of Wales, Aberystwyth, United Kingdom

R. E. Wetton, Department of Chemistry, Loughborough University,
    Loughborough, United Kingdom

Graham Williams, Edward Davies Chemical Laboratory, University
    College of Wales, Aberystwyth, United Kingdom

R. N. Work, Physics Department, Arizona State University, Tempe,
    Arizona

H. Yu, Department of Chemistry, Dartmouth College, Hanover,
    New Hampshire

Bruno H. Zimm, Department of Chemistry, Revelle College, University
    of California (San Diego), La Jolla, California

CONTENTS

SPATIAL CORRELATIONS IN COPOLYMERS FROM DIELECTRIC MEASUREMENTS

F. H. Smith, Benger Laboratory, E. I. duPont deNemours

and Company, Waynesboro, Virginia; L. C. Corrado,

University of Wisconsin-Green Bay, Manitowoc Campus,

Manitowoc, Wisconsin; and R. N. Work, Physics Department,

Arizona State University, Tempe, Arizona

INTRODUCTION

Debye and Bueche (1) have shown that a comparison of the dipole moment of a polymeric molecule with that of one of its structural units can yield information concerning the average chain configuration. Greater detail may be obtained by comparing measured values of dipole moment with those calculated by means of the rotational isomeric state model (2) (RISM), providing there is adequate knowledge of the statistical weights and geometric parameters needed for this model.

Our purpose here is to show that structural information can also be obtained from the compositional dependence of molecular dipole moments of appropriate copolymers. No geometric model or information on statistical weights is needed for this method. It depends instead on the occurrence of typal correlations, which may be defined in terms of non-randomness in the sequences of the types of structural units which make up the copolymer chain.

The theory given here generalizes and in a sense reinterprets the findings of an earlier attempt at the same problem (3) and it provides somewhat different insights than those obtained by Birshtein, Burshtein, and Ptitsyn (4) from their treatment of the problem in the rotational isomeric state approximation. No attempt is made here either to explore the relationship between typal correlations

1

and other properties of copolymers (5) or to treat the problem of obtaining information on copolymerization kinetics from the dependency of molecular dipole moments on copolymer composition (6).

THEORY

We begin by observing that the mean-square electric dipole moment of a polymeric molecule composed of N dipolar segments is given by the equation

$$<\mu_m^2> = \sum_i \sum_j <\mu_i \cdot \mu_j> \tag{1}$$

where μ_i is the dipole moment of the ith structural unit and the serial indices i and j count all of the structural units in the chain. Most of the terms in the summation of Equation 1 are negligible for flexible molecules, and hence they may be omitted. This may be accomplished systematically by including only those terms for which $k = j-i \leqslant M$, where the integer M represents the maximum separation of pairs of structural units over which angular correlation persists; that is, M may be interpreted as an angular correlation length. Neglecting all terms for which k is larger than M, Equation 1 becomes after dividing by N

$$\mu_e^2 \equiv <\mu_m^2>/N = \mu_i^2 + 2\sum_{j=i+1}^{i+M} <\mu_i \cdot \mu_j> - (2/N) \; x$$

$$\sum_{i=1}^{M} \sum_{j=2i}^{i+M} <\mu_i \cdot \mu_j> \tag{2}$$

where μ_e^2 is the effective square moment per structural unit in the polymer. This equation includes all significant correlation terms for any chain as long as $N \geqslant 2M+1$. The last sum in Equation 2 corrects the second term for having included correlations involving nonexistent structural units beyond the ends of the chain. The correction is approximate in that it does not account for the possibility of differences in angular relationships between pairs of structural units that are located close to the ends of the chain. For chains of even moderate length, this approximation is of little consequence (7). The form of Equation 2, incidentally, displays clearly the reason for the slow convergence, that is usually observed, of $<\mu_e^2>$ toward its assymptotic value as N increases.

In what follows, we assume for simplicity that M << N so that the last term in Equation 2 may be dropped. With this approxima- tion, only the relative locations, k = j-i, of the dipole pairs are of importance; and, because we will be treating copolymer systems, it will be convenient to redefine the subscripts and also to intro- duce superscripts in order to designate the types of structural units involved. In this notation, the average cosine of the angle between pairs of dipoles separated by k-1 structural units which are arranged in a particular ordering s may be defined to be

$$\eta_{ks}^{\alpha\beta} \equiv <\underset{\sim}{\mu}_\alpha \cdot \underset{\sim}{\mu}_\beta>_{ks}/\mu_\alpha\mu_\beta \qquad (3)$$

where μ_α and μ_β are the dipole moments of the structural units at the ends of the particular sequence. The ordering parameter s might be expressed, for instance, as a sequence of binary digits as suggested by Price (8).

With approximations mentioned above, Equation 2 may be written in a form suitable for copolymers

$$\mu_e^{2} = \sum_{\alpha=0}^{1} \sum_{\beta=0}^{1} \sum_{k=0}^{M} \sum_{\{s_k\}} \mu_\alpha\mu_\beta A_{ks}^{\alpha\beta} \eta_{ks}^{\alpha\beta}, \qquad (4)$$

where $A_{ks}^{\alpha\beta}$ (k > 0) is twice the probability of occurrence of a sequence that is terminated with elements of types α and β and that has the k-1 intervening structural units arranged in the particular ordering designated by s. The notation $\{s_k\}$ denotes the set of all orderings of the k-1 intermediate structural units. The value of $\eta_0^{\alpha\alpha}$ is, of course, unity and $\eta_0^{\alpha\beta}$ is undefined. Coefficients for the first few values of k are given in Table I.

If the composition x_α and reactivity ratio product (9) R = r_1r_2 are known for the copolymer, the probability $P_{\alpha\alpha}$ can be found from the quadratic equation

$$X(R-1)P_{\alpha\alpha}^{2} - [1+(2R-1)X]P_{\alpha\alpha} + RX = 0 \qquad (5)$$

where X = x_α/x_β is the molar ratio of the two types of structural units.

TABLE I

Probability Coefficients $A_{ks}^{\alpha\beta}$ in Equation 4[a]

k	s	A_{ks}^{00}	$A_{ks}^{01}+A_{ks}^{10}$	A_{ks}^{11}
0	-	x_0	0	x_1
1	-	$2x_0 P_{00}$	$2x_0 P_{01}+2x_1 P_{10}$	$2x_1 P_{11}$
2	1	$2x_0 P_{00}P_{00}$	$2x_0 P_{00}P_{01}+2x_1 P_{10}P_{00}$	$2x_1 P_{10}P_{01}$
2	2	$2x_0 P_{01}P_{10}$	$2x_0 P_{01}P_{11}+2x_1 P_{11}P_{10}$	$2x_1 P_{11}P_{11}$

[a]The mole fraction of type α structural units in the copolymer is x_α, and $P_{\alpha\beta}$ is the probability that a structural unit in the copolymer of type α is immediately followed by one of the type β.

The quantity R and hence also $P_{\alpha\alpha}$ contains the information on typal correlations within the copolymer chain. Values of R which are less than unity correspond to a tendency toward excess alternation of types of neighboring structural units. Values of R which are greater than unity indicate a tendency toward the occurrence of sequences of similar types of structural units. The probabilities for the occurrence of the different types of pairs are related by the equations $P_{\alpha\beta} = 1-P_{\alpha\alpha}$, and $x_\alpha P_{\alpha\beta} = x_\beta P_{\beta\alpha}$. Using these relations, it is possible to express all of the coefficients $A_{ks}^{\alpha\beta}$ in terms of $P_{\alpha\alpha}$ and x_α only.

By inserting experimental values of μ_e^2 for each of several copolymer compositions into Equation 4 together with the corresponding calculated values of $A_{ks}^{\alpha\beta}$, one can generate a set of simultaneous algebraic equations with the $\eta_{ks}^{\alpha\beta}$ being unknowns. In general, however, it is not possible to find unique solutions to such a set of equations because the coefficients $A_{ks}^{\alpha\beta}$ are not linearly independent functions of composition. It is thus necessary to group together those terms which contain linearly dependent coefficients and then to solve for the corresponding linear combinations of the η_{ks}'s. Further dissection of these combinations requires special approximations or assumptions.

The simplest useful special case arises when the dipole moment of one of the structural units of the copolymer is negligible with respect to the other. Accordingly, by setting $\mu_0 = 0$, Equation 4 becomes

$$\mu_e^2 = \mu^2 \sum_{k=0}^{M} \sum_{\{s_k\}} A_{ks} \, \eta_{ks} \tag{6}$$

where the labels $\alpha = \beta = 1$ have been dropped. All coefficients A_{ks} for $k \leq 3$ are linearly independent functions and it is possible to evaluate them separately.

Longer sequences, which are related through permutations in the ordering of their structural units have linearly dependent coefficients, and it is possible to solve only for linear combinations of the corresponding η_{ks}'s.

Another approximation may be useful whenever there is sufficient similarity between the structural units of the copolymer to assume that the values of η_{ks} are independent of the types of structural units of which the sequence is composed. This approximation is useful when the dipole moment of neither structural unit is negligible. We later test such an assumption for the case of poly(4-chlorostyrene, 4-methylstyrene) copolymers, referred to here as (4CS-4MS). In this copolymer, substantially the only difference between the two structural units is in the magnitudes of their dipole moments.

With the assumption that $\eta_k \equiv \eta_{ks}^{\alpha\beta}$ depends only on the length of the sequence, Equation 4 becomes

$$\mu_e^2 = \sum_{\alpha=0}^{1} \sum_{\beta=0}^{1} \sum_{k=0}^{M} \mu_\alpha \mu_\beta A_k^{\alpha\beta} \, \eta_k \tag{7}$$

where $A_k^{\alpha\beta} \equiv \sum_{\{s_k\}} A_{ks}^{\alpha\beta}$. Summation of $A_{ks}^{\alpha\beta}$ over all sequences of length k leads to the relatively simple equations for the typal coefficients

$$A_k^{\alpha\alpha} = 2x_\alpha \{x_\alpha + x_\beta \, [(P_{\alpha\alpha} - x_\alpha)/x_\beta]^k\}, \qquad k > 0 \tag{8a}$$

and

$$A_k^{\alpha\beta} = 2 - A_k^{\alpha\alpha} - A_k^{\beta\beta}. \tag{8b}$$

Insertion of Equations 8a and 8b into Equation 4 yields the equation

$$\mu_e^2 = x_0\mu_0^2 + x_1\mu_1^2 + 2(x_0\mu_0 + x_1\mu_1)^2 \sum \eta_k$$

$$+ 2x_0x_1(\mu_1 - \mu_0)_2^2 \sum [(P_{11} - x_1)/x_0]^k \eta_k \qquad (9)$$

where now, all non-zero coefficients of the η_k's in the last term are linearly independent functions of the composition. It is clear from the form of Equation 9 that typal correlations must be present to evaluate any η_k separately from the others.

Now, the terms containing the typal correlations become negligible either as r_1r_2 approaches unity or as k becomes large. The former reflects the necessity of nonrandomness in the copolymerization kinetics for typal correlations to exist, and the latter reflects the decrease in typal correlations as the separation between pairs of structural units increases. This suggests that a typal correlation length L may be defined in terms of the separation parameter k such that any k > L implies that $|[(P_{11} - x_1)/x_0]^k| < \varepsilon$, where ε is an arbitrary small number. The correlation length is, of course, a function of ε as well as of $R = r_1r_2$.

The necessary condition for the separability of the η_k's can be stated quite simply in terms of the typal correlation length and of the angular correlation length M, which may be defined with precision in a manner similar to L. Only those η_k's can be found for which k is less than the smaller of either L or M, and to find every significant value of the η_k's requires that L be greater than M.

Now the effective dipole moment μ_e is a convenient and familiar concept to use for explaining the development of the theory. It need not, however, be evaluated explicitly for our purposes. In fact, suppression of the dipole moment in favor of a ratio of dipole moments $r = \mu_0/\mu_1$ may result in a substantial reduction of uncertainties owing to approximations used to represent the local field.

To illustrate this point, consider the Onsager equation (10) for the dipole moment of a spherical polar molecule in a pure polar liquid,

$$(\varepsilon_s - \varepsilon_\infty) = \frac{3\varepsilon_s}{2\varepsilon_s + \varepsilon_\infty} \frac{4\pi}{3kT} N_o \left[\frac{\varepsilon_\infty + 2}{3}\right]^2 \mu_v^2 \qquad (10)$$

where ε_s and ε_∞ are the low- and high-frequency dielectric constants respectively, k is the Boltzmann constant, T is the temperature, and N_0 is the number of dipoles per unit volume. In this model the dipole moment of the molecule is represented by a rigid dipole of moment μ at the center of a sphere of dielectric constant ε_∞. The factor $3\tilde\varepsilon_s/(2\varepsilon_s+\varepsilon_\infty)$ is the ratio of the local field strength at the dipole to the macroscopic field strength, and the moment μ_V is the dipole moment of an equivalent molecule in free space. It is related to μ through the expression $\mu = \mu_V(\varepsilon_\infty+2)/3$.

Both factors $3\varepsilon_s/(2\varepsilon_s+\varepsilon_\infty)$ and $(\varepsilon_\infty+2)/3$ derive from assumptions of sphericity of both the molecule and of the cavity in which it is located (at least on the average), and of isotropic polarizabilities of the molecule and its surroundings. Onsager's equation is based on the further assumption of the absence of short range forces which would lead to angular correlations between a molecule and its surrounding dipoles. Failure to satisfy these assumptions leads, typically, to differences of up to $\pm$ 20% between dipole moments found from measurements of polar liquids and those from measurements on vapors.

In the approach being considered here, directional correlations which arise from short range forces within a given molecule are included in the expression for the effective square moment, given in Equation 4. Insertion of μ_e^2 from Equation 4 into Onsager's equation thus carries with it the tacit assumption that effects of any directional correlations which might originate from other short range forces, such as forces between chains, are insignificant compared to effects of forces within a given chain.

To show how the use of dipole moment ratios reduces the effects of unsatisfied assumptions regarding the local field and short range forces, consider the expression for the ratio of the effective square dipole moments $\mu_e^2(x_1)/\mu_e^2(1)$, where $\mu_e^2(x_1)$ is the square moment per structural unit of a copolymer containing a mole fraction x_1 of structural units of type 1, and $\mu_e^2(1)$ is the corresponding moment squared of the homopolymer. This ratio can be written in terms of experimentally obtainable quantities by using Equation 10. It is given by the expression

$$\frac{\mu_e^2(x_1)}{\mu_e^2(1)} = \frac{(\varepsilon_s-\varepsilon_\infty)_x\ (2\varepsilon_s+\varepsilon_\infty)_x\ (\varepsilon_s)_1\ \overline{M}_x\ \rho_1}{(\varepsilon_s-\varepsilon_\infty)_1\ (2\varepsilon_s+\varepsilon_\infty)_1\ (\varepsilon_s)_x\ M_1\ \rho_x} \qquad (11)$$

where the subscripts x and 1 refer to the copolymer and homopolymer
respectively. It is clear that the term $3/(\varepsilon_\infty+2)$ is irrelevant and
therefore questions concerning the method of converting dipole moments to the equivalent values in the vapor phase do not enter the
discussion. It is also seen that any error due to uncertainty in
the local field correction $3\varepsilon_s/(2\varepsilon_s+\varepsilon_\infty)$ is reduced by the fractional
change in the field factor that occurs with changing composition.
For example, in the case of P(4CS-4MS) copolymers, the maximum
change in the local field term is about 10% of its value; and hence
any uncertainty from this term is reduced by a factor of about 10
when dipole moment ratios are used rather than the moments themselves.

Using Equation 10 together with the dipole moment ratio
$r = \mu_0/\mu_1$, one finds for the ratio of the effective square moments

$$\mu_e^2(x_1)/\mu_e^2(1) = \{x_0 r^2 + x_1 + 2(x_0 r + x_1)^2 \sum \eta_k \tag{12}$$

$$+ 2x_0 x_1 (1-r)^2 \sum [(P_{11}-x_1)x_0]^k \eta_k \}/(1 + 2 \sum \eta_k)$$

The denominator of the RHS of Equation 12 is just the familiar
Kirkwood g-factor (11) which accounts for those angular correlations
that arise from short range forces, and which relates the effective
dipole moment of a structural unit to that of an equivalent isolated
unit through the relation $u_e^2 = g \mu^2$.

EXPERIMENTAL

Values of dipole moments have been obtained from a set of
P(4-CS, 4-MS) copolymers covering the whole range of compositions.
The copolymers were prepared by polymerizing monomer mixtures in
the absence of air, at 403°K using thermal initiation. Polymeriza-
tions were carried to about 40% conversion for alternate composition
ratios and to about 80% for the others. No effect has been found
which might have originated from the differences in the extent of
conversion. Conventional techniques of precipitation were used to
separate the polymer from the residual monomer mixture, and in each
case the polymer was freeze dried from a benzene solution prior to
use. Specimens 1 mm thick and 38 mm in diameter were molded in
vacuum, and three-terminal measurements (12-14) of the complex
dielectric constants $\varepsilon'-j\varepsilon''$ were carried out in the temperature
range of 380°K to 450°K, within an evacuated chamber. Frequencies
were varied from 10 Hz to 100 kHz. Values of ε_∞ and $\varepsilon_s-\varepsilon_\infty$ were

adjusted to obtain 100 kHz superposition of the experimental data, normalized in the form $\varepsilon''/(\varepsilon_s-\varepsilon_\infty)$ vs. $(\varepsilon'-\varepsilon_\infty)/(\varepsilon_s-\varepsilon_\infty)$, when plotted in the complex plane. This method yielded values of ε_∞ which are self-consistent to about $\pm$ 0.2%, and values of $\varepsilon_s-\varepsilon_\infty$ which have a reliability of about $\pm$ 0.5%. Values of dipole moment ratios were then calculated by means of Equation 11 using densities derived from hydrostatic weighings at room temperature with corrections for thermal expansion. Composition ratios were obtained from chlorine analyses of the copolymers.

Dipole moment ratios have also been obtained from three-terminal measurements at 298° of 10% solutions of the same set of copolymers dissolved in p-xylene (15).

Details of experimental procedures will be published later.

ANALYSIS OF EXPERIMENTS

Values of the dipole moment ratios obtained from the solid specimens of P(4CS-4MS) are plotted vs composition in Figure 1 and compared with curves calculated, using only the first two terms of Equation 12, for each of the indicated values of the g-factor. The absence of angular correlation effects for g=1 is evident by the linear variation of the effective square moment with changing composition.

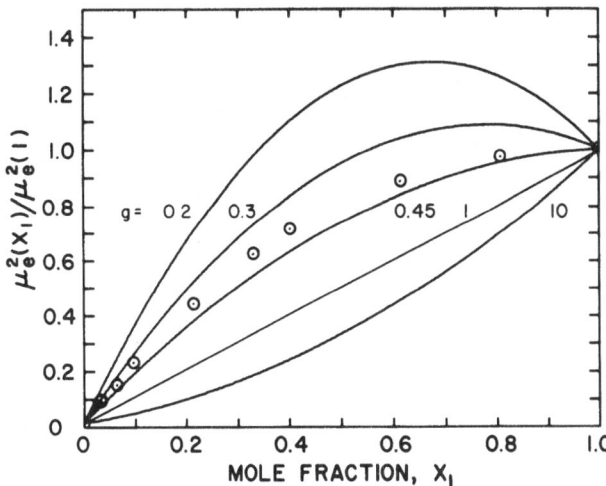

Figure 1.   Ratio of effective square dipole moments of a copolymer vs mole fraction x_1 for random copolymerization (r_1r_2=1), and with g = $\mu_e^2(1)/\mu_1^2$ having the indicated values. Dipole moment ratio μ_1/μ_0 of structural units is 0.13. Circles represent experimental values for P(4CS, 4MS) copolymer at 428°K.

Adjustment of g and of the dipole moment ratio r to minimize
the differences between experimental and calculated values yields
values of g=0.39 and of r=0.13. Differences between experiment
and calculated values of the dipole moment ratios obtained by using
these values for g and r are plotted vs composition in Figure 2.
Any attempt to further reduce the residuals for this case by including the last term in Equation 12 and adjusting values of η_k
will not be effective because, it turns out, each typal coefficient
is a symmetric function of composition with the value $x_1=0.5$ being
the center of symmetry. The coefficients for k=1, 2, and 3 are
plotted in Figure 3, again for the special case of P(4CS-4MS),
with $r_1 r_2=0.75$, g=0.39 and r=0.13. The maximum contributions to
the dipole moment ratio of the typal correlation terms are seen
to be about 0.08 and 0.01 for k equal to 1 and 2, respectively.
The contribution of the term for k=3 is indiscernable. It is clear
that for the copolymer systems such as P(4CS-4MS), unrealistically
high precision of experimental values would be required in order
to evaluate η_k for k $\geq$ 3.

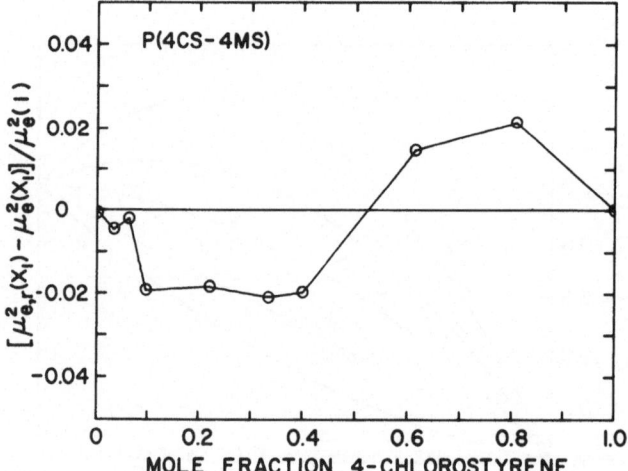

Figure 2.   Residual differences between calculated $\mu_{e,r}^2(x_1)$ and
             experimental $\mu_e^2(x_1)$ values of dipole moments for random
             copolymerization ($r_1 r_2=1$), assuming $\eta_k = \eta_{ks}^{\alpha\beta}$ for
             P(4CS-4MS). Points are connected to aid in
             visualization.

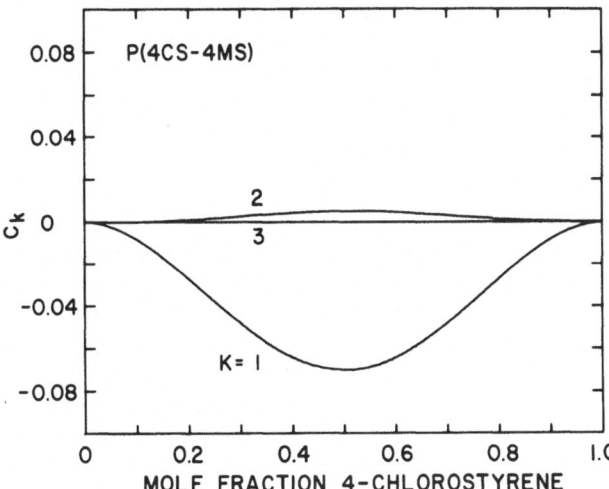

Figure 3.   Typal correlation coefficients $C_k = 2 g^{-1} x_0 x_1 (1-r)^2$
$[(P_{11}-x_1)/x_0]^k$ (see Equation 12) with g =
0.39, $r_1 r_2 = 0.75$, $r = \mu_0/\mu_1 = 0.13$, and k as indicated.

Now, the systematic nature of the residuals shown in Figure 2
suggests that they do, in fact, result from typal correlations. By
relaxing the requirement that all $\eta_{ks}^{\alpha\beta} = \eta_k$ it is indeed possible to
match the calculated curve to experimental values to within a
standard deviation of about 0.05 by using typal correlation terms
for k equal to one and two. Higher values of k are unnecessary at
this level of precision.

Interpretation of the values of these fitting parameters must
await further analysis of the forms and relationships among the
coefficients. It is clear, however, that the assumption that η_k
depends only on k, and not on the types of structural units involved,
fails even for structural units as nearly similar as 4-chlorostyrene
and 4-methylstyrene. This suggests that differences in the dipole-
dipole interaction energies for the two types of structural units
are responsible for the differences which must exist in the values
of $\eta_{ks}^{\alpha\beta}$ for each k < 3.

A preliminary analysis of dipole moment ratios obtained from
10 mole percent solutions of these copolymers in p-xylene yields
similar results. There is, however, a small but insignificant increase
in the g factor for the copolymers in solution. Further study is
needed to see whether this difference indicates a need to take into

account such things as the solvent-polymer interaction energy, or possibly the effects of correlations of structural units which are located on different but adjacent chains.

DISCUSSION

Basically, the method described here for calculating angular correlation parameters η_k depends only on a knowledge of non-randomness in sequences of structural units, together with experimental values of dipole moment ratios obtained from an adequately large set of copolymer compositions. We have assumed that the correlation parameters $\eta_{ks}^{\alpha\beta}$ themselves are independent of copolymer composition, that short range forces between structural units located on different chains make a negligible contribution to the η_k's, and finally that the chains and branches, if any, are long. There is no assumption regarding chain geometry, stereoregularity, or rotational states. Thus, to the extent that the basic assumptions are valid, the values of the different η_k's may be regarded as being essentially independent, experimental quantities. Hence, they may at least, in principle, be used to test the internal consistency of theoretical models such as the rotational isomeric state model.

Testing the RISM through use of the parameters η_k rather than through the dipole moment ratios themselves may result in a substantial reduction in the effort needed for RISM calculations simply because it is possible to deal with chain segments having lengths of the order k rather than N. Such a saving is particularly evident if the dipole-dipole interaction energy is of importance, as it seems to be, or if stereoregularity is to be included in the RISM calculation.

Finally, it seems appropriate to discuss a ramification of this method of evaluating spatial correlation parameters in copolymers, with respect to temporal correlations and relaxation processes.

Discussion of dielectric relaxation in copolymers may be given in terms of the time-dependent correlation functions (16).

$$\eta_{ks}^{\alpha\beta}(t) \equiv \langle \underset{\sim}{\mu_i^\alpha}(0) \cdot \underset{\sim}{\mu_{i+k}^\beta}(t)\rangle_s / \mu_i^\alpha \, \mu_{i+k}^\beta \qquad (13)$$

where $\mu_i^\alpha(0)$ is the dipole moment vector of the ith structural unit of the copolymer at an arbitrary origin of time, and $\underset{\sim}{\mu_{i+k}^\beta}(t)$ is the moment of the (i+k)th structural unit as a time t later, and the

average is taken over all possible initial states, each weighted
according to its probability of occurrence. Other symbols retain
their previously defined meanings. Using this notation, Equation 12
may be immediately adapted to give the decay function for the
special case of a copolymer in which all $\eta_{ks}^{\alpha\beta} = \eta_k$. It is given by
the equation

$$D(t) = \frac{U(x_1)\eta_0(t) + V(x_1) \sum_k \eta_k(t) + \sum_k W_k(x_1)\ \eta_k(t)}{U(x_1)\eta_0(0) + V(x_1) \sum_k \eta_k(0) + \sum_k W_k(x)\ \eta_k(0)} , \qquad (14)$$

where $\eta_0(t)$ is the autocorrelation function $\langle\mu_i(0) \cdot \mu_i(t)\rangle/\mu_i^2$
and the correlation parameter η_k as used previously is now
written $\eta_k(0)$ to emphasize its stationary nature. The functions
U, V, and W represent the linear, quadratic, and higher power
terms in x_1 respectively as given in Equation 12. All previous
arguments applying to the separability of the η_k's apply to their
time dependent analogues $\eta_k(t)$, and because $\eta_k(t) \leq \eta_k(0)$ the
same angular correlation length M applies to the time dependent as
well as the steady state case. Again, the last term contains the
information on typal correlations, and its presence is essential
to the decomposition of the decay function into separate contributions of the η_k's.

   In order to consider the effects of chain stiffness on the
relaxation spectrum, it is instructive to begin with an examination
of Equation 14 for two limiting cases. Firstly, if the chain is
composed of freely jointed segments, no angular correlation terms
are present, and D(t) is just the autocorrelation function $\eta_0(t)$.
The function $\eta_0(t)$ for such a polymer will in general not be the
same as for independently orienting structural units because of
the possibility of motional correlations (17). On the other hand,
all three terms may be present in Equation 14, but if the copolymer
is rigid, then all η_k again have the same time dependence, namely
$\eta_k(t) = \eta_0(t)$, though, of course, this is not the same function as
for the freely jointed polymer. A real polymer chain may be considered to be made up of segments which are locally stiff such that
the configuration at each such segment, which may change during long
intervals of time, does not change significantly within a few
relaxation times. This suggests the need for introducing a stiffness correlation length, in order to treat relaxation phenomena in
polymers. The stiffness correlation length S may be defined in a
manner directly analogous to the definition of typal correlation
lengths, but with the size of the quantity $\langle d\eta_k(t)/dt\rangle$ being the

criterion rather than the size of $(|P_{11}-x_1|/x_0)$. The stiffness correlation length used here is essentially the same as the angular velocity correlation length introduced by Clark and Zimm (17).

In terms of the three correlation lengths, typal, angular and stiffness, the necessary criteria for the separability of the cross-correlation functions $\eta_k(t)$ is that k be greater than S but less than either L or M whichever is smaller. To find all $\eta_k(t)$ for k greater than S, it is necessary that L be larger than M.

These criteria provide a basis for explaining the observation by Leffingwell and Bueche (18) that the shapes of the loss peak for poly(styrene, 2-chlorostyrene) are independent of the copolymer composition. As we have seen, such independence is obtained either if the chain is freely jointed so that all η_k vanish for k > 0, or if the stiffness correlation length is greater than either the angular correlation length or the typal correlation length, which-ever is less. A choice between these possibilities cannot be made without further information. To this end we have examined the relaxation spectra of the copolymer system P(4CS-4MS), and we find the same independence of shape on composition. Now, we know that angular correlations exist in P(4CS-4MS) copolymers because of the necessity of introducing a value of 0.39 for the g-factor, and we also know that the angular correlation length is at least as great as the typal correlation length L=2. Hence, we can conclude that the copolymer chain P(4CS-4MS) is locally stiff over a length which includes at least three structural units.

CONCLUSIONS

Measurements of dipole moment ratios of copolymers of variable composition can be used effectively to determine the angular corre-lation length M as well as the values of the angular correlation parameters η_k, k < M, providing the typal correlation length L is greater than M. Further, there is a possibility of determining separately the contributions to the relaxation spectrum of the cross correlation terms $\eta_k(t)$ for S < k < M providing the typal correla-tion length L is, in fact, greater than M.

The shortness of the typal correlation length in the system P(4CS-4MS) precludes a detailed analysis of the correlation para-meters of this system beyond k=2. Such an analysis could, however, be carried out with a different copolymer system having a longer typal correlation length. Copolymers with greater typal correlation lengths could be obtained either by choosing a system with an $r_1 r_2$ value which is substantially less than unity, or by the choice of a set of condensation polymers such as poly(oxy-n-methylene),

representing several values of n, in which the separation of the
polar units is well defined over the whole chain length.

ACKNOWLEDGMENTS

The authors would like to thank David G. Shaw for his part in
summing the series used in Equation 8, Susan L. Kreuser for repre-
cipitating and freeze-drying the materials used for the solid
polymer specimens, and John D. Irvine for help in programming calculations and for assistance in the dielectric measurements.

Support by the U.S. Atomic Energy Commission of that part of
this work which was carried out at the Pennsylvania State University, and support by the National Science Foundation of the later
portions done at Arizona State University are gratefully acknowledged.

REFERENCES

1. P. Debye and F. Bueche, J. Chem. Phys. 19, 589 (1951).
2. See for instance: P. J. Flory, STATISTICAL MECHANICS OF
    CHAIN MOLECULES, Interscience, New York, 1969.
3. R. N. Work and Y. Trehu, J. Appl. Phys. 27, 1003 (1956).
4. T. M. Birshtein, L. L. Burshtein and O. B. Ptitsyn, Zhur.
    Tekh. Fiz. 29, 7 (1959); Engl. Trans.: Soviet Phys. - Tech.
    Phys. 4, 810 (1959).
5. H. J. Harwood, Angew. Chem. (Int. Ed. Engl.) 4, 1051 (1965).
6. M. Shima, J. Polymer Sci. 56, 213 (1962).
7. Ref. 2, p. 80 ff.
8. F. P. Price, J. Chem. Phys. 36, 209 (1962).
9. See for instance: G. E. Ham in COPOLYMERIZATION, G. E. Ham,
    Ed., Interscience, New York, 1964.
10. L. Onsager, J. Am. Chem. Soc. 58, 1486 (1936); See also:
    H. Frohlich, THEORY OF DIELECTRICS, Oxford, London, 1949.
11. J. G. Kirkwood, J. Chem. Phys. 7, 911 (1939).
12. L. C. Corrado, Ph.D. Thesis, Arizona State University, 1969.
13. R. D. McCammon and R. N. Work, Rev. Sci. Instrum. 36, 1169
    (1965).
14. L. C. Corrado and R. N. Work, Rev. Sci. Instrum. 41, 598 (1970).
15. F. H. Smith, Ph.D. Thesis, The Pennsylvania State University,
    1964.
16. L. D. Landau and E. M. Lifshitz, STATISTICAL PHYSICS, Pergamon,
    London, 1958, p. 374 ff. See also: R. H. Cole, J. Chem.
    Phys. 42, 637 (1965), and M. Cook, D. C. Watts and G. Williams,
    Trans. Faraday Soc. 66, 2503 (1970).
17. M. B. Clark and B. H. Zimm, this volume, p. 45.
18. J. Leffingwell and F. Bueche, J. Appl. Phys. 39, 5910 (1968).

MULTIPLE DIELECTRIC RELAXATION PROCESSES IN AMORPHOUS POLYMERS AS A FUNCTION OF FREQUENCY, TEMPERATURE AND APPLIED PRESSURE

Graham Williams and David C. Watts

Edward Davies Chemical Laboratory, University College of Wales, Aberystwyth, United Kingdom

INTRODUCTION

It is well known (1,2) that most dipolar amorphous polymers exhibit multiple dielectric relaxation processes. At low temperatures the α and β processes are observed which, on raising the temperature at a given pressure, tend to coalesce to form the (αβ) relaxation process. The application of pressure at a given temperature may resolve an (αβ) process into the α and β processes. The general features of the α, β and (αβ) processes have become clarified as comprehensive data, obtained over a wide range of frequency and temperature (1,2) and applied pressure (3-15), are obtained for polymers of different chemical structure and stereoregularity. It is the aim of the present work to describe the behavior which is characteristic of α, β and (αβ) processes, with a special reference to the effect of an applied hydrostatic pressure. This behavior is then discussed in molecular terms with the aid of the dipole correlation function approach of Glarum (16) and Cole (17) as developed for polymer systems by Cook, Watts and Williams (18). An attempt is made to give a unified approach to the α, β and (αβ) processes, and it is suggested that the broad loss curves obtained for the α process arise from a non-exponential correlation function, which is characteristic of a cooperative relaxation process.

THE α RELAXATION

The α relaxation has been studied in many polymers over a wide range of frequency and temperature (1,2). In addition, several studies have been made using applied pressure as an additional variable (3-15). Table I indicates some of the amorphous polymers whose dielectric properties have been studied as a function of frequency, temperature and applied pressure. It would not be appropriate to consider the α process in these polymers in detail. It suffices to describe the results for a model case - the α process as observed in polyethyl acrylate (14,15).

Figure 1 shows the plot of the dielectric permittivity ε' and loss factor ε'' against log frequency at 20.3°C and four applied pressures for amorphous polyethyl acrylate (14,15). The single absorption observed is the α process, whose mechanism is the Brownian motions of the chains. As the pressure is raised, the curves move rapidly to lower frequencies without a change in shape. This is made clear in Figure 2 which shows that the curves for the four pressures have the same shape. Very similar behavior was observed (14,15) for this polymer over the temperature range -6.5° to +40°C and 1 to 2400 atm. In fact, the shape of the reduced dispersion and absorption curves are essentially independent of pressure at a given temperature, and narrow only slightly as the temperature is increased. Figure 3 shows log f_{max} against pressure at given temperatures, and two features should be noted. At a given temperature, the plot is curved, the slope increasing with increasing pressure. In addition, at a given pressure, the slope increases with decreasing temperature. These results, Figures 1 - 3 for polyethyl acrylate, are similar to those obtained for the α process in polyvinyl acetate (4,12), polymethyl acrylate (6,9), polypropylene oxide (7) and polyethylene terephthalate (12). The results for the α processes for the other polymers listed in Table I are not so well defined - due in most cases to the proximity of a substantial β process.

The data, Figures 1 - 3, may be considered in three parts. (i) The magnitude ($\varepsilon_0 - \varepsilon_\infty$) for the process, where ε_0 and ε_∞ are the limiting low and high frequency permittivities respectively for the process. This magnitude is proportional to $N\mu_\alpha^2$, where N is the number of dipoles per unit volume, and μ_α^2 is the square of the effective dipole moment involved in the α process. (ii) The shape of the dispersion and absorption curves (e.g. Figure 2) which contains information on the molecular mechanisms for Brownian motion. This will be considered in greater detail below. (iii) The plot of log f_{max} against pressure (Figure 3) and the cross plot of log f_{max} against (1/T) at given pressures, gives values of the apparent activation volume $\Delta V(T,P)$ and constant pressure apparent activation energy $Q_p(T,P)$ respectively (6). It is of interest to consider (ii) and (iii) in a little more detail at this point. First, the asymmetry of the plot of loss factor against log f (Figure 2) is characteristic of α processes (1,4,6,7,9,12). It will be seen below that this relates to the form of the dipole correlation function, but an alternative approach is to express the dispersion and absorption curves in terms of the macroscopic time-frequency relation (1,14,18,21,22).

TABLE I

Polymer	Process Observed and Reference	ΔV (cc/mole)			Q_v/Q_p	
		α	β	(αβ)	α	β
Polyvinyl Acetate	α (4,12)	140 (12)	—	—	—	—
Polymethyl Acrylate	α (6,9)	80-150 (6)	—	—	0.71-0.78 (6)	—
Polyethyl Acrylate	α (14,15)	117-176 (14)	—	—	0.7 (14)	—
Polymethyl Methacrylate	α, β, (αβ) (5,13)	—	21 (13)	53 (13)	—	—
Polyethyl Methacrylate	α, β, (αβ) (9,11,13)	—	22 (13)[a]	63 (13)	—	—
Poly-n-butyl Methacrylate	α, β, (αβ) (8,13)	105 (13)	12 (13)[b]	44 (13) 75 (8)	—	—
Polyoctyl Methacrylate	α, β, (αβ) (13)	95 (13)	38 (13)	95 (13)	—	—
Polynonyl Methacrylate	α, β, (αβ) (19)	—	—	150-220 (19)	0.60 (19)	—

TABLE I
(Continued)

Polymer	Process Observed and Reference	ΔV (cc/mole)			Q_v/Q_p	
		α	β	(αβ)	α	β
Polylauryl Methacrylate	(αβ) (19)	—	—	37–52 (19)	0.80 (19) (19)	—
Polypropylene Oxide	α (7)	90–120 (7)	—	—	0.7 (7)	—
Polyethylene Terephthalate	α, β (9,10,12)	520 (12)	35 (12)	—	—	0.86 (10)
Polychlorotri-fluoroethylene	β (12)	—	31 (12)	—	—	—
Polyvinyl Chloride	α, β (3,12,15,20)	320 (12)	25 (12) 21 (15)	—	—	0.88 (15,20)

[a]See also Reference 11.

[b]See also Reference 8.

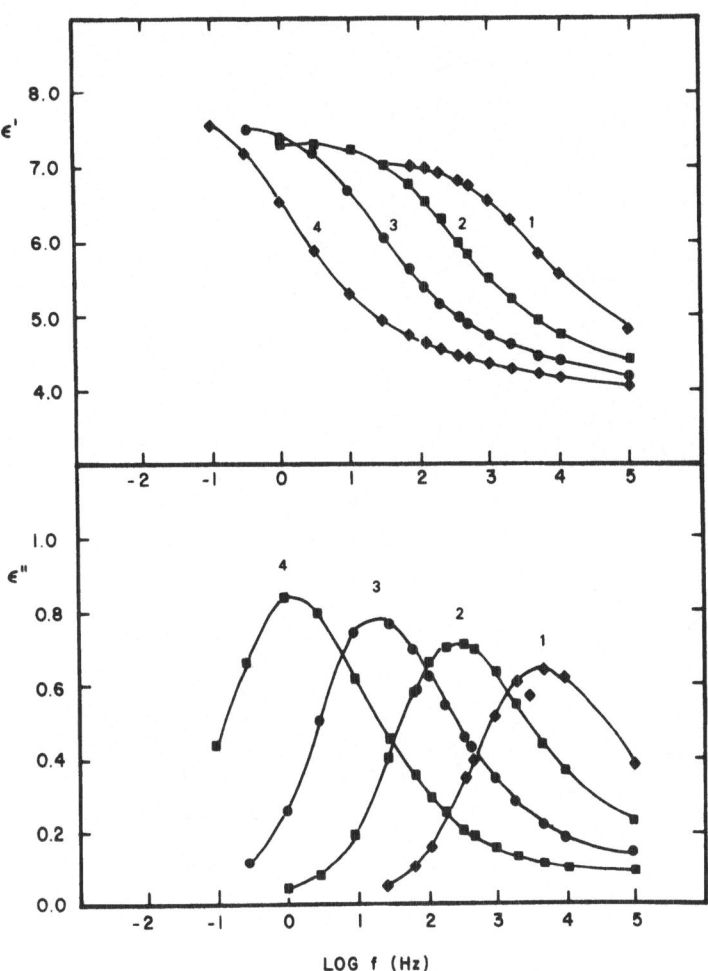

Figure 1. ϵ' and ϵ'' against log (frequency) (Hz) for the α process
            in amorphous polyethyl acrylate at 20.3°C and four applied
            pressures.   Curves 1, 2, 3 and 4 correspond to 550, 1100,
            1560 and 2060 atm respectively.

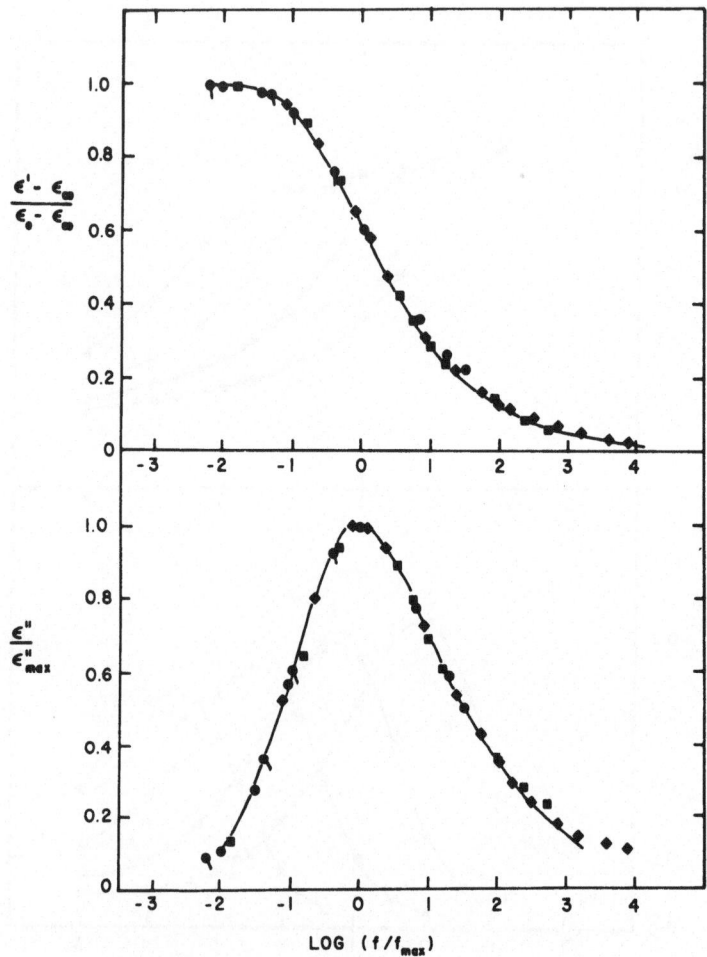

Figure 2.   The normalized permittivity $(\varepsilon' - \varepsilon_\infty)/(\varepsilon_0 - \varepsilon_\infty)$ and
normalized loss factor $\varepsilon''/\varepsilon''_{max}$ against $\log(f/f_{max})$
for the α process in amorphous polyethyl acrylate at
20.3°C. $\bullet$, $\bullet$, $\blacksquare$ and $\blacklozenge$ correspond to 550, 1100, 1570
and 2060 atm respectively.   The continuous curves cor-
respond to $\bar{\beta} = 0.41$.

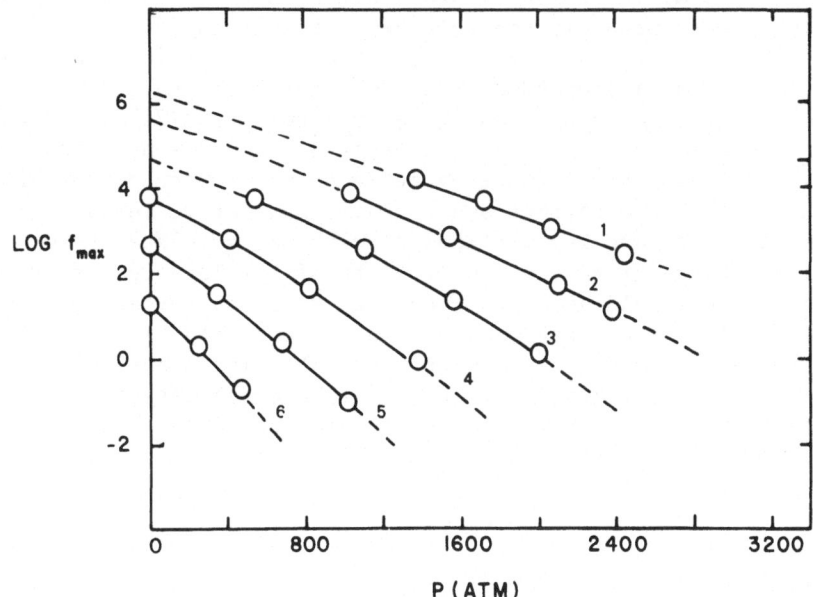

Figure 3. log f_{max} against applied pressure at different temperatures for the α process in amorphous polyethyl acrylate. Curves 1, 2, 3, 4, 5 and 6 correspond to 40.0, 30.0, 20.3, 10.3, 0.0 and −6.5°C respectively.

$$\frac{\epsilon^*(i\omega) - \epsilon_\infty}{\epsilon_0 - \epsilon_\infty} = \int_0^\infty \left[\frac{-d\phi(t)}{dt} \right] [\exp - i\omega t]\, dt \qquad (1)$$

$\phi(t)$ is the normalized charge decay function obtained when a steady electric field is removed from a specimen. We have found (21,22) that the shape of the α relaxation curves (Figure 2) may be fitted quite well using the empirical form for the decay function

$$\phi(t) = \exp - (t/\tau_0)^{\bar\beta} \; ; \qquad 0 < \bar\beta \le 1 \qquad (2)$$

where τ_0 is some effective relaxation time and $\bar{\beta}$ is a parameter.
The continuous curves in Figure 2 were calculated from Equations 1
and 2 for $\bar{\beta} = 0.41$, and are seen to give a very satisfactory representation of the α process at this temperature. Table II shows the
values of $\bar{\beta}$ for the α process in several polymers (22). The asymmetry is predicted in a quantitative manner, and we note that the
Davidson-Cole empirical representation of relaxation curves (23-25),
although it has the correct sense of the asymmetry, leads to loss
curves that are more asymmetrical than those calculated from
Equation 2 for a given total half width of the loss factor plot in
Figure 2. It should be emphasized that Equation 2 does not involve
a distribution of relaxation times, but represents $\phi(t)$ in a form
that is non-exponential in linear time. Now it is possible to fit
the present experimental data (Figure 2) and the data for other
α relaxations to a distribution of relaxation times (1,26) $H(\tau)$
where

$$\phi(t) = \int_0^\infty H(\tau) \exp(-t/\tau) \, dt \qquad (3)$$

and a combination of Equations 1 and 2 yields the familiar relation

$$\frac{\varepsilon^*(i\omega) - \varepsilon_\infty}{\varepsilon_0 - \varepsilon_\infty} = \int_0^\infty \frac{H(\tau)}{1+i\omega\tau} \, dt \qquad (4)$$

It is to be expected that Equation 2 would correspond to a particular form for $H(\tau)$. However, such a procedure may not be necessary
to account for the experimental results. It is quite possible that
the molecular motions responsible for the α process lead quite
naturally to a $\phi(t)$ which has a 'non-exponential' dependence on
linear time. This will be discussed further below.

As indicated above, the plots of log f_{max} against pressure at
given temperatures, and of log f_{max} against $(1/T)$ at given pressures
yield $\Delta V(T,P)$ and $Q_p(T,P)$ respectively, where

$$\left(\frac{\partial \log f_{max}}{\partial P}\right)_T = -\frac{\Delta V(T,P)}{2.3RT} \; ; \left(\frac{\partial \log f_{max}}{\partial T}\right)_P = \frac{Q_p(T,P)}{2.3RT^2} \qquad (5)$$

TABLE II

Polymer	$\bar{\beta}$	Reference to Experimental Data
Polyvinyl Acetate	0.56 at 62.5°C	40
Polymethyl Acrylate	0.40 (master curve)	6
Polyethyl Acrylate	0.38 at 0.0°C 0.41 at 20.0°C	14,15
Polypropylene Oxide	0.45 (master curve)	7
Polyvinyl Octanoate	0.56 (master curve)	41

It is clear from Figure 3 for polyethyl acrylate that both $\Delta V(T,P)$ and $Q_p(T,P)$ vary with temperature and pressure. This is true for other α relaxations (1,6,7). From Figure 3, Q_p (20°, 1 atm) = 39 kcal/mole rising to Q_p (-6.5°, 1 atm) = 61 kcal/mole. Also ΔV (20°, 1 atm) = 117 cm^3/mole rising to ΔV (-6.5°, 1 atm) = 176 cm^3/mole. Data for the α relaxations in several polymers are shown in Table I. First, great caution must be applied in interpreting values for $Q_p(T,P)$ and $\Delta V(T,P)$. It should not be thought that $Q_p(T,P)$ is the barrier required for a unit in a chain to overcome in reorientation, since an α process is a cooperative process. (Remember that bond dissociation energies lie in the range 50 to 100 kcal/mole!) Similarly, $\Delta V(T,P)$ should not be interpreted as the actual volume required to reorient a chain unit. The quantities $Q_p(T,P)$ and $\Delta V(T,P)$ may be interpreted in terms of several models for relaxation (1,6,7,26). Further considerations of their molecular significance will be made in the section below which is concerned with correlation functions. A further macroscopic quantity is the constant volume apparent activation energy (6,7) $Q_V(T,V)$, which is given by (6)

$$Q_V(T,V) = 2.3RT^2 \left(\frac{\partial \log f_{max}}{\partial T}\right)_V = Q_p(T,P) - T \left(\frac{\partial P}{\partial T}\right)_V \Delta V(T,P)$$

$$(6)$$

The thermal pressure coefficient $(\partial P/\partial T)_V$ is equal to (6) the ratio
of the isobaric expansion coefficient to the isothermal compressi-
bility of the medium. Values of $[Q_V(T,V)/Q_p(T,P)]$ may be obtained
from Equation 6 knowing $Q_p(T,P)$, $\Delta V(T,P)$ and $(\partial P/\partial T)_V$. For the
process in polyethyl acrylate, this ratio is near (14) to 0.7.
Table I gives $[Q_V(T,V)/Q_p(T,P)]$ at 1 atm for several polymers,
and the values are seen to lie in the range 0.6 to 0.8 for the α
process. This clearly means that $Q_V(T,V)$ forms a large part of
$Q_p(T,P)$, and has important implications for the free volume
theories of relaxation (6,26). If we write (6,26),

$$f_{max} = f_o \exp(-B/v_f) \tag{7}$$

where v_f is a suitably defined (6) free volume, then from Equation 7

$$\left(\frac{\partial \log f_{max}}{\partial T}\right)_V = \frac{B}{2.3 v_f^2}\left(\frac{\partial v_f}{\partial T}\right)_V > 0 \tag{8}$$

where > 0 indicates the experimental result. Hence $(\partial v_f/\partial T)_V > 0$.
A similar relation is given by Ferry (see Reference 26, page 326).
If v_f is defined as the actual volume V minus the 'occupied volume'
V_o, then $(\partial v_f/\partial T)_V > 0$ requires that $(\partial V_o/\partial T)_V$ is negative – which
is very difficult to understand (27).

THE β RELAXATION

Any description of the β process in polymers must note that
there are two distinct types of polymer chains, and these may give
different dielectric β processes. The first type has the dipoles
rigidly attached to the polymer chain. Examples are polyvinyl
chloride, polyethylene terephthalate and polychlorotrifluoroethylene.
In these cases the β process must involve the motion of the chain
backbones. The second type has dipoles contained in the side groups
off the chain. Examples are the polyalkyl methacrylates. In these
cases, a portion of the mean square dipole moment may be relaxed by
the reorientation of the side groups.

Figure 4 shows the effect of pressure (15,20) on the β process
in polyvinyl chloride at 36.5°C. As the pressure is increased the
loss curve shifts only slightly to lower frequencies, with

$(\partial \log f_{max}/\partial P)_T \approx 0.35$ decades per 1000 atm, and shows a marked
reduction in its magnitude. Figure 5 shows $\log f_{max}$ against pressure at given temperatures. The contrast between the effect of
pressure in the β process in polyvinyl chloride (Figures 4 and 5)
and on the α process in polyethyl acrylate (Figures 1 and 3) is
obvious. Results similar to those for polyvinyl chloride have been
obtained for the β process in amorphous polyethylene terephthalate
(10,12). The β process in the 'flexible side group' polymers has
been extensively studied (1,2,8,9,11,13). The magnitude of the
β process is greater than that of the α process for the conventional
lower polyalkyl methacrylates (1,2,13), which is the reverse of the
situation for polyvinyl chloride and polyethylene terephthalate.
Williams (8,9,11) and Sasabe and Saito (13) found that the magnitude
of the β process in polyethyl methacrylate and poly-n-butyl methacrylate decreased very rapidly with increasing pressure, and
(like polyvinyl chloride) $\log f_{max}$ was only slightly dependent upon
pressure at a given temperature. This is a quantitative difference
between the β processes in the two types of polymer. Williams and
Watts (20) give[+] $[1/\varepsilon''_m(\partial\varepsilon''_m/\partial P)_T]$ in the range (0.5 to 1.0) x
10^{-4} atm^{-1} for polyvinyl chloride (20) and polyethylene terephtha-
late (10), and give (3 to 5) x 10^{-4} atm^{-1} for polymethyl methacrylate (13), polyethyl methacrylate (9,11,13) and poly-n-butyl methacrylate (8,13). Thus it appears that the magnitude for the β process decreases more rapidly with increasing pressure for the flexible side group polymers. Table I shows $\Delta V(T,P)$ obtained for several
polymers and it is clear that $\Delta V(\beta) << \Delta V(\alpha)$ in all cases. The
ratio $[Q_V(T,V)/Q_p(T,P)]$ at 1 atm is 0.86 and 0.88 for the two
polymers in Table I and is indicative of a barrier mechanism for
the β process in these polymers (10,20). The actual mechanism for
the β process will be discussed further below. Note that the decrease in the magnitude of the β process in polyethyl methacrylate
(9,11,13) and in poly-n-butyl methacrylate (8,13) as pressure is
increased is accompanied by a corresponding increase in the magnitude of the β process, indicating a connection in their origins even
when they are well resolved on the log frequency axis (18).

[+]ε''_m is the maximum loss factor in the plot ε'' against log f at a
given (T,P) condition.

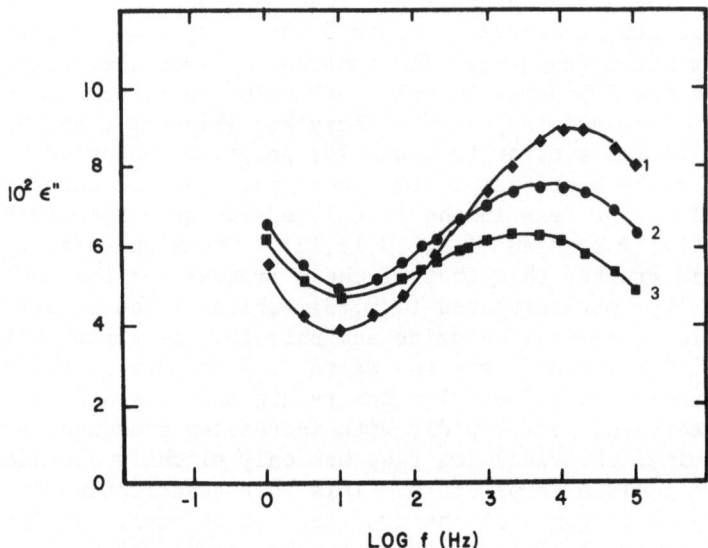

Figure 4. The effect of applied pressure on the β process in amorphous polyvinyl chloride at 36.5°C. Curves 1, 2 and 3 correspond to 1, 1040 and 2120 atm respectively.

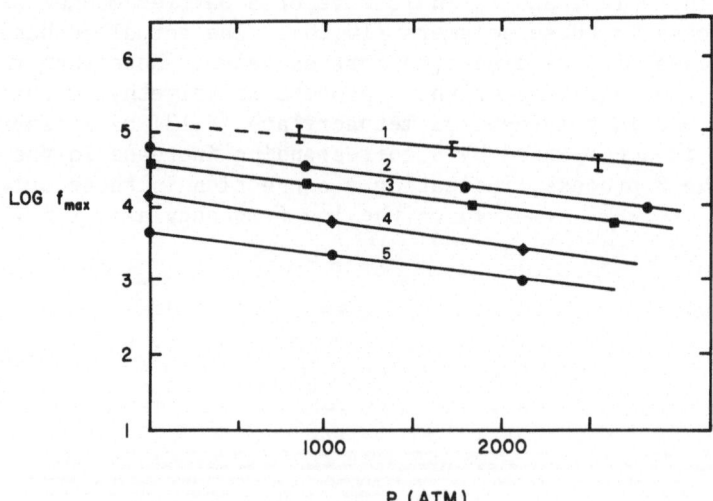

Figure 5. log f_{max} against applied pressure at different temperatures for the β process in amorphous polyvinyl chloride. Curves 1, 2, 3, 4,and 5 correspond to 80, 64, 51, 36.5 and 23.5°C respectively

THE ($\alpha\beta$) RELAXATION

The coalescence of the α and β processes to form the ($\alpha\beta$) process has been demonstrated for a number of polyalkyl methacrylates (8,9,11,13,19). Table I indicates those which have been studied over a range of frequency, temperature and pressure. It was mentioned above that the β process decreases in magnitude as the alkyl side chain is increased in length for conventional polyalkyl methacrylates (13). Thus although the decomposition of the ($\alpha\beta$) process into α and β processes on raising the pressure is well defined for the lower polyalkyl methacrylates [e.g. polyethyl methacrylate (9,11,13) and poly-n-butyl methacrylate (8,13)], it may not be clear if the single absorption in long side chain polyalkyl methacrylates is an α or an ($\alpha\beta$) process, since the resolved β process may be very small. Figure 6 shows the absorption observed[+] in a polynonyl methacrylate (15,19) as a function of frequency at different pressures. At this temperature of 62°C the single absorption may be an α or an ($\alpha\beta$) process. However, Figure 7 shows that at 24.7°C, application of pressure resolves the single process into an α process and the β process. The β process is not well defined since, as it is resolved from the α process, the increase of pressure reduces its magnitude [as in the lower polyalkyl methacrylates (8, 9,11,13)]. Figure 8 shows log f_{max} against pressure at different temperatures for the ($\alpha\beta$) process in this polynonyl methacrylate (19). These plots are quite similar to Figure 3 and emphasize the similar behavior of α and ($\alpha\beta$) processes. The experimental behavior of the ($\alpha\beta$) process in the lower polyalkyl methacrylates (8,9,11,13) is quite similar to an α process, and this will be discussed further below. It should be noted that the constant pressure apparent activation energy for the ($\alpha\beta$) process falls with increasing length of the n-alkyl side chain (1,13). It has been suggested (19) that there is an essential difference between the ($\alpha\beta$) process in the lower polyalkyl methacrylates and in the higher polyalkyl methacrylates. In the lower polyalkyl methacrylates, at high temperatures the ($\alpha\beta$) process corresponds primarily to the microbrownian motions of the chain backbone. In the higher polyalkyl methacrylates at high temperatures, the long side chains play an important part in determining the microbrownian motions of the dipoles, and it is this cooperation that leads to a lowering of the apparent activation energy.

[+]Note the alkyl group in this polynonyl methacrylate is
R = $(CH_2)_2$ $CH(CH_3)CH_2$ $C(CH_3)_2$ CH_3.

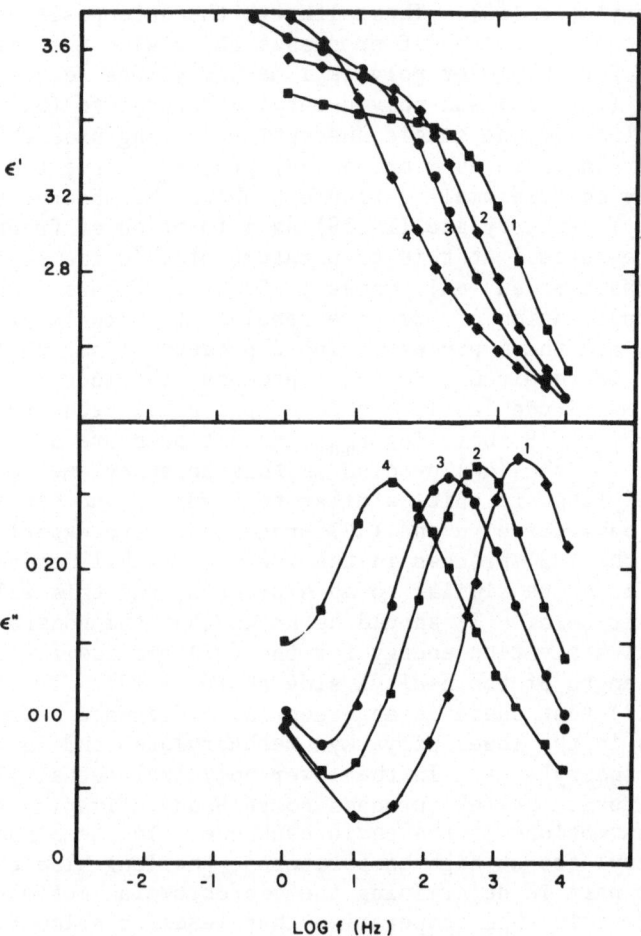

Figure 6.   ε" against log (frequency) (Hz) for a polynonyl meth-
            acrylate at 62°C and different applied pressures.
            Curves 1, 2, 4 and 4 correspond to 480, 830, 1030 and
            1380 atm respectively.

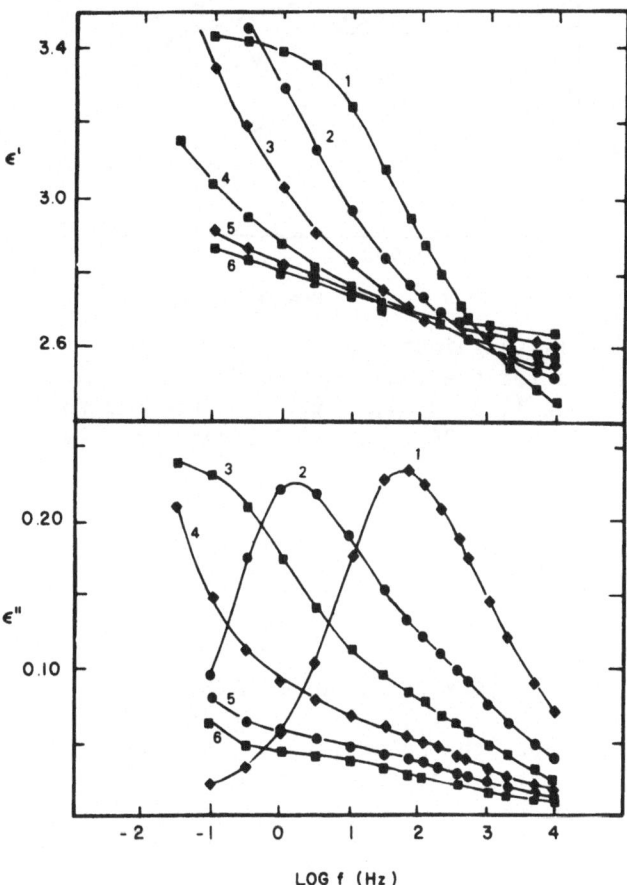

Figure 7.   ε" against log (frequency) (Hz) for a polynonyl methacrylate at 24.7°C and different applied pressures.
Curves 1, 2, 3, 4, 5 and 6 correspond to 1,407, 696,
1034, 1393 and 1750 atm respectively.

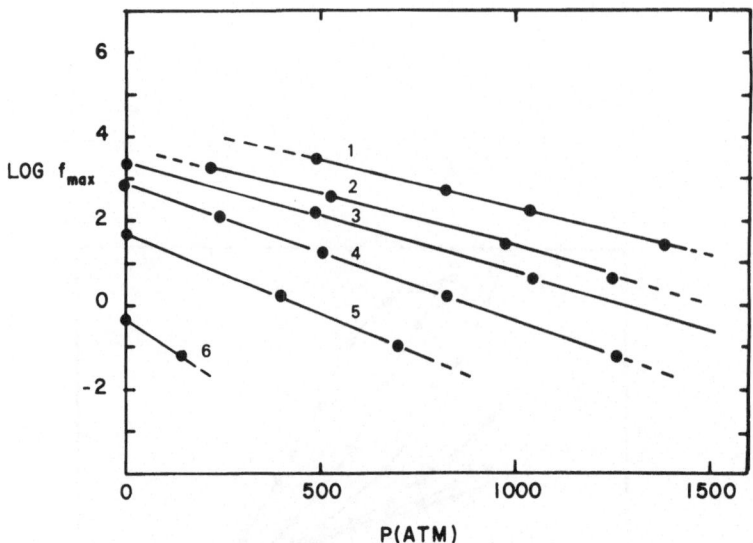

Figure 8. log f_{max} against applied pressure at different temperatures for the ($\alpha\beta$) process in a polynonyl methacrylate.
Curves 1, 2, 3, 4, 5 and 6 correspond to 62.0, 51.6,
45.7, 34.8, 24.7 and 10.0°C respectively.

THEORY OF DIPOLE RELAXATION

We now consider the molecular approach to the dielectric α,
β and ($\alpha\beta$) relaxation processes. In doing so, it is appropriate
first to outline the dipole correlation function method for independent dipoles in the liquid state (e.g. liquid chlorobenzene)
and then to indicate how this is generalized in order to include
dipolar polymeric chains (15,18). The dielectric theory using
correlation functions was developed by Glarum (16) and Cole (17),
using the Kubo method (28,29) for handling the external force
problem in time dependent statistical mechanics. The general
approach has been reviewed by Zwanzig (30).

Consider first non-correlated dipole molecules in the liquid
state where each molecule has a dipole moment of magnitude μ.
The dipole correlation function for a reference dipole i may be
written as

$$\Gamma_{ii}(t) = \frac{<\underset{\sim}{\mu_i}(0) \cdot \underset{\sim}{\mu_i}(t)>}{\mu^2} \qquad (9)$$

The physical meaning of $\Gamma_{ii}(t)$ is illustrated in Figure 9. Consider the reference dipole at $t = 0$, in an orientation $\underset{\sim}{\mu_i}(0)$.
As time develops, the dipole moves in Brownian motion, and we
require $<\mu_i(0) \cdot \mu_i(t)>$. Now all initial directions in the liquid
state are equally probable, so it follows that $<\mu_i(0) \cdot \mu_i(t)> = \mu_i(0) \cdot <\mu_i(t)>$. The quantity $<\mu_i(t)>$ is the average value of $\mu_i(t)$
given that the dipole had the value $\mu_i(0)$ at $t = 0$. This may be
obtained by averaging over a very large number of trials as shown
in Figure 9. At $t = t_1$ the quantity $<\mu_i(t_1)>$ is strongly correlated with $\mu_i(0)$, but as time develops $<\mu_i(t)>$ falls, so that in
the limit $t \to \infty$, $<\mu_i(t \to \infty)> \to 0$, as indicated in the figure. The
normalized correlation function $\Gamma_{ii}(t)$ is indicated schematically
in Figure 10a. If there is one general reorientation process,
$\Gamma_{ii}(t)$ falls to zero along a single decay curve [curve (i)]. If,
however, there are two reorientation mechanisms, this will be
reflected in $\Gamma_{ii}(t)$ and is indicated schematically as curve (ii)
in Figure 10a. The complex permittivity is given by the relation,

$$\left[\frac{\varepsilon^*(i\omega) - \varepsilon_\infty}{\varepsilon_0 - \varepsilon_\infty} \right] \overline{f}(i\omega) = \int_0^\infty \left[-\frac{d\Gamma_{ii}(t)}{dt} \right] [\exp -i\omega t] \, dt \qquad (10)$$

where $\overline{f}(i\omega)$ depends upon the details of local field considerations
(16,17,31). For approximate work $\overline{f}(i\omega)$ may be taken to be unity
(17,18) and we see Equation 10 and the macroscopic relation
Equation 1 are very similar in form. A single decay in time
[Figure 10a, curve (i)], Equation 10, leads to a single loss process, as indicated schematically in Figure 10b, curve (i). Similarly, the two well-separated decays in time [Figure 10a, curve (ii)]
lead to two well-separated loss curves as indicated in Figure 10b,
curve (ii). If we write $\Gamma_{ii}(t) = \exp - t/\tau$, then Equation 10 gives
the usual single relaxation time expression if $\overline{f}(i\omega) = 1$. Note
that the exponential decay is not an acceptable correlation function.
One feature that is unacceptable is that its derivative $[d\Gamma_{ii}(t)/dt]$
at $t = 0$ is not zero. This arises due to the neglect of inertia
terms, and has received much attention for non-polymeric systems
(32).

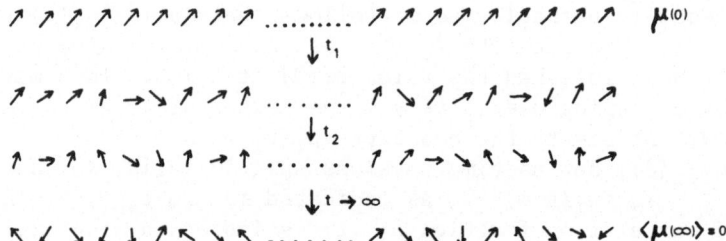

Figure 9.   Schematic illustration of the randomization of a dipolar
            vector via Brownian motion.

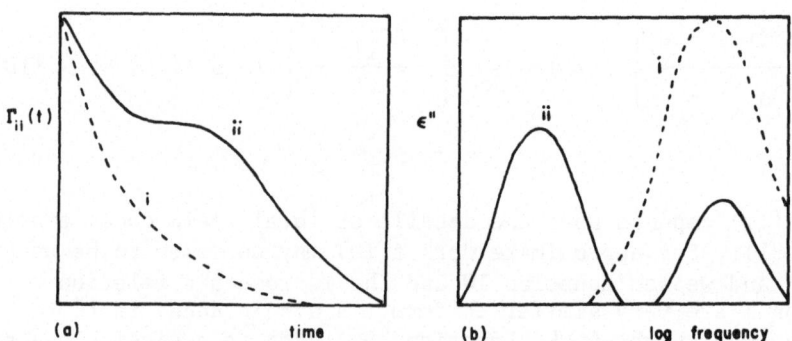

Figure 10a. $\Gamma_{ii}(t)$ (schematic) against time for (i) a single
             decay and (ii) two well-separated decays.

     10b.   The loss factor (schematic) against log frequency for
            (i) the single decay and (ii) the two well-separated
            decays of Figure 10a.

Thus the dielectric relaxation of small molecule systems in the liquid state may be considered in terms of the dipole correlation function and this has been done for various systems (16,17, 33), including internal reorientation processes in small molecules (34). The essential difference between the small non-associated molecules and polymer molecules lies in the presence of the dipole orientation correlations along a polymer chain. The equilibrium theory is well known (1,18) and $(\varepsilon_0 - \varepsilon_\infty)$ is proportional to $N<\mu^2>$, where N is the number of dipoles per unit volume. For a homopolymer chain containing only one kind of dipole along the chain (e.g. polyvinyl chloride), $<\mu^2>$ may be written as (1,18)

$$<\mu^2> = \mu^2 + \sum_j <\underset{\sim}{\mu}_i(0)\cdot\underset{\sim}{\mu}_j(0)> \tag{11}$$

$\mu_i(0)$ is the dipole moment of a reference dipole i along the chain, and $\mu_j(0)$ is the dipole moment of a further dipole j contained in the same polymer chain as dipole i, see Figure 11. The sum is taken over all the dipoles which correlate with the reference dipole i. Clearly for flexible polymer chains $<\mu_i(0)\cdot\mu_j(0)>$ decreases with an increase in the separation of the dipoles along the chain. Cook, Watts and Williams (18) have evaluated the magnitude of these cross correlation terms for a series of model polyethers and have found that for flexible chains only the two or three neighboring dipoles give terms which contribute significantly to $<\mu^2>$. The details of equilibrium correlations are considered for a range of polymer chains by Volkenstein (35) and Flory (36).

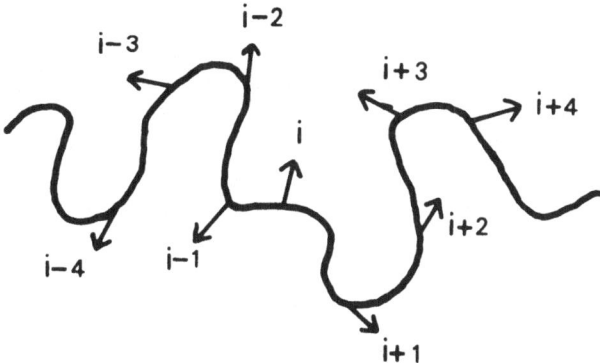

Figure 11.   The arrangement of dipoles along a polymer chain (schematic).

The dynamic theory for the dielectric behavior of dipolar polymers has been outlined by Cook, Watts and Williams (18) using the dipole correlation function $\Lambda(t)$ for a homopolymer chain containing only one kind of dipole.

$$\Lambda(t) = \frac{<\mu_i(0) \cdot \mu_i(t)> + \sum_j <\mu_i(0) \cdot \mu_j(t)>}{<\mu^2>} \tag{12}$$

Equation 12 is seen to be (18) the generalization of the Kirkwood equilibrium theory involving correlations to the dynamic situation. The autocorrelation term $<\mu_i(0) \cdot \mu_i(t)>$ refers to the reorientation of a reference dipole i in the chain, while the $<\mu_i(0) \cdot \mu_j(t)>$ are the dynamic cross correlations between the dipole j and the reference dipole i. The magnitude of these cross correlation terms are $<\mu_i(0) \cdot \mu_j(0)>$. Denoting $<\mu_i(0) \cdot \mu_j(0)>$ as ξ_{ij} say, we may define a time dependent decay function

$$\lambda_{ij}(t) = \frac{<\mu_i(0) \cdot \mu_j(t)>}{<\mu_i(0) \cdot \mu_j(0)>} \tag{13}$$

so that Equation 12 becomes

$$\Lambda(t) = \frac{\xi_{ii}\lambda_{ii}(t) + \sum_j \xi_{ij}\lambda_{ij}(t)}{\xi_{ii} + \sum_j \xi_{ij}} \tag{14}$$

Note $\lambda_{ij}(0) = 1$, $\lambda_{ij}(\infty) = 0$ and ξ_{ij} may be positive or negative for $i \neq j$. The relation between $\Lambda(t)$ and the complex permittivity is made complicated by local field factors, but under certain assumptions we may write (18)

$$\left[\frac{\varepsilon^*(i\omega) - \varepsilon_\infty}{\varepsilon_o - \varepsilon_\infty} \right] p(i\omega) = \int_o^\infty \left[-\frac{d\Lambda(t)}{dt} \right] [\exp -i\omega t] \, dt \tag{15}$$

where p(iω) is given by $[\varepsilon_0(2\varepsilon^*+\varepsilon_\infty)]/[\varepsilon^*(2\varepsilon_0+\varepsilon_\infty)]$. For cases of practical interest, p(iω) may be taken to be unity, so that Equation 15 is very similar in form to Equation 1 above. Equations 11, 12, 14 and 15 relate the observed (macroscopic) complex permittivity to the dipole correlation function $\Lambda(t)$, where $\Lambda(t)$ necessarily involves the dipole correlations along a polymer chain. Equations 12 and 15 may be regarded as a restatement of the dielectric theory for polymers, but where the role of correlations may possibly be assessed (18) in a more obvious manner than in theories which do not employ the correlation function approach.

DISCUSSION

The experimental behavior of the α, β and (αβ) processes with changing temperature and pressure has been outlined above, and the theory of dielectric relaxation has also been outlined. It is now necessary to attempt to explain the α, β and (αβ) processes in terms of the general theory. In general terms, we may explain the α, β and (αβ) processes as follows (14,15,20). At low temperatures the α and β processes occur, where $\Lambda(t)$ is partially relaxed by local motions (as yet undefined) and the remainder is relaxed by Brownian motion. As the temperature is raised, the frequency of the Brownian motions (α process) increases more rapidly than that for the β process, so that at higher temperatures there is a tendency for the two to cross, as indicated schematically in Figure 12. However, when the α process becomes faster than the extrapolated β process, then all of the available $<\mu^2>$ will be relaxed by the α process. This means that the (αβ) process has the same mechanism as the α process - i.e. large scale Brownian motions of chains (13-15). There is a difference between their magnitudes since the α relaxation relaxes only a part of $<\mu^2>$, whereas all of $<\mu^2>$ is relaxed by the (αβ) process. This explanation suffices to account for the general features of the α, β and (αβ) processes given above. Thus (i) the α and β processes coalesce to form the (αβ) process at high temperature and (ii) since the total magnitude of the relaxation is shared between α and β processes at lower temperatures, it necessarily follows that if pressure decreases the magnitude of the β process, there must be a corresponding increase in the magnitude of the α process. This is the experimental result for the polyalkyl methacrylates (8,9,11,13-15), as was indicated above, and is particularly striking for poly-n-butyl methacrylate (13) where the relative magnitudes of the α and β processes are reversed by application of pressure.

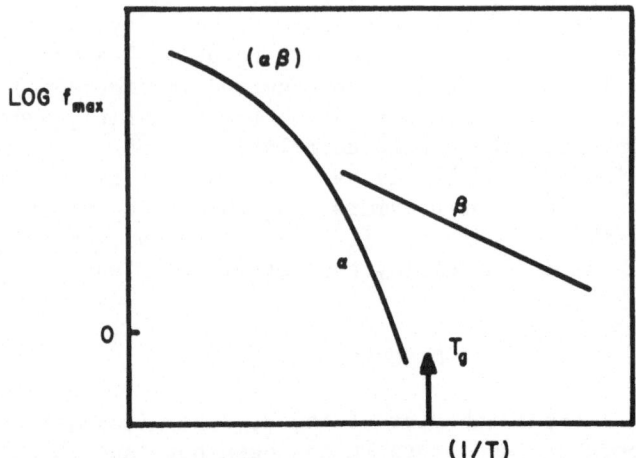

Figure 12. log f_{max} against $(1/T)$ (schematic) for the α, β and ($\alpha\beta$) relaxation process.

The shape of the dielectric α relaxation may be considered in terms of Equations 14 and 15. The shape of the plots of $(\varepsilon'-\varepsilon_\infty)/(\varepsilon_0-\varepsilon_\infty)$ and $(\varepsilon''/\varepsilon''_{max})$ against $\log(f/f_{max})$ are determined by the ξ_{ij} and $\lambda_{ij}(t)$ in Equation 14. Consider the experimental results for certain copolymers (37,38) where the shape of the α relaxation is independent of the dipole concentration in the chains. This suggests that in these cases the autocorrelation terms $\lambda_{ii}(t)$ and cross correlation terms $\lambda_{ij}(t)$ have approximately the same dependence on time. Thus if $\lambda_{ii}(t) = \lambda_{ij}(t) = \lambda(t)$ say, then from Equation 14, $\Lambda(t) = \lambda(t)$. This is of special interest since it would mean that the cross correlations are of importance in determining the magnitude of the relaxation process - via $<\mu^2>$, but may not affect the shape of the dispersion and absorption process - since auto and cross correlation terms may have the same time dependence. This may also be the case for the ($\alpha\beta$) process. The exact shape of the α and ($\alpha\beta$) processes (for examples see Figures 2 and 6) is well represented by the empirical macroscopic decay function Equation 2 with $\bar{\beta}$ as a parameter. As discussed above, such behavior may be expressed in terms of a distribution of relaxation times Equation 3. However, we would suggest that this may not be the case and instead would suggest that the cooperative molecular motions involved in α and β processes naturally lead to a 'nonexponential' correlation function $\Lambda(t)$ in time. It is usually considered that the apparent 'distribution of relaxation times' in polymers arises due to the modes of motion of the <u>polymer</u> chain. Clearly the chain connectivity would play an important role in determining such a distribution. Johari and Goldstein have observed both α and β dielectric processes in a number of small

molecule glass forming systems (42) and have remarked on the similarity between these systems and polymer systems. We have recently
obtained (39) very broad, non-symmetrical loss curves for the reorientation of small polar molecules in a supercooled liquid solvent. Figure 13 shows ε'' against log frequency for nitrobenzene
and for anthrone in the essentially non-polar solvent ortho-
terphenyl (39). The loss process is nearly all due to the solute
molecule, and note the low frequency at which the process is
observed. Since there is no permanent association between molecules in such simple systems, clearly the broad <u>asymmetric</u> loss
curves and their low frequency are due to the cooperative motions
of the small molecules. These cooperative motions give a non-
exponential correlation function which may be approximately represented as the empirical decay function Equation 2. The similarity
between the curves of Figure 13 for the small molecules and those
for the α and $(\alpha\beta)$ relaxations in polymers (Figures 2 and 8 are
examples) is so striking that we are led to the following conclusion. It would appear that chain connectivity, although clearly
a factor in allowing chain segments to undergo Brownian motion,
is not the major factor leading to the broad loss curves for α and
$(\alpha\beta)$ relaxations. The major factor is that chain segments relax
in cooperation with their environment, and it is this cooperative
factor which leads to a low frequency process, of high apparent
activation energy and having a 'non-exponential' correlation function in linear time. This has implications with regard to the
$Q_p(T,P)$, $\Delta V(T,P)$ and $Q_v(T,V)$ values for α and $(\alpha\beta)$ relaxations
(see Table I). Since the processes are cooperative, it is not
possible to give a simple physical interpretation for these apparent
activition quantities. Thus $Q_p(T,P)$, for example, should not be
interpreted as the average barrier a unit has to overcome in reorientation. It is now necessary to consider the mechanism for the
dielectric β process. Williams and Watts (14,15,20) simplified the
general theory, Equations 12 and 15, by attempting to generate α,
β and $(\alpha\beta)$ processes in terms of the autocorrelation function $\Gamma_{ii}(t)$
(see Equation 10). They assumed that a reference dipole i may find
itself in a number of different environments. For a given environment s say, it is assumed that the dipole may be <u>partially</u> relaxed
by local motions (β process). The environments may themselves be
relaxed by microbrownian motions [α process, $(\alpha\beta)$ process]. If
the dipole decay function for the α process is $\phi_\alpha(t)$ and the dipole
decay function for the β process in environment s is $\phi_{\beta_s}(t)$ then
(14,15,20)

$$\Gamma_{ii}(t) = \phi_\alpha(t) \left[\sum_s P_s q_{\alpha_s} + \sum_s P_s q_{\beta_s} \phi_{\beta_s}(t) \right] \qquad (16)$$

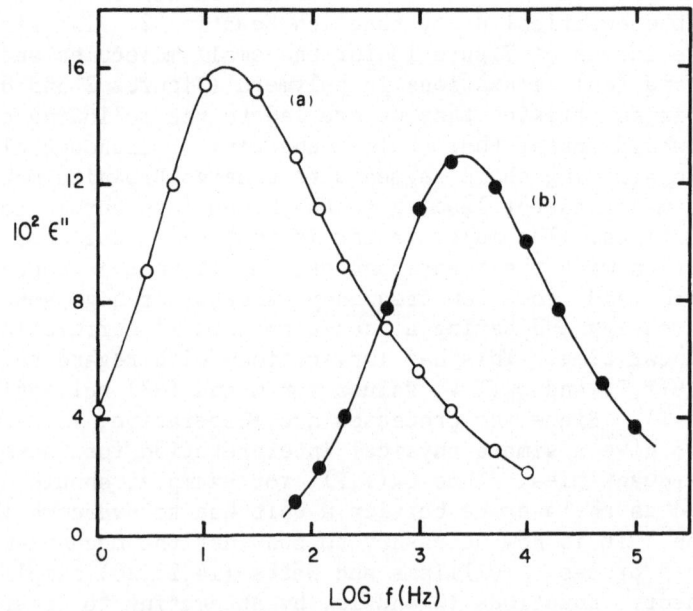

Figure 13. ε'' against log frequency for (a) 3.14 percent (w/w)
of nitrobenzene in ortho-terphenyl at $-21.3°C$, and
(b) 4.17 percent (w/w) of anthrone in ortho-terphenyl
at $-6.2°C$.

p_s is the probability of finding the dipole in environment 's' and $q_{\alpha_s} = 1 - q_{\beta_s} = [<\underset{\sim}{\mu}>_s^2/\mu^2]$ where $<\underset{\sim}{\mu}>_s$ is the net dipole moment obtained for environment s by averaging over a time scale long compared with the local motions but short compared with the Brownian motions. The sum extends over all the available environments. This simple model predicts: (i) at low temperatures $\phi_{\beta_s}(t)$ decays to zero faster than $\phi_\alpha(t)$ so that two separate relaxation regions will be obtained. The higher frequency process is characterized by the $\phi_{\beta_s}(t)$, and, since there will be a range of these, with their effective weighting factors $p_s q_{\beta_s}$, then a broad β loss process will result. The lower frequency process will be due to the cooperative Brownian motions, characterized by $\phi_\alpha(t)$. The relative magnitude of the overall β process to that of the α process will be

$$\sum_s p_s q_{\beta_s} / [\sum_s p_s q_{\alpha_s}]$$

(ii) At high temperatures the local environments relax so quickly that the local motions are not detected, i.e. $\phi_\alpha(t)$ decays to zero far faster than the $\phi_{\beta_s}(t)$ decays to zero. Thus $\Gamma_{ii}(t)$ in Equation 16 is equal to $\phi_\alpha(t)$ and the dipole is randomized by the cooperative (Brownian) motions, giving an α process only. This process is called $(\alpha\beta)$ to distinguish it from the α process at lower temperatures since it carries with it all the available magnitude. This simple model is consistent with Figure 12, which has been discussed above with regard to the α and $(\alpha\beta)$ processes. The question that now arises is what are the local motions which give rise to the β process? Williams and Watts (14,15,20) suggest that in flexible side group polymers the local motion may involve the reorientation of the side groups while in the other polymers the local motion is restricted to relatively free environments. However, in both cases it is the local environment which is a key factor in allowing local relaxation to occur, and this provides the common factor for the processes in the two types of polymer. Note that Johari and Goldstein (42) have observed a very broad β process in a number of small molecule glass forming systems which appears to be quite similar to the β process observed in polymers where the dipoles are rigidly attached to the main chain.

SUMMARY

The effect of temperature and pressure on the dielectric α, β and $(\alpha\beta)$ relaxations in amorphous polymers has been discussed with the aid of the experimental results for polyethyl acrylate, polyvinyl chloride and a polynonyl methacrylate. The theory of dielectric relaxation in polymers has been outlined using the dipole correlation function approach, and its application to the α, β and $(\alpha\beta)$ processes is discussed in general terms and with the aid of a simple model for multiple relaxations. It is suggested that the α and $(\alpha\beta)$ dielectric relaxations are due to the large scale Brownian motions of chains and that those motions naturally lead to a dipole correlation function which is non-exponential in linear time and hence to broad loss curves in the frequency domain. The observation of very similar behavior in small molecule glass forming systems (39,42) would suggest that chain connectivity is not the major factor involved in the non-exponential form for the dipole correlation function and the large apparent activation energy for the α and $(\alpha\beta)$ processes. It would appear that the major factor is that the dipole moves cooperatively with its environment. It is suggested that the β process involves the partial reorientations of dipoles in a range of local environment and a distinction should be made between the β process for polymers containing flexible dipolar side groups and those which contain dipoles rigidly attached to the main chain.

ACKNOWLEDGMENT

The authors wish to thank Mr. P. J. Hains for material prior to publication.

REFERENCES

1. N. G. McCrum, B. E. Read and G. Williams, ANELASTIC AND DIELECTRIC EFFECTS IN POLYMERIC SOLIDS, John Wiley, New York and London, 1967.
2. Y. Ishida, J. Polymer Sci. A-2, 7, 1835 (1969).
3. J. Koppelmann and J. Gielessen, Z. Elektrochem. 65, 689 (1961).
4. J. M. O'Reilly, J. Polymer Sci. 59, 429 (1962).
5. P. Heydemann, Koll. Z. 195, 122 (1964).
6. G. Williams, Trans. Faraday Soc. 60, 1548, 1556 (1964).
7. G. Williams, Trans. Faraday Soc. 61, 1564 (1965).
8. G. Williams and D. A. Edwards, Trans. Faraday Soc. 62, 1329 (1966).
9. G. Williams in MOLECULAR RELAXATION PROCESSES, The Chemical Society Special Publ. No. 20, p. 21, The Chemical and Academic Press, 1966.
10. G. Williams, Trans. Faraday Soc. 62, 1321 (1966).

11. G. Williams, Trans. Faraday Soc. 62, 2091 (1966).
12. S. Saito, H. Sasabe, I. Nakajima and K. Yada, J. Polymer Sci. A-2, 6, 1297 (1968).
13. H. Sasabe and S. Saito, J. Polymer Sci. A-2, 6, 1401 (1968).
14. G. Williams and D. C. Watts in NMR, BASIC PRINCIPLES AND PROGRESS, Volume 4, Springer Verlag, Heidelberg, 1971, pp. 271-285.
15. G. Williams and D. C. Watts, Polymer Preprints (American Chemical Society) 12, No. 1, 79 (1971).
16. S. H. Glarum, J. Chem. Phys. 33, 1371 (1960).
17. R. H. Cole, J. Chem. Phys. 42, 637 (1965).
18. M. Cook, D. C. Watts and G. Williams, Trans. Faraday Soc. 66, 2503 (1970).
19. G. Williams and D. C. Watts, Trans. Faraday Soc., 1971, in press.
20. G. Williams and D. C. Watts, Trans. Faraday Soc., 1971, in press.
21. G. Williams and D. C. Watts, Trans. Faraday Soc. 66, 80 (1970).
22. G. Williams, D. C. Watts, S. B. Dev and A. M. North, Trans. Faraday Soc., 1971, in press.
23. D. W. Davidson and R. H. Cole, J. Chem. Phys. 18, 417 (1950).
24. D. W. Davidson and R. H. Cole, J. Chem. Phys. 19, 1484 (1951).
25. D. W. Davidson, Canad. J. Chem. 31, 571 (1961).
26. J. D. Ferry, VISCOELASTIC PROPERTIES OF POLYMERS, Second Ed., John Wiley, New York, 1971.
27. J. D. Hoffman, G. Williams and E. Passaglia, J. Polymer Sci. C, 14, 173 (1966).
28. R. Kubo, J. Phys. Soc. (Japan) 12, 570 (1957).
29. R. Kubo, LECT. IN THEOR. PHYS., Volume 1, Interscience, New York, 1961, p. 120.
30. R. Zwanzig, Ann. Rev. Phys. Chem. 16, 67 (1965).
31. T. W. Nee and R. Zwanzig, J. Chem. Phys. 52, 6353 (1970).
32. G. Birnbaum and E. R. Cohen, J. Chem. Phys. 53, 2885 (1970).
33. S. H. Glarum, J. Chem. Phys. 33, 639 (1960).
34. G. Williams, Trans. Faraday Soc. 64, 1219, 1934 (1968).
35. M. V. Volkenstein, CONFIGURATIONAL STATISTICS OF POLYMERIC CHAINS, Interscience, New York, 1963.
36. P. J. Flory, STATISTICAL MECHANICS OF CHAIN MOLECULES, Interscience, New York, 1969.
37. J. Leffingwell and F. Bueche, J. Appl. Phys. 39, 5910 (1968).
38. G. P. Mikhailov and L. V. Krasner, Vysokomol. Soed. 5, 144, 151 (1963).
39. G. Williams and P. J. Hains, manuscript in preparation.
40. Y. Ishida, M. Matsuo and K. Yamafuji, Kolloid Z. 180, 108 (1962).
41. A. J. Bur, Ph.D. Thesis, Pennsylvania State University, 1962, p. 56, Figure 11.
42. G. P. Johari and M. Goldstein, J. Chem. Phys. 53, 2372 (1970).

Figures 1 to 8 appears in Polymer Preprints 12, No. 1, March, 1970, page 79 et. seq. and are reprinted by permission of the copyright owner, the American Chemical Society.

Figure 1 and 3 appeared also in G. Williams and D. C. Watts, "Natural and Synthetic High Polymers," NMR, BASIC PRINCIPLES AND PROGRESS, Volume 4, Springer Verlag, Heidelberg, 1971, pp. 271-285.

Figures 4 to 8, inclusive, appeared in the Transactions of the Faraday Society, 1971, in press, and we thank the Council of the Faraday Society for permission to reproduce them.

A LINEARIZED CHAIN MODEL FOR DIELECTRIC LOSS IN POLYMERS

Mary Barkley Clark* and Bruno H. Zimm

Department of Chemistry, Revelle College, University of

California (San Diego), La Jolla, California

ABSTRACT

In order to treat dielectric dispersion arising from the
motions of main-chain dipoles perpendicular to the polymer chain
axis, the polymer molecule is approximated by a linear mechanical
model consisting of a chain of torsional springs, dashpots, and
beads bearing dipoles. The equations of rotational motion can be
solved analytically for this model. We derive an expression for
the correlation length, a measure of the extent to which rotational
motion is correlated along the chain. We obtain equations for the
dielectric dispersion.

INTRODUCTION

Anomalous dielectric dispersion in liquids containing polar
molecules is explained by the classical theory of Debye (1,2).
However, the corresponding phenomenon in high polymers is less
clearly understood from a theoretical viewpoint (3). In polymers,
the loss peak is frequently asymmetric, the peak height is depressed,
and the dispersion extends over a wider frequency range than observed
with small molecules. Relaxation times also are longer, varying
from 10^{-4} to 10^{-11} sec, in contrast to the times characteristic of
small polar molecules of 10^{-11} to 10^{-12} sec. In some cases, several
distinct dispersion regions are seen.

*Present Address: Salk Institute for Biological Studies, La Jolla,
California.

Kirkwood and Fuoss (4) presented an early theoretical treatment of dielectric loss in polymers. Their model had a high degree of verisimilitude, but in the mathematical development, which was otherwise quite elegant, they were forced into pre-averaging techniques that made the accuracy of their results questionable.

In this paper, we suggest a new approach to the problem of anomalous dispersion in polymers. Like Kirkwood and Fuoss, we treat diffusion in multi-dimensional chain space, but we have simplified the model to a point where the dynamical equations are compatible with exact techniques of solution.

The model that we use here is closely related to the damped torsional oscillator (DTO) model of Tobolsky and coworkers (5,6); in fact, we were led to this model by discussions with Drs. Tobolsky and Du Pre. We depart from the DTO model in two ways: (1) we ignore the inertial terms for reasons discussed later, and (2) we introduce a second damping element (ρ' in Figure 1b) in order to allow complete rotation around bonds. Variation of the ρ' element allows the symmetry of the loss peak to be more widely varied. We feel that it also introduces a greater degree of realism into the model. Moreover, there are significant differences in the mathematical formulations of the DTO and of our model; for example, we have found it desirable to go beyond the mere calculation of the normal frequencies.

Work and Fujita (7) and Anderson (8) have discussed the polarization behavior of the one-dimensional Ising lattice as a model for the dielectric relaxation of a polymer chain. Orwoll and Stockmayer (9) have also developed a somewhat different version of the dynamic Ising model. The relation of these models to our work is not clear to us. We shall make some superficial comparisons at the end of the second paper.

As has been discussed previously (3), the dielectric behavior of polymer chains depends greatly on the geometry of the attachment of the dipole moments to the chain, i.e. whether the dipoles have components parallel to or perpendicular to the chain axis. When the dipoles have moments parallel to the chain axis, their behavior can be described satisfactorily by the bead-spring model of Rouse (3,10,11,12). In this case, the dipoles have non-vanishing average projections on the chain end-to-end vector, so that the total dipole moment of the molecule is correlated with the end-to-end vector. As is well known, the relaxation behavior of the latter vector can be described by a set of normal coordinates whose relaxation times vary markedly with the number of segments in the chain. The dielectric relaxation of chains with "parallel" dipole moments shows the same size-dependent relaxation times. The theory for this case is already well worked out.

On the other hand, the case of "perpendicular" dipoles, exemplified by the substituted vinyl polymers, for instance, has never been treated theoretically in a satisfactory way. The motions of such dipoles do not correlate well with those of the end-to-end vector. Experimentally it is found that the relaxation times are not usually dependent on the number of segments in the chain, in contrast to the "parallel" case.

We hope that the theory presented in this paper, even though it treats a highly idealized model of perpendicular dipoles, will help to throw light on the differences in behavior between the two cases. The development is very different in structure from that associated with the bead-spring model. In particular, the collective, or normal, motions so prominent in the Rouse theory appear in our treatment only as a means of facilitating the computations and not as objects of imposing physical significance. What appears in their place, as the central concept in the theory, is the correlation length for angular velocity, a measure of the extent to which the motions of nearby dipoles are correlated. This feature is completely foreign to the bead-spring model. Its presence in our model leads to a ready explanation of the lack of dependence of relaxation times on chain length, since for long chains the total chain length is many times greater than the correlation length, and the dipolar motions are affected only by their immediate environment. The correlation length concept also provides a simple rationalization of the diverse types of relaxation behavior shown by our model when the parameters are varied, as will be discussed in detail later.

THE MODEL

The response of a polar molecule to an electric field was described by Debye (1) as rotational diffusion retarded by the viscous resistance of the surrounding solvent. In a chain polymer, the individual dipoles are further constrained by the interactions with their neighbors through the intervening skeletal bonds. The possible motions are effectively reduced, at least in chains of carbon and oxygen atoms, to rotations about the bonds. These internal rotations are opposed by a hindering potential, usually with three minima.

We propose a simple method of treating bond rotations in terms of a roughly equivalent mechanical model (the Maxwell model), which has long been used in the theory of viscoelasticity. The Maxwell model contains two linear elements: a spring (the energy storage element) and a "dashpot" (the frictional or energy dissipative element). The spring allows the small-amplitude torsional oscillations of bond angles near the minima in the potential functions, and the dashpot allows the occasional large-amplitude rotations

that take the bond angle over the rotational barriers from one
minimum to another.

The formal equivalence of relaxation processes arising from
barrier-hopping between discrete states and from viscous retarda-
tion in a continuous medium has been recognized by numerous authors
(1,13). The rationale for relating passage over barriers to rotation of a dashpot is reviewed here briefly. Consider an ensemble
of single beads bearing dipoles at constant temperature and hence
subject to Brownian motion. The dipoles are constrained to rotate
about one axis only. We shall compare the two mechanisms for
internal rotation: (1) the barrier or discrete model and (2) the
dashpot or continuum model. For simplicity, we shall consider the
barrier model to have three equivalent orientational sites separated
by barriers of energy ΔF^+ per mole. At equilibrium, the dipoles
in the barrier model will be equally distributed among the three
sites. For the dashpot model, we find a uniform distribution over
all angles. Under the influence of a static electric field applied
perpendicular to the axis of rotation, both models lead to a Boltzmann distribution of orientations, biased along the direction of
the applied field and proportional to the field strength.

Moreover, the dynamic response to a fluctuating electric field
or to abrupt removal of a perturbing field exhibits first-order
kinetics for both the barrier and the dashpot models, for fields
small compared kT, where k is Boltzmann's constant and T is absolute
temperature. Hoffman (14) has analyzed the relaxation process for
a single-axis rotor having three equivalent sites, and finds that
it is exponential in time. He obtains a single relaxation time τ_H,

$$\tau_H = \frac{1}{3k_r}$$

where k_r is the rate constant for passage over the barrier. According to Eyring's absolute reaction rate theory (15),

$$k_r = \frac{kT}{h} e^{-\Delta F^+/RT} \tag{1}$$

where h is Planck's constant, ΔF^+ is the barrier height per mole,
and R is the gas constant. Similarly, for the continuum model,
Debye (1) obtains exponential relaxation with time constant τ_D,

$$\tau_D = \frac{\rho'}{kT}$$

for a one-dimensional treatment. Here ρ' is the frictional coefficient of the dashpot.

In summary, passage over barriers and rotation of a dashpot display similar properties. For weak applied fields, both models show a displacement of the equilibrium linear in the field strength, and both obey first-order kinetics. Therefore, to a good approximation, we can assume that the two models adequately describe the same phenomenon and that the relaxation times τ_H and τ_D for the two models are equal. Thus, we can write,

$$k_r = \frac{kT}{3\rho'} \tag{2}$$

We turn now to the treatment of the dynamical equations of the model chosen to represent torsional oscillations and internal rotation. A torsional spring of spring constant c and a rotational dashpot of frictional coefficient ρ' are connected in series; the resulting mechanical model is a Maxwell element (16,17) (Figure 1a). Imagine for the moment a bead attached at one end of the Maxwell element; the other end of the Maxwell element is fixed in an immovable wall. Since the relaxation processes in dielectric experiments are observed at radio and microwave frequencies, and since inertial motion is normally seen in the infrared, those contributions to the equations of motion from inertial terms may be ignored. The mutual torque $L_{b,w}$ acting between the bead and the wall due to motion of the dashpot is

$$L_{b,w} = \rho' v_d$$

where v_d is the angular velocity $\dot{\theta}_d$ of the dashpot. The torque $L_{b,w}$ is also the linear restoring torque arising from the torsional spring,

$$L_{b,w} = c\theta_s$$

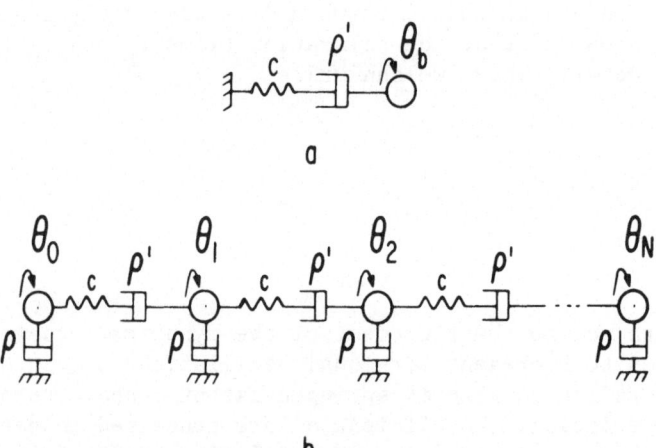

Figure 1.   Schematic Drawings of Model.

   (a) Maxwell element. Rotational dashpot of frictional
        coefficient ρ' represents barriers to internal
        rotation. Torsional spring of spring constant c
        represents parabolic potential minimum.

   (b) Chain of Maxwell elements. Rotational dashpot of
        frictional coefficient ρ represents viscous drag
        exerted by fluid.

where θ_s is the angular displacement of the spring. If we take
the time derivative of the latter expression and combine it with
the former, we obtain the dynamical equations of the Maxwell
element:

$$v_b = v_s + v_d$$

$$= \frac{1}{\rho'} L_{b,w} + \frac{1}{c} \dot{L}_{b,w} \tag{3}$$

Here v_s is velocity of the spring; v_b is the velocity of the bead.
We shall use the Maxwell element to construct a simple mechanical
model of a polymer molecule.

Consider a linear array of (N+1) beads each bearing a dipole,
numbered from 0 to N. The beads are a fixed, regular distance
apart. The angle θ_ℓ designates the position of the ℓth dipole
about the chain axis with respect to an arbitrary reference coordinate. Note that these beads represent individual repeating segments having permanent dipole moments, and hence should be thought
of as being the size of one or so monomer units. The chain is
immersed in a viscous fluid, and the polymer-solvent system is
assumed to be at constant temperature. For dilute solutions,
polymer intermolecular interactions are unimportant. As mentioned
above, we shall neglect inertial terms. The (N+1) beads are sup-
posed linked by N identical Maxwell elements (Figure 1b). The
dipoles rotate about the chain axis. Their motion is constrained
by internal dashpots of frictional coefficient ρ', torsional springs
of spring constant c, and external dashpots of frictional coefficient ρ. From Equation 3, we can write down dynamical equations
for the Maxwell elements immediately.

$$v_\ell - v_{\ell+1} = \frac{1}{\rho'} L_{\ell,\ell+1} + \frac{1}{c} \dot{L}_{\ell,\ell+1} \qquad \text{for all } \ell \tag{4}$$

Here v_ℓ is the angular velocity $\dot{\theta}_\ell$ of the ℓth bead; $L_{\ell,\ell+1}$ is the
mutual torque acting between the ℓth and $(\ell+1)$th beads. The exter-
nal resistance is equal to the resistance ρ times the velocity $\dot{\theta}_\ell$
of each bead; it represents the viscous drag exerted by the solvent
on the rotating dipole. In terms of molecular properties, the chain
of Maxwell elements substitutes damped rotary motion for internal
rotation between discrete states, and simple harmonic motion for
small oscillations within a given rotational potential minimum.

We now follow a development similar to that of Debye for a single dipole. The motion of the chain is also influenced by Brownian motion. The torque at the ℓth dipole will be the sum of three torques: an external torque T_ℓ, a torque arising from the frictional resistance ρ of the fluid on the ℓth bead, and an effective torque resulting from Brownian motion of the ℓth element.

$$L_{01} = - T_0 - \rho v_0 - kT \frac{\partial \ln\psi}{\partial \theta_0} \tag{5a}$$

$$L_{\ell,\ell+1} - L_{\ell-1,\ell} = - T_\ell - \rho v_\ell - kT \frac{\partial \ln\psi}{\partial \theta_\ell} \qquad 0 < \ell < N \tag{5b}$$

$$-L_{N-1,N} = - T_N - \rho v_N - kT \frac{\partial \ln\psi}{\partial \theta_N} \tag{5c}$$

where $\psi(\theta_0, \theta_1, \ldots, \theta_\ell, \ldots, \theta_N; t)$ is the multivariant time-dependent distribution function of the θ_ℓ's. It is convenient to rewrite Equations 5, expressing the velocities of the beads in terms of the unknown torques $L_{01}, \ldots, L_{\ell,\ell+1}, \ldots, L_{N-1,N}$.

$$\rho v_0 = - T_0 - L_{01} - kT \frac{\partial \ln\psi}{\partial \theta_0} \tag{6a}$$

$$\rho v_\ell = - T_\ell + L_{\ell-1,\ell} - L_{\ell,\ell+1} - kT \frac{\partial \ln\psi}{\partial \theta_\ell} \qquad 0 < \ell < N \tag{6b}$$

$$\rho v_N = - T_N + L_N - kT \frac{\partial \ln\psi}{\partial \theta_N} \tag{6c}$$

The mutual torques are determined by combining Equations 4 and 6, eliminating the velocities, and then solving the resulting set of first-order differential equations. It is straightforward to show that for weak applied fields sinusoidal in time,

$$\frac{\partial L_{\ell,\ell+1}}{\partial t} = i\omega L_{\ell,\ell+1} \qquad \text{for all } \ell \tag{7}$$

Substituting Equation 7 into Equation 4, we obtain the following
expression for the torques,

$$L_{\ell,\ell+1} = \zeta(v_\ell - v_{\ell+1}) \qquad\qquad \text{for all } \ell \qquad (8)$$

where ζ is defined as the complex impedance of the Maxwell element,
that is,

$$\zeta \equiv \frac{1}{1/\rho' + i\omega/c} \qquad\qquad (9)$$

The equations of motion are obtained by putting Equation 8 into
Equations 6:

$$\rho v_0 + \zeta(v_0 - v_1) = -T_0 - kT\frac{\partial \ln\psi}{\partial\theta_0} \qquad\qquad (10a)$$

$$\rho v_\ell + \zeta(-v_{\ell-1} + 2v_\ell - v_{\ell+1}) = -T_\ell - kT\frac{\partial \ln\psi}{\partial\theta_\ell} \qquad 0<\ell<N \qquad (10b)$$

$$\rho v_N + \zeta(-v_{N-1} + v_N) = -T_N - kT\frac{\partial \ln\psi}{\partial\theta_N} \qquad\qquad (10c)$$

At this juncture, it may be useful to point out the limiting
properties of this mechanical model. If we let either ρ' or $c \to 0$,
$\zeta \to 0$ and the problem degenerates to the Debye case. The consequence
is obvious from purely physical arguments. When the impedance to
internal rotation disappears, because either the ρ' dashpots or the
torsional springs are weak enough to permit essentially uncorrelated
motion, each bead rotates independently in the viscous fluid. The
individual beads will exhibit Debye dispersion with the character-
istic relaxation time, ρ/kT. For the case where $\rho' \to \infty$, $\zeta \to c/i\omega$, the
internal dashpots are frozen and one has a chain of torsional
springs. On the other hand, if $c \to \infty$, $\zeta \to \rho'$, one has a chain of dashpots.
Finally, we consider the limit where ρ' and $c \to \infty$ simultaneously. All
internal motion is prevented, and the chain rotates about its long
axis like a rigid rod. Again, we find Debye behavior, but in this

case with relaxation time $(N+1)\rho/kT$. It is clear that there are
two extreme cases: the one of very loose, the other of very tight
coupling between adjacent dipoles. Both approach the Debye result,
but with different time constants. On the basis of this preliminary
discussion, we can make two general statements. First, the cases
of interest from the standpoint of duplicating experiment are those
involving intermediate values of the internal parameters, ρ' and c.
And second, we may suspect that the degree to which motions are
correlated along the chain will bear directly on those aspects of
the problem we seek to understand. Toward this end, then, we shall
attempt to derive a measure of these correlations.

ROTATIONAL CORRELATION LENGTH

To define a correlation length for angular velocity, we apply
an arbitrary torque to one bead and ask at what distance along the
chain the velocity has dropped to the fraction 1/e of the velocity
of the perturbed bead; this distance we call the rotational correlation length. The calculation is done for long chains in order
that the correlation length may be independent of end effects. The
correlation length thus estimated will be used to interpret dielectric dispersion curves.

Imagine a chain of $(N+1)$ Maxwell elements suspended in a viscous liquid. In this derivation, we omit Brownian motion. The
zeroth bead is driven by a weak sinusoidally varying torque $Le^{i\omega\tau}$.
There are no external torques acting on the remaining N beads of
the chain. The equations of motion for this hypothetical situation
can be written down directly (compare Equations 10):

$$\rho v_0 + \zeta(v_0 - v_1) = Le^{i\omega t} \qquad\qquad (11a)$$

$$\rho v_\ell + \zeta(-v_{\ell-1} + 2v_\ell - v_{\ell+1}) = 0 \qquad 0 < \ell < N \qquad (11b)$$

$$\rho v_N + \zeta(-v_{N-1} + v_N) = 0 \qquad\qquad (11c)$$

It is convenient to put these equations into vector notation.
Let $\underset{\sim}{v}$ be the $(N+1)$-dimensional column vector of the velocities,
whose components are v_ℓ, $\ell = 0, 1, \ldots, N$. Let $\underset{\sim}{A}$ be the $(N+1) \times (N+1)$
symmetric matrix

$$A \equiv \begin{bmatrix} 1 & -1 & 0 & 0 \ldots 0 & 0 & 0 \\ -1 & 2 & -1 & 0 \ldots 0 & 0 & 0 \\ 0 & -1 & 2 & -1 \ldots 0 & 0 & 0 \\ 0 & 0 & -1 & 2 \ldots 0 & 0 & 0 \\ \cdot & & & & & \\ \cdot & & & & & \\ \cdot & & & & & \\ 0 & 0 & 0 & 0 \ldots -1 & 2 & -1 \\ 0 & 0 & 0 & 0 \ldots 0 & -1 & 1 \end{bmatrix} \qquad (12)$$

And let L be the (N+1)-dimensional column vector,

$$L \equiv \begin{bmatrix} Le^{i\omega t} \\ 0 \\ \cdot \\ \cdot \\ \cdot \\ 0 \end{bmatrix} \qquad (13)$$

We can write the set of Equations 11 as

$$[I + \frac{\zeta}{\rho} A]v = \frac{1}{\rho} L \qquad (14)$$

Where I is the identity matrix, and

$$\frac{\zeta}{\rho} = \frac{1}{r + i\omega\rho/c} \qquad (15)$$

with

$$r \equiv \frac{\rho}{\rho'} \tag{16}$$

The formal solution for the velocities is then:

$$\underset{\sim}{v} = \frac{1}{\rho} [1 + \frac{\zeta}{\rho} \underset{\sim}{A}]^{-1} \underset{\sim}{L} \tag{17}$$

At this time, we shall digress briefly to invert the matrix $[\underset{\sim}{I} + \frac{\zeta}{\rho} \underset{\sim}{A}]$. Define the eigenvalues of $[\underset{\sim}{I} + \frac{\zeta}{\rho} \underset{\sim}{A}]^{-1}$ as χ_k, where $k=0, 1, 2, \ldots, N$; $\underset{\sim}{X}$ is the diagonal matrix composed of the $(N+1)$ χ_k's. The matrix $\underset{\sim}{A}$ can be diagonalized by a similarity transformation $\underset{\sim}{Q}$.

$$\underset{\sim}{Q}^{-1}\underset{\sim}{Q} = \underset{\sim}{I} = \underset{\sim}{Q}\underset{\sim}{Q}^{-1}$$
$$\underset{\sim}{Q}^{-1}\underset{\sim}{A}\underset{\sim}{Q} = \underset{\sim}{M} \tag{18}$$

Here $\underset{\sim}{M}$ is a diagonal matrix whose $(N+1)$ diagonal elements μ_k (18) are the eigenvalues of the $\underset{\sim}{A}$ matrix,

$$\mu_k = 2(1-\cos \frac{k\pi}{N+1}) \qquad k=0, 1, 2, \ldots, N \tag{19}$$

Since $\underset{\sim}{A}$ is symmetric, $\underset{\sim}{Q}$ is an orthogonal matrix,

$$\underset{\sim}{Q}^{-1} = \underset{\sim}{Q}^{T}$$

Moreover, Q is formed by the set of $(N+1)$ column vectors which are the $(N+1)$-dimensional eigenvectors α_k of A. Let $\alpha_k(\ell)$ be the ℓth component of the kth eigenvector. Then the normalized eigenvectors (18) of the A matrix are

$$\alpha_k(\ell) = \left(\frac{2}{N+1}\right)^{1/2} \begin{cases} \left(\frac{1}{2}\right)^{1/2} & k = 0 \\ \\ \cos \frac{k\pi}{N+1} \left(\ell + \frac{1}{2}\right) & k > 0 \end{cases} \tag{20}$$

$$\ell = 0, 1, 2, \ldots, N$$

Now $[I + \frac{\zeta}{\rho} A]^{-1}$ can be diagonalized by the same transformation:

$$Q^T [I + \frac{\zeta}{\rho} A]^{-1} Q = X \tag{21}$$

or,

$$[I + \frac{\zeta}{\rho} A]^{-1} = QXQ^T \tag{22}$$

To show this, invert Equation 21 and apply Equation 18; the result is diagonal:

$$X^{-1} = I + \frac{\zeta}{\rho} M \tag{23}$$

Hence, the inverse matrix $[I + \frac{\zeta}{\rho} A]^{-1}$ is in fact given by Equation 22, where X_k is

$$X_k = \frac{1}{1 + \frac{\zeta}{\rho} \mu_k} \qquad k = 0, 1, 2, \ldots, N \tag{24}$$

Returning to the expression for the velocity vector $\underset{\sim}{v}$ (Equation 17), we can use Equation 22 to get

$$\underset{\sim}{v} = \frac{1}{\rho} \underset{\sim}{Q} \underset{\sim}{X} \underset{\sim}{Q}^T \underset{\sim}{L} \tag{25}$$

By carrying out the matrix multiplications in the last equation, and taking the ratio of the velocities of the ℓth and 0th beads, we find:

$$\frac{v_\ell}{v_0} = \frac{\displaystyle\sum_{j=0}^{N} X_j \, \alpha_j(\ell) \, \alpha_j(0)}{\displaystyle\sum_{j=0}^{N} X_j \, \alpha_j^{\,2}(0)} \tag{26}$$

$$\ell = 0, 1, 2, \ldots, N$$

We note several properties of these equations describing the velocities of the beads. First, the velocities are complex functions; the real and imaginary parts represent motion in phase and out of phase with the external torque, respectively. The amplitude of the motion is given by the magnitude of this complex function. Second, Equation 26 exhibits the appropriate properties for the special cases. If $c \to \infty$,

$$X_k \to \frac{1}{1 + \mu_k/r}$$

and one obtains the solution for a chain of dashpots only. In this case the velocity is real in accord with expectation, since the only motion possible in the absence of inertial term is along the direction of the external field. In addition, one finds the expected behavior for the two limiting cases. If $\rho' \to 0$ or if $c \to 0$, then $\zeta \to 0$ and

$$X_k \to 1, \quad \text{for all } k.$$

Using the orthogonality properties of the eigenvectors, it is then easy to show that

$$v_0 = \frac{Le^{i\omega t}}{\rho}$$

and

$$v_\ell = 0 \qquad \text{for } \ell > 0$$

This is also the anticipated result, since there is no mechanism for transmitting torque along the chain. On the other hand, if $\rho' \to \infty$ and $c \to \infty$, then $\zeta \to \infty$ and we find that

$$\chi_0 \to 1,$$

$$\chi_k \to 0, \qquad \text{for } k > 0$$

From Equations 20 and 25 it follows that

$$v_0 = \frac{Le^{i\omega\tau}}{(N+1)\rho}$$

and

$$v_\ell = v_0$$

for all other ℓ. The elements of the rigid chain rotate with a single velocity.

For large N, the summations in Equation 26 can be expressed in closed form. Starting with the eigenvectors from Equation 20 and applying simple trigonometric identities, we write the v_ℓ summation,

$$\sum_{j}^{N} x_j \alpha_j(\ell) \, \alpha_j(0) = \left(\frac{1}{N+1}\right) \, [1 + \sum_{j=1}^{N} x_j \, \cos \frac{j\pi}{N+1} (\ell+1)$$

$$+ \sum_{j=1}^{N} x_j \, \cos \frac{j\pi}{N+1} \, \ell] \tag{27}$$

This summation is evaluated by the following method. Substitute Equation 19 for μ_k into Equation 24 for x_k,

$$x_k = \frac{\rho/\zeta}{(\rho/\zeta+2)} \, [\frac{1}{1 - \dfrac{2\cos k\pi/(N+1)}{(\rho/\zeta+2)}}], \tag{28}$$

and introduce this expression into Equation 27. The Euler-Maclaurin summation formula (19) approximates the discrete sum from 1 to N by a definite integral from 0 to N+1.

$$\sum_{j=1}^{N} x_j \, \cos \frac{j\pi}{N+1} \, n \approx \frac{\rho/\zeta}{(\rho/\zeta+2)} \left(\int_0^{N+1} \frac{1}{1 - \dfrac{2\cos j\pi/(N+1)}{(\rho/\zeta+2)}} \cos \frac{j\pi}{N+1} \, k \, dj \right.$$

$$\left. - \frac{1}{2} \, \frac{1}{1-2/(\rho/\zeta+2)} + \frac{(-1)^n}{1+2/(\rho/\zeta+2)} \right) \quad n = 0, 1, 2, \ldots, N$$

With the variable change $x = j\pi/N+1$, the integral of the above equation is transformed into a known definite integral from 0 to π (20). One obtains,

$$\sum_{j=1}^{N} x_j \, \cos \frac{j\pi}{N+1} \, n \approx \frac{(N+1) \, \rho/\zeta}{\sqrt{(\rho/\zeta+2)^2-4}} \left[\frac{(\rho/\zeta+2) - \sqrt{(\rho/\zeta+2)^2 - 4}}{2} \right]^n$$

$$- \frac{1}{(\rho/\zeta+4)} \begin{cases} (\rho/\zeta+2), & n - \text{even} \\ 2, & n - \text{odd} \end{cases} \tag{29}$$

$$n = 0, 1, 2, \ldots, N$$

Equation 29 is introduced into Equation 27, yielding an approximation in closed form for Equation 26 for large N:

$$\frac{v_\ell}{v_0} = \left[\frac{(\rho/\zeta+2) - \sqrt{(\rho/\zeta+2)^2 - 4}}{2} \right]^\ell \tag{30}$$

It can be shown that the bracketed term on the right-hand side of
Equation 30 lies between 1 and 0, and hence that v_ℓ/v_0 for non-zero
$|\rho/\zeta|$ drops off with increasing ℓ. Let us rewrite Equation 30 for
v_ℓ/v_0 as an explicit exponential function of ℓ,

$$\frac{v_\ell}{v_0} = e^{-\ell y} \tag{31}$$

where

$$y = \ln \left[\frac{2}{(\rho/\zeta+2) - \sqrt{(\rho/\zeta+2)^2 - 4}} \right]$$

$$= - \ln [1+\rho/2\zeta - \sqrt{(1+\rho/2\zeta)^2 - 1}],$$

$$= \text{Arch } (1+\rho/2\zeta) \tag{32}$$

Arch $(1+\rho/2\zeta)$ denotes the principal value of the inverse hyperbolic
cosine of $(1+\rho/2\zeta)$.

We are now in a position to extract a correlation length λ from
Equation 31 by determining the distance along the chain where the
magnitude of the velocity ratio has decayed to $1/e$, that is

$$\text{Re } \lambda y = 1$$

Then λ^{-1} is the real part of y in Equation 32. After considerable manipulation, one uncovers the following general results, with $r = \rho/\rho'$;

$$\lambda^{-1} = \frac{1}{2} \text{Arch} \left[(1+r/2)^2 + (\omega\rho/2c)^2 + \{ (1+r/2)^4 + 2(1+r/2)^2 \right.$$

$$\left. (\omega\rho/2c)^2 + (\omega\rho/2c)^4 - 2(1+r/2)^2 + 2(\omega\rho/2c)^2 + 1 \}^{1/2} \right]$$

$$(33a)$$

It should be mentioned that λ may assume non-integer values. For $|\rho/\zeta| \ll 1$, Equation 33a simplifies considerably (Equation 32),

$$\lambda \simeq \text{Re}(\rho/\zeta)^{-1/2}$$

$$= \left(\frac{2}{r + \sqrt{r^2 + \omega^2\rho^2/c^2}} \right)^{1/2} \qquad (33b)$$

If ρ' and $c \to 0$, $\zeta \to 0$, all correlations vanish and $\lambda \to 0$. For the rigid rod, ρ' and $c \to \infty$, $\zeta \to \infty$, and the correlation length becomes very large. We notice, moreover, that Equations 33 describe the correlation length as a function of the angular frequency ω for intermediate values of the internal parameters ρ' and c. For very low frequencies, such that $\omega\rho/c \ll 1$ and $\omega\rho/c \ll r$, we find that λ becomes independent of both frequency and spring constant,

$$\lambda = \frac{1}{\ln 2 - \ln[(r+2) - \sqrt{(r+2)^2 - 4}]} \qquad (34a)$$

and for $r < 1$ additionally,

$$\lambda \simeq \frac{1}{\sqrt{r}} \qquad (34b)$$

Equations 34 are also the result for the chain of simple dashpots ($c \to \infty$, $\zeta \to \rho'$). Hence at low frequencies, the motion of a chain of Maxwell elements is coordinated to about the same extent as that of a chain of dashpots without springs. At very high frequencies, $\omega\rho/c \gg 1$, and λ approaches zero quite rapidly, with the motion of the beads becoming essentially uncorrelated. We have then a picture in which our chain of Maxwell elements behaves at low frequencies like a chain of beads and dashpots and at high frequencies like a collection of independent beads. In the middle frequency range, the correlation length is changing as an explicit function of frequency from its maximum value given by the chain-of-dashpots-limit of Equation 34 to zero.

DIELECTRIC DISPERSION

We shall now proceed with our treatment of the problem of dielectric relaxation. As before, we assume a chain of (N+1) beads, joined by N Maxwell elements, diffusing in a viscous liquid. We consider the polarization arising from orientation of the permanent dipoles along the direction of the applied sinusoidally varying field. Since the dispersion processes occurring at radio and microwave frequencies have times characteristic of those for molecular motions, we need not include polarization due to deformation of the electronic structure of the molecule. That is to say, the polymer dielectric constant will be that of our model plus the optical value, ε_∞.

Imagine a weak oscillating field $\underset{\sim}{E}e^{i\omega t}$ applied perpendicular to the chain axes. A torque T_ℓ acts on the ℓth dipole and tends to align it along the field. This torque is the vector product of the electric moment with the external field:

$$T_\ell = E\mu e^{i\omega t} \sin\theta_\ell \qquad \text{for all } \ell \qquad (35)$$

where μ is the permanent dipole moment and θ_ℓ is the angle between the ℓth dipole and the external field. The dynamical equations can be written down immediately by combining Equation 35 and Equations 10:

$$\rho\underset{\sim}{v} + \zeta A\underset{\sim}{v} = -E\mu e^{i\omega t}\underset{\sim\sim\sim}{\sin\theta} - kT\frac{\partial \ln\psi}{\underset{\sim\sim}{\partial\theta}} \qquad (36)$$

Here $\sin\theta$ is the (N+1)-dimensional column vector, whose components
are $\sin\tilde{\theta}_{\ell}$, and $\partial/\partial\theta$ is the (N+1)-dimensional differential operator,
whose components are $\partial/\partial\theta_{\ell}$, $\ell=0, 1, \ldots, N$.

Applying the method of the previous section, we derive an
expression for the velocity v from Equation 36,

$$v = - \frac{E\mu}{\rho} e^{i\omega t} QXQ^{T} \cdot \sin\theta - \frac{kT}{\rho} QXQ^{T} \cdot \frac{\partial\ln\psi}{\partial\theta} \tag{37}$$

The matrices Q and X are defined by Equations 18-24. The distribution function ψ must also satisfy the continuity equation,

$$\text{div}\psi v = - \frac{\partial\psi}{\partial t} \tag{38}$$

The divergence operator in angular coordinates is

$$\text{div} \equiv \frac{\partial^{T}}{\partial\theta} \tag{39}$$

where the superscript T denotes the transpose of a vector. By substituting Equation 37 for v into the equation of continuity, we
generate the diffusion equation,

$$\frac{\partial\psi}{\partial t} = \frac{E\mu}{\rho} e^{i\omega t} \{\psi \frac{\partial^{T}}{\partial\theta} \cdot QXQ^{T} \cdot \sin\theta + \frac{\partial\psi^{T}}{\partial\theta} \cdot QXQ^{T} \cdot \sin\theta\}$$

$$+ \frac{kT}{\rho} \frac{\partial^{T}}{\partial\theta} \cdot QXQ^{T} \cdot \frac{\partial\psi}{\partial\theta} \tag{40}$$

Execution of the indicated matrix operations yields a linear second-
order partial differential equation for the unknown function ψ.

$$\frac{\partial \psi}{\partial t} = \frac{E\mu}{\rho} e^{i\omega t} \{\psi \sum_{m=0}^{N} (QXQ^T)_{mm} \cos\theta_m + \sum_{n=0}^{N} \sum_{m=0}^{N} (QXQ^T)_{mn}$$

$$\frac{\partial \psi}{\partial \theta_m} \sin\theta_n\} + \frac{kT}{\rho} \sum_{n=0}^{N} \sum_{m=0}^{N} (QXQ^T)_{mn} \frac{\partial^2 \psi}{\partial \theta_m \partial \theta_n} \qquad (41)$$

where $(QXQ^T)_{mn}$ is the (m,n) element of the product matrix QXQ^T. Moreover, in order that ψ have physical meaning, we must stipulate that ψ and its first derivatives be periodic in 2π for all θ_ℓ,

$$\psi(\theta_0, \theta_1, \ldots, \theta_\ell - \pi, \ldots, \theta_N; t)$$

$$= \psi(\theta_0, \theta_1, \ldots, \theta_\ell + \pi, \ldots, \theta_N; t), \qquad \text{for all } \ell \qquad (42a)$$

$$\frac{\partial \psi(\theta; t)}{\partial \theta_\ell}\bigg|_{\theta_\ell - \pi} = \frac{\partial \psi(\theta; t)}{\partial \theta_\ell}\bigg|_{\theta_\ell + \pi} \qquad \text{for all } \ell \qquad (42b)$$

and that ψ be normalized,

$$\int_{-\pi}^{\pi} \psi(\theta; t)d\theta = 1 \qquad (43)$$

We suggest that the solution of Equation 41 can be expanded in a Fourier series,

$$\psi(\theta; t) = \psi_0 \sum_{\nu=0}^{\infty} \psi_\nu e^{i\omega\nu t} \qquad (44)$$

where ψ_0 is equal to unity. By collecting just the time-independent terms in Equation 41, we obtain an equation for ψ_0, the solution in the absence of applied field,

$$\sum_{n=0}^{N} \sum_{m=0}^{N} (QXQ^T)_{mn} \frac{\partial^2 \psi_0}{\partial\theta_m \partial\theta_n} = 0 \tag{45}$$

In angular coordinates, the solution to Laplace's equation satisfying periodic boundary conditions is a constant,

$$\psi_0 = \frac{1}{(2\pi)} N+1 \tag{46}$$

All derivatives of ψ_0 vanish. If we equate coefficients of $e^{i\omega\nu t}$ for $\nu \geq 1$ in Equation 41, we generate a set of recursion relations for the remaining ψ_ν's,

$$i\omega\nu\psi_\nu = \frac{E\mu}{\rho} \{ \psi_{\nu-1} \sum_{m=0}^{N} (QXQ^T)_{mm} \cos\theta_m + \sum_{n=0}^{N} \sum_{m=0}^{N} (QXQ^T)_{mn}$$

$$\frac{\partial\psi_{\nu-1}}{\partial\theta_m} \sin\theta_n \} + \frac{kT}{\rho} \sum_{n=0}^{N} \sum_{m=0}^{N} (QXQ^T)_{mn} \frac{\partial^2\psi_\nu}{\partial\theta_m \partial\theta_n} \tag{47}$$

$$\nu \geq 1$$

In particular, the equation for ψ_1 is

$$i\omega\psi_1 = \frac{E\mu}{\rho} \sum_{m=0}^{N} (QXQ^T)_{mm} \cos\theta_m$$

$$+ \frac{kT}{\rho} \sum_{n=0}^{N} \sum_{m=0}^{N} (QXQ^T)_{mn} \frac{\partial^2\psi_1}{\partial\theta_m \partial\theta_n} \tag{48}$$

Equation 48 is a non-homogeneous differential equation. Since ψ_0 is the distribution for the field-free case, ψ_1 must vanish when $E = 0$; the coefficients of the solutions to the corresponding homogeneous

equation must also vanish. Hence, Equation 48 is solved by a
particular solution satisfying the non-homogeneous differential
equation. The solution, except for a normalizing factor, is

$$\psi_1 = \frac{E\mu}{kT} \sum_{m=0}^{N} \frac{1}{1+i\omega\rho/kT(QXQ^T)_{mm}} \cos\theta_m \qquad (49)$$

Inspection of Equation 47 indicates that ψ_2 will be proportional to
$(E\mu/kT)^2$. For weak applied fields, we can exclude these higher
order terms from our expression for ψ. Therefore, the solution of
Equation 41 that is sufficient for our purposes is

$$\psi(\underset{\sim}{\theta}; t) = \frac{1}{(2\pi)^{N+1}} \{1 + \frac{E\mu}{kT} e^{i\omega t} \sum_{m=0}^{N}$$

$$\frac{1}{1 + i\omega\rho/kT (QXQ^T)_{mm}} \cos\theta_m\} \qquad (50)$$

We use this distribution function to calculate ensemble
averages of the requisite molecular parameters. The polarization
per molecule $\underset{\sim}{p}$ is the average electric moment arising from the
(N+1) dipoles; only the component p parallel to the external field
is non-vanishing. The ℓth dipole has a component $\mu\cos\theta_\ell$ in the
direction of the external field. Thus

$$p = \mu \sum_{m=0}^{N} <\cos\theta_m> \qquad (51)$$

These averages are computed in the usual manner.

$$<\cos\theta_\ell> = \int_{-\pi}^{\pi} \cos\theta_\ell \underset{\sim}{\psi} \, d\underset{\sim}{\theta},$$

$$= \frac{1}{2} \frac{E\mu}{kT} e^{i\omega t} [\frac{1}{1+i\omega\rho/kT(QXQ^T)_{\ell\ell}}] \qquad \text{for all } \ell \qquad (52)$$

Inserting this result into Equation 51, we determine the polarization. We now define the polarizability per unit dipole, α, as the ratio of p to the applied field:

$$\alpha \equiv \frac{p}{(N+1)} \, Ee^{i\omega t}$$

$$= \frac{\mu^2}{2(N+1)kT} \sum_{m=0}^{N} \frac{1}{1+i\omega\rho/kT(QXQ^T)_{mm}} \tag{53}$$

Now, if we consider the formula for the diagonal elements $(QXQ^T)_{\ell\ell}, \ell = 0, 1, 2, \ldots, N$, of the matrix product QXQ^T,

$$(QXQ^T)_{\ell\ell} = \sum_{j=0}^{N} X_j \alpha_j^2(\ell) \tag{54}$$

and if we recall Equation 20 for the set of $\alpha_k(\ell)$'s, it is trivial to show that

$$\alpha_j^2(N-\ell) = \alpha_j^2(\ell) \qquad \text{for all } j$$

and it follows immediately that

$$(QXQ^T)_{N-\ell, \, N-\ell} = (QXQ^T)_{\ell\ell} \tag{55}$$

We expect symmetry about the center of the chain on the basis of physical arguments. Consequently, we can rewrite Equation 53 for the polarizability,

$$\alpha = \frac{\mu^2}{(N+1)kT} \begin{cases} \displaystyle\sum_{m=0}^{(N-1)/2} \frac{1}{1+i\omega\rho/kT(QXQ^T)_{mm}} \quad , \quad (N+1) - \text{even} \\[2em] \displaystyle\sum_{m=0}^{(N-2)/2} \frac{1}{1+i\omega\rho/kT(QXQ^T)_{mm}} \end{cases} \tag{56}$$

$$+ \frac{1}{2} \frac{1}{1+i\omega\rho/kT(QXQ^T)_{N/2,N/2}} \quad , \quad (N+1) - \text{odd}$$

The polarizability α can be related to the complex dielectric constant ε through the relation,

$$\varepsilon - 1 = 4\pi\alpha/v'$$

where v' is the volume per monomer unit. Thus, the forms of the polarizability dispersion and of the dielectric dispersion are the same. It suffices to leave our results in terms of the polarizability. In order to facilitate comparison between polymers having different kinds of dipoles, we define, in analogy to the dielectric case, a reduced polarizability α_r,

$$\alpha_r \equiv \frac{\alpha - \alpha_\infty}{\alpha_0 - \alpha_\infty} = \frac{\varepsilon - \varepsilon_\infty}{\varepsilon_0 - \varepsilon_\infty}$$

such that α_r lies between unity and zero. α_0 will be the static polarizability, observed at zero frequency. α_∞ would denote the high frequency or optical polarizability; in our model α_∞ is of course zero. Accordingly,

$$\alpha_r = \frac{2}{N+1} \begin{cases} \displaystyle\sum_{m=0}^{(N-1)/2} \frac{1}{1+i\omega\rho/kT(QXQ^T)_{mm}}, & (N+1) - \text{even} \\[2em] \displaystyle\sum_{m=0}^{(N-2)/2} \frac{1}{1+i\omega\rho/kT(QXQ^T)_{mm}} \end{cases} \tag{57}$$

$$+ \frac{1}{2} \frac{1}{1+i\omega\rho/kT(QXQ^T)_{N/2,N/2}}, \quad (N+1) - \text{odd}$$

Equation 57 for the polarizability indicates a complicated relaxation process. The "apparent relaxation times," $\rho/kT(QXQ^T)_{\ell\ell}$, are summations containing complex functions of the angular frequency ω. However, there are a few remarks we can make for the two limiting cases of vanishing coupling and of tight coupling. Referring to the previous section, we find that, in the absence of internal interactions, $\chi_k \to 1$ for all k. Substitution of this result into Equation 54 yields

$$(QXQ^T)_{\ell\ell} \to 1 \tag{58}$$

This leads to the Debye result for the polarizability α_r,

$$\alpha_r = \frac{1}{1+i\omega\tau} \tag{59}$$

where τ is the relaxation time for a single bead in a one-dimensional treatment:

$$\tau = \frac{\rho}{kT} \tag{60}$$

In a plot of α_r versus $\log\omega\tau$, the loss curve maximum appears at $\omega\tau = 1$. For the other extreme of very tight coupling, $\chi_0 \to 1$ and $\chi_k \to 0$ for k>0, and one finds

$$(QXQ^T)_{\ell\ell} \rightarrow \frac{1}{N+1} \tag{61}$$

$$\alpha_r = \frac{1}{1+i\omega(N+1)\tau} \tag{62}$$

Equation 62 also predicts Debye behavior, but with the time constant $(N+1)\tau$. The loss curve maximum occurs at $\omega\tau = 1/N+1$. In words, a rigid rod relaxes by the same mechanism as a single element in a one-dimensional treatment; but, because of its increased frictional resistance $(N+1)\rho$, the dispersion is shifted to lower frequency. It is convenient to rewrite Equation 57 in terms of the Debye relaxation time τ of Equation 60.

$$\alpha_r = \frac{2}{N+1} \begin{cases} \displaystyle\sum_{m=0}^{(N-1)/2} \frac{1}{1+i\omega\tau/(QXQ^T)_{mm}}, & (N+1) - \text{even} \\[3em] \displaystyle\sum_{m=0}^{(N-2)/2} \frac{1}{1+i\omega\tau/(QXQ^T)_{mm}} \\[3em] + \frac{1}{2}\frac{1}{1+i\omega\tau/(QXQ^T)_{N/2,N/2}}, & (N+1) - \text{odd} \end{cases} \tag{63}$$

For intermediate cases, evaluation of Equations 54 and 63 is readily handled numerically. Discussions of these calculations and interpretations of the results with the aid of the correlation length are contained in the following paper.

ACKNOWLEDGMENTS

We are indebted to Professor Arthur V. Tobolsky for correspondence, beginning in 1966, which motivated the present work. We are also indebted to Professor Walter H. Stockmayer for discussions and for sharing unpublished material with us. This research was funded from the following grants from the U.S. Public Health Service: GM 11916 and GM 01045.

REFERENCES

1. P. Debye, POLAR MOLECULES, Dover Publications, New York, 1945.
2. C. P. Smyth, DIELECTRIC BEHAVIOR AND STRUCTURE, McGraw-Hill, New York, Toronto and London, 1955.
3. W. H. Stockmayer, Pure Appl. Chem. 15, 539 (1967).
4. J. G. Kirkwood and R. M. Fuoss, J. Chem. Phys. 9, 239 (1949).
5. A. V. Tobolsky and J. J. Aklonis, J. Phys. Chem. 68, 1970 (1964).
6. A. V. Tobolsky and D. B. Du Pre, Adv. Polymer Sci. 6, 103 (1969).
7. R. N. Work and S. Fujita, J. Chem. Phys. 45, 3779 (1966).
8. J. E. Anderson, J. Chem. Phys. 52, 2821 (1970).
9. R. A. Orwoll and W. H. Stockmayer, Adv. Chem. Phys. 15, 305 (1969).
10. P. E. Rouse, Jr., J. Chem. Phys. 21, 1272 (1953).
11. B. H. Zimm, J. Chem. Phys. 24, 269 (1956).
12. M. E. Baur and W. H. Stockmayer, J. Am. Chem. Soc. 86, 3485 (1964).
13. H. Frohlich, THEORY OF DIELECTRICS, Oxford University Press, London, 1958.
14. J. D. Hoffman, J. Chem. Phys. 23, 1331 (1955).
15. S. Glasstone, K. J. Laidler and H. Eyring, THE THEORY OF RATE PROCESSES, McGraw-Hill, New York and London, 1941.
16. T. Alfrey, Jr., MECHANICAL BEHAVIOR OF HIGH POLYMERS, Interscience Publishers, New York, 1948.
17. J. D. Ferry, VISCOELASTIC PROPERTIES OF POLYMERS, John Wiley, New York and London, 1961.
18. (a) D. E. Rutherford, Proc. Roy. Soc. (Edinburgh) $\underset{\sim}{A}$ 62, 229 (1947); (b) R. A. Orwoll and W. H. Stockmayer, Adv. Chem. Phys. 15, 305 (1969); (c) R. Ullman, J. Chem. Phys. 43, 3161 (1965). There is some confusion, ranging from errors to misprints, in the literature regarding the eigenvalues and eigenvectors of $\underset{\sim}{A}$. See Orwoll and Stockmayer (Equation 12a) for the correct expression for the eigenvalues μ_k of $\underset{\sim}{A}$. That our equations for the eigenvectors α_k are correct can be verified by direct substitution.
19. M. Abramowitz and I. A. Stegun, HANDBOOK OF MATHEMATICAL FUNCTIONS, Dover Publications, New York, 1965.
20. I. S. Gradshteyn and I. M. Ryzhik, TABLE OF INTEGRALS, SERIES, AND PRODUCTS, Academic Press, New York and London, 1965.

COMPUTATIONS BASED ON A CHAIN MODEL FOR DIELECTRIC LOSS

Mary Barkley Clark[*]

Department of Chemistry, Revelle College, University of

California (San Diego), La Jolla, California

ABSTRACT

Equations developed in the preceding paper are used to calculate the dielectric loss curves for polymers. Results are computed
for infinite chains and for finite chains. The correlation length
is used to interpret the dispersion curves. For reasonable values
of the parameters, considerable depression of the peak of maximum
loss and extension of the dispersion region can be obtained, even
for very short chains. For long chains, the expressions for the
dielectric dispersion are independent of the number of segments, in
contrast to the findings for the Rouse model for viscoelastic
properties.

INTRODUCTION

In the previous paper, hereafter referred to as I, we discussed
a chain of Maxwell models in order to treat the dielectric behavior
of polymers having dipoles perpendicular to the chain contour. We
derived an expression for the correlation length, a measure of the
extent to which rotatory motion is propagated along the chain. We
also obtained an expression for the polarizability as a function
both of frequency and of the parameters of our model. In this paper,
we shall examine in detail the properties of the Maxwell model in
the limit of long chains. We shall then investigate the effects of
the conditions at the ends of short chains on the dielectric relaxation. Finally, we shall mention the manner in which this model might
be applied to real polymers.

[*]Present Address: Salk Institute for Biological Studies, La Jolla,
California.

As in I, the model is a linear chain of (N+1) beads bearing dipoles joined by N Maxwell elements, of spring constant c and of frictional coefficient ρ'. The chain is suspended in a fluid of viscous resistance ρ. The reader is reminded of the following notation: The relaxation time τ of a single bead in a one-dimensional treatment,

$$\tau = \rho/kT,$$

the ratio of the frictional coefficients r,

$$r = \rho/\rho',$$

and the complex impedance ζ of a Maxwell element in the presence of a weak electric field of angular frequency ω,

$$\zeta = \frac{1}{1/\rho' + i\omega/c}$$

Equations from I are designated Equation I.1, etc.

Before proceeding, we should mention orders of magnitude to be expected for the molecular parameters ρ', c, and ρ. We extract ρ' by combining Equation I.1 and I.2. The spring constant c is computed, for small angles θ, by equating the curvatures of a Hooke's law potential and a symmetrical, rotational barrier potential $V(\theta)$ of the form

$$V(\theta) = \frac{V_0}{2}(1-\cos 3\theta)$$

where V_0 is taken as equal to the energy of the barrier, ΔF^+. From the barrier heights for small molecules, we may guess a reasonable average barrier for the polymer. We obtain ρ from the Stokes' law expression for the rotational diffusion coefficient of a sphere of the same volume as the monomer unit:

$$\rho = 6\eta \frac{m}{d}$$

where η is the solvent viscosity, m is the mass of a monomer, and d is the density of the polymer. At room temperatures, for a polymer such as polyvinyl chloride with barriers to internal rotation of about 4 Kcal/mole, we estimate $c/kT \simeq 1.8$, $\rho/\rho' \simeq 1$. For a polymer with barriers of about 3 Kcal/mole, we find $c/kT \simeq 1$ and $\rho/\rho' \simeq 5$. The calculated relaxation time for a monomer, ρ/kT, is approximately 6×10^{-11} sec. The corresponding relaxation times for small molecules in dilute solution are $\rho/2kT$ (2). For example, Smyth (3) gives relaxation times for ethyl bromide and chlorobenzene in benzene solution at 20°C of 0.3×10^{-11} and 0.8×10^{-11} sec, respectively. The discrepancy between the observed and the calculated values of the rotational frictional coefficients of small molecules has been discussed in the recent literature (4). It is apparent that c/kT and ρ' are quite sensitive to changes in the height of the barrier, and that estimates of ρ/kT are of low accuracy. Therefore, in our calculations, we shall choose values one or two orders of magnitudes on either side of the above estimated values.

RESULTS FOR LONG CHAINS

In its present form, our expression (Equation I.63) for the frequency dependence of the polarizability α_r defies ready interpretation. For long chains, though, some simplifications are possible. We begin with Equation I.54 for the $(QXQ^T)_{\ell\ell}$'s,

$$(QXQ^T)_{\ell\ell} = \left(\frac{1}{N+1}\right) \{1 + \sum_{j=1}^{N} x_j \cos \frac{j\pi}{N+1} (2\ell+1)$$

$$+ \sum_{j=1}^{N} x_j\} \tag{1}$$

and proceed in the manner used to evaluate Equation I.27,

$$(QXQ^T)_{\ell\ell} = \frac{\rho/\zeta}{\sqrt{(\rho/\zeta+2)^2-4}}$$

$$x \left\{1 + \left[\frac{(\rho/\zeta+2) - \sqrt{(\rho/\zeta+2)^2-4}}{2}\right]^{2\ell+1}\right\}$$

$$= \frac{\rho/\zeta}{\sqrt{(\rho/\zeta+2)^2-4}} \left\{1+e^{-(2\ell+1)y}\right\} \qquad (2)$$

where y has the previous meaning of Equation I.32. Equation 2 does not exhibit the symmetry property demanded by Equation I.55. This anomalous behavior can be blamed on the phase correlations of the eigenvectors $\alpha_k(\ell)$. At the ends, ℓ very small and ℓ very large, the phases of the k eigenvectors are strongly correlated; in the middle, they are uncorrelated. As $\ell \to N/2$, the approximation of the first sum on the right-hand side of Equation 1 by an integral breaks down and we lose sight of the fact that the correlations reappear as $\ell \to N$. However, since we know that the behavior at the far end of the chain must be the same as that at the near end, we can make use of Equation 2 with the restriction $\ell < N/2$.

Now, we note from an earlier discussion (see I) that y lies between zero and infinity. For non-zero $|\rho/\zeta|$, the exponential term in Equation 2 dies out rapidly with increasing ℓ; consequently we may neglect the ℓ-dependence of the $(QXQ^T)_{\ell\ell}$'s for values of ℓ not near the end of the chain. Reference to Equation I.63 shows that the contribution to the polarizability dispersion from the few end terms is insignificant compared to the contribution from the many interior elements. Therefore, we conclude that for large N, the $(QXQ^T)_{\ell\ell}$'s may be considered to be effectively independent of ℓ, as far as the calculation of the polarizability is concerned.

$$(QXQ^T) = \sqrt{\frac{\rho/\zeta}{(\rho/\zeta+4)}} \qquad (3a)$$

Here we have removed the subscripts from $(QXQ^T)_{\ell\ell}$ in order to emphasize that we have neglected end effects. If $|\rho/\zeta| \ll 1$, Equation 3a reduces further,

$$(QXQ^T) = \frac{\sqrt{\rho/\zeta}}{2} \tag{3b}$$

In either case, we can immediately compute the summation in Equation I.63 for the polarizability,

$$\alpha_r = \frac{1}{1+i\omega\tau/(QXQ^T)} \tag{4}$$

Inspection of Equation 3a shows that for values of the constant r large compared to one,

$$(QXQ^T) \to 1.$$

In this case, then, we will observe Debye dispersion arising from the relaxation of single beads against the viscous fluid. Likewise, if $kT\omega\tau/c$ is very much greater than one, i.e., $\omega\tau \gg c/kT$, one obtains the Debye result. Consequently, we shall be most concerned with values of r and $kT\omega\tau/c$ less than and of the order of unity. For the majority of cases, Equation 3b for (QXQ^T) will apply.

We propose a simple calculation on the basis of the above conditions. We expect at least its qualitative features to be useful in further discussions of the dielectric dispersion predicted both by the general result, Equation I.63, and by the long chain result, Equation 4. For large N and non-vanishing $|\rho/\zeta|$, the correlation length λ is, of course, less than the chain length. We can rewrite Equation 3b for (QXQ^T) as a function of the correlation length defined in Equation I.33b. After separating the real and imaginary parts, we get

$$(QXQ^T) = \frac{1+i(1-\lambda^2 r)^{1/2}}{2} \tag{5}$$

Substitution into Equation 4 and rearrangement leads to

$$\alpha_r = \frac{2-\lambda^2 r}{2-\lambda^2 r+2\omega\tau\lambda(1-\lambda^2 r)^{1/2}}$$

$$\left\{ \frac{i}{1+\dfrac{2i\omega\tau\lambda}{2-\lambda^2 r+2\omega\tau\lambda(1-\lambda^2 r)^{1/2}}} \right\} \tag{6}$$

At very low frequencies, such that $r \gg kT\omega\tau/c$, the correlation length is at its maximum value λ_{max}. Equation I.34b predicts that λ_{max} goes as $1/\sqrt{r}$, that is, $\lambda_{max}^2 r = 1$. The above expression for the polarizability α_r assumes a simple form,

$$\alpha_r = \frac{1}{1+2i\omega\tau\lambda_{max}} \tag{7a}$$

Or in terms of the molecular constants ρ and ρ',

$$\alpha_r = \frac{1}{1+2i\omega\tau(\rho'/\rho)^{1/2}} \tag{7b}$$

Equation 7b is exactly the dispersion relation derived for a long chain of dashpots. The relaxation process is Debye-like with time constant $2\tau/(\rho/\rho')^{1/2}$ or $2\left(\frac{\rho}{kT}\frac{\rho'}{kT}\right)^{1/2}$, a sort of geometric mean of the relaxation times of the individual dashpots. We can think of the low frequency dispersion as arising from the relaxation of the internal ρ' dashpots and the external ρ dashpots. At these frequencies, the springs do not participate in the relaxation; the motion is too slow to develop appreciable deformation of the springs before the internal dashpots relax. The low-frequency component of the dispersion is attributed to motion of chain segments of average length $2\lambda_{max}$. The loss curve is the usual Debye shape, with the maximum occurring at $\omega\tau = (\rho/\rho')^{1/2}/2$.

For intermediate frequencies, $r \leq kT\omega\tau/c$, the correlation length becomes a decreasing function of frequency. Within this range, we can write Equation 6 as follows

$$\alpha_r = \frac{1}{1+i\omega\tau\lambda} \left\{ \frac{1}{1 + \dfrac{i\omega\tau\lambda}{1+\omega\tau\lambda}} \right\} \qquad (8)$$

If we further assume that $\omega\tau\lambda < 1$, or $\omega\tau < kT/2c$, we obtain an approximate relation for the low-frequency side of the polarizability,

$$\alpha_r \simeq \frac{1}{1+i\omega\tau\lambda} \qquad (9)$$

Qualitatively, the dispersion at a particular frequency can be thought of as Debye-like dispersion with relaxation time $\tau\lambda$. It is recalled that λ is the correlation length at that frequency. In other words, we observe relaxation of segments of length $\lambda < \lambda_{max}$, where λ now depends on both ρ/ρ' and c. Presumably, we could reconstruct the dispersion curves by joining segments of Debye curves with maxima located at the appropriate $\omega\tau = 1/\lambda$. We can also expect that the greater the change in λ over the proper frequency range, the broader will be the dispersion curve.

As mentioned above, at high frequencies $\omega\tau > c/kT$, our approximate relations for (QXQ^T) and α_r do not apply. Rather, the high frequency behavior is extracted directly from Equation 3a. We stated earlier that $(QXQ^T) \to 1$ and that we obtain

$$\alpha_r = \frac{1}{1+i\omega\tau} \qquad (10)$$

the Debye equation for individual dipoles with the usual relaxation time τ. Thus, at higher frequencies in the dispersion, we see the external ρ dashpots relaxing alone. The amplitude of the motion is, of course, small, since the velocity of the bead, damped by the fluid, times the period of a cycle, is small. Consequently, almost no torque is put on the Maxwell element, and neither the springs nor the internal ρ' dashpots participate in the relaxation mechanism.

We can summarize our analysis of the mechanism of dielectric dispersion for long chains as expressed by Equation 4. We find a Debye-like relaxation process at very low frequencies, occurring

primarily through the ρ and ρ' dashpots. This low-frequency behavior
is visualized as rotation of segments of length $2\lambda_{max}$. As the frequency is raised to intermediate values, the springs become increas-
ingly important and the relaxation mechanism becomes a complicated
function involving both rotational and oscillatory motions of segments λ beads long. Again, at high frequencies, we see Debye relaxation, this time from the ρ dashpots alone. It is apparent that the
presence of the springs is responsible for the changing correlation length, and thus for the broadening of the dispersion curve.

Figure 1 shows some results for long chains. The dielectric
dispersion curves were actually calculated from Equations I.54 and
I.63, although curves calculated from Equations 2 and I.63 or
Equations 3a and 4 are indistinguishable from the ones presented
here. The solid lines indicate the real α_r' and imaginary α_r'' parts
of the complex polarizability α_r. For convenience, we have also
displayed the relative change in the correlation length as a function of frequency. The correlation length λ was computed from
Equation I.33a; λ/λ_{max} is denoted by a broken line. Figures 1i,
1ii, and 1iii refer to calculations for a chain of Maxwell elements
having weak, moderate, and stiff springs, respectively. The subtitles (a), (b), and (c) designate computations for a single spring
constant but for increasing values of the ratio of frictional
coefficients r, in order. We have already remarked that we will
not be concerned with values of $r > 1$, since larger values of r, or
smaller values of ρ', in effect imply the Debye result.

There are several general points which we wish to enumerate
before discussing specific examples. First, for long chains we
find a single loss peak. Second, the loss curve maximum does not
necessarily occur at the frequency at which $\alpha_r' = 0.5$, as it does in
the Debye theory; the complexities of the "apparent relaxation times"
$\tau/(QXQ^T)_{\ell\ell}$ destroy the simple relation between the real and imaginary
parts of the polarizability α_r. Thirdly, the broadening of the loss
curve and depression of the peak height depend on two factors: 1)
the magnitude of the change in the correlation length in going from
low to high frequencies, and 2) the location of the frequency range
in which this change takes place with respect to the dielectric dispersion process. For a given value of c/kT, loss curves for smaller
values of r and thus larger values of λ_{max} will generally be broader
and flatter than those for greater values of r; however in some cases
the change may be too small to be apparent. Also, for a given value
of r, loss curves for different values of c/kT will be broader and
flatter where λ/λ_{max} is varying rapidly with frequency in the middle
of the dispersion region.

This last point is illustrated by reference to the dispersion
curves themselves. Comparison of the three curves in Figure 1ii for
springs of intermediate spring constant reveals clearly the relation

between the broadening of the loss peak and the magnitude of change
in the correlation length. For Figures 1ii(a), 1ii(b), and 1ii(c)
the corresponding values of λ_{max} are 10, 3.2, and 1, respectively;
in all cases, λ for the highest frequencies is about 0.3. Figures
1i(a), 1ii(a), and 1iii(a) for $r = .01$ exhibit the effect of the
amount of overlap between the dispersion of the polarizability and
the change in the correlation length on the shape of the dispersion
curves. In these examples, the correlation length drops to 20% of
its low frequency value in two decades. However, in Figure 1i(a),
we see that the change in the correlation length occurs at frequencies considerably lower than those at which significant dispersion is observed. The outcome is a somewhat Debye-shaped peak
at $\omega\tau = 1.5$; most of the relaxation arises from the motion of individual dipoles. In Figure 1ii(a), the correlation length and the
polarizability are changing rapidly in the same frequency region.
We see the profound effect of this coincidence in the very broad
and flat dispersion curve. Figure 1iii(a), on the other hand, shows
the correlation length varying at higher frequencies than the dispersion. The result is, again, reminiscent of Debye relaxation,
but with loss peak at $\omega\tau = .05$. The chain is behaving primarily
like a chain of dashpots with correlation length $\lambda_{max} = 10$ and is
relaxing in segments $2\lambda_{max} = 20$ beads long.

The remaining curves can be interpreted along the same lines.
We feel the kind of analysis performed above serves to verify some
of our previously tentative conclusions. We hope it demonstrates
the application of the general results of our theory to particular
cases.

RESULTS FOR SHORT CHAINS

The foregoing discussion applies to long chains; generally
speaking, the chain length is very much greater than the correlation length, and the ends do not contribute appreciably. We wish
now to direct our attention to the behavior of short chains, for
which end effects will be important. We can imagine extreme situations at the ends of our linear array: 1) free ends, where the
motion of the end beads is restricted only by the usual resistance
ρ; and 2) fixed ends, where the end beads have zero velocity. The
first case merely refers to short linear chains suspended in a
viscous fluid. In itself, this model with free ends is not particularly relevant to experiments on high polymers. The second
case, on the other hand, might represent short linear segments of
a coiled polymer (Figure 2). If the chain is twisted or kinked in
such a way that rotation about certain bonds is severely hindered,
then the angular velocities of the restricted dipoles will be very
small or zero. We designate the beads whose motion is prohibited
as the fixed ends of the short segment joining any two such immovable

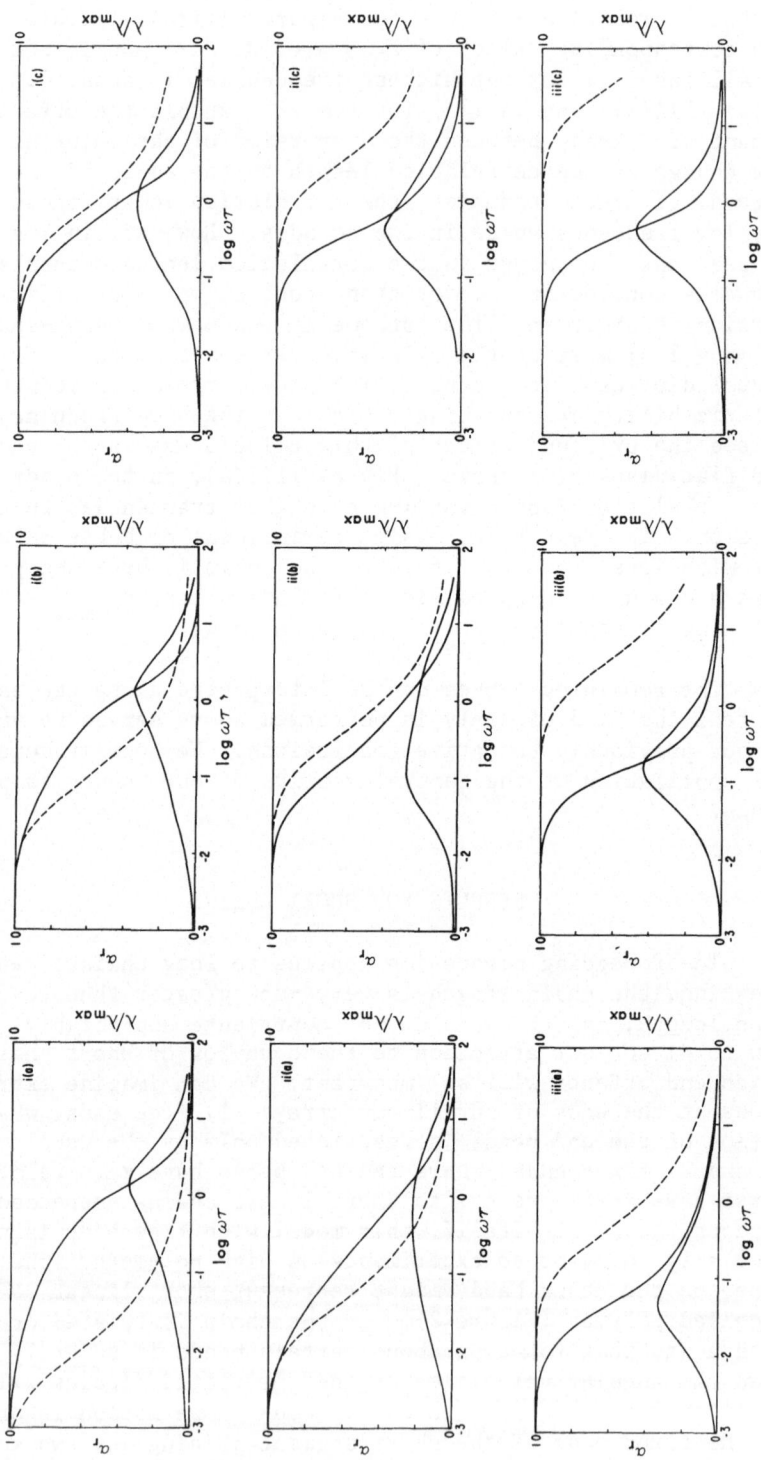

Figure 1. Theoretical Dispersion Curves of the Complex Polarizability and the Correlation Length for Long Chains.

The real part of the reduced polarizability α'_r and the reduced loss factor α''_r are plotted in solid lines. The relative correlation length λ/λ_{max} is plotted in broken lines. Calculations are for an 100-bead chain.

i(a). $c/kT = .24$, $r = .01$; $\lambda_{max} = 10.0$
i(b). $c/kT = .24$, $r = .1$; $\lambda_{max} = 3.2$
i(c). $c/kT = .24$, $r = 1.0$; $\lambda_{max} = 1.0$
ii(a). $c/kT = 1.0$, $r = .01$; $\lambda_{max} = 10.0$
ii(b). $c/kT = 1.0$, $r = .1$; $\lambda_{max} = 3.2$
ii(c). $c/kT = 1.0$, $r = 1.0$; $\lambda_{max} = 1.0$
iii(a). $c/kT = 10.0$, $r = .01$; $\lambda_{max} = 10.0$
iii(b). $c/kT = 10.0$, $r = .1$; $\lambda_{max} = 3.2$
iii(c). $c/kT = 10.0$, $r = 1.0$; $\lambda_{max} = 1.0$

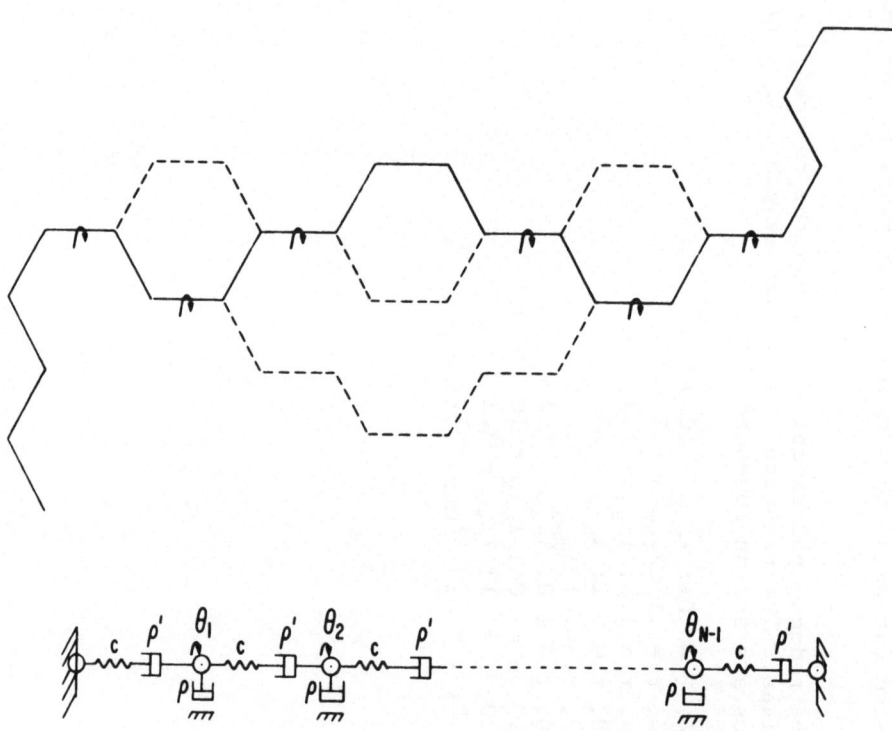

Figure 2.   Drawing of Short Linear Chain with Fixed Ends.

A simple example of a short linear segment of a real
polymer between two kinks in the polymer chain.

regions. A short-chain model with restricted ends may thus be more
realistic than our long linear chain model. Finally, we remark that
for short chains, the chain length may be less than the correlation
length λ.

For free ends, the derivation of the previous section applies.
The polarizability is given by Equation I.63, where the summations
in Equations I.54 and I.63 must be computed directly. Such calculations will be valuable for comparison with the results for fixed
ends and with the long chain result.

For fixed ends, we require that the velocities, v_0 and v_N, of
the end beads vanish. The mutual torques between the 0th and first
beads and between the $(N-1)^{th}$ and N^{th} beads become

$$L_{0,1} = -\zeta v_1 \tag{11a}$$

$$L_{N-1,N} = \zeta v_{N-1} \tag{11b}$$

The $L_{\ell,\ell+1}$ for $0<\ell<N-1$ are given by Equation I.8. Since the end beads
cannot move, their dynamical equations drop out. We are left with
a set of $(N-1)$ equations of motion.

$$\rho v_1 + \zeta(2v_1 - v_2) = -E\mu e^{i\omega t}\sin\theta_1 - kT \frac{\partial\ln\psi}{\partial\theta_1} \tag{12a}$$

$$\rho V_\ell + \zeta(-v_{\ell-1} + 2v_\ell - v_{\ell+1}) = -E\mu e^{i\omega t}\sin\theta_\ell - kT \frac{\partial\ln\psi}{\partial\theta_\ell} \quad 1<\ell<N-1 \tag{12b}$$

$$\rho v_{N-1} + \zeta(-v_{N-2} + 2v_{N-1}) = -E\mu e^{i\omega t}\sin\theta_{N-1} - kT \frac{\partial\ln\psi}{\partial\theta_{N-1}} \tag{12c}$$

Transcribe the dynamical equations into vector notation,

$$\rho\underset{\sim}{v} + \zeta\underset{\sim}{B}\underset{\sim}{v} = -E\mu e^{i\omega\tau}\underset{\sim}{\sin\theta} - kT \frac{\partial\ln\psi}{\partial\underset{\sim}{\theta}} \tag{13}$$

where $\underset{\sim}{v}$, $\underset{\sim}{\sin\theta}$, and $\partial/\partial\theta$ are now the appropriate (N-1)-dimensional vectors. $\underset{\sim}{B}$ is an (N-1) x (N-1)-dimensional symmetric matrix, the cyclical analogue of the $\underset{\sim}{A}$ matrix,

$$
\underset{\sim}{B} \equiv
\begin{bmatrix}
2 & -1 & 0 & 0 \dots 0 & 0 & 0 \\
-1 & 2 & -1 & 0 \dots 0 & 0 & 0 \\
0 & -1 & 2 & -1 \dots 0 & 0 & 0 \\
0 & 0 & -1 & 2 \dots 0 & 0 & 0 \\
\cdot & \cdot & \cdot & \cdot & \cdot & \cdot & \cdot \\
\cdot & \cdot & \cdot & \cdot & \cdot & \cdot & \cdot \\
\cdot & \cdot & \cdot & \cdot & \cdot & \cdot & \cdot \\
0 & 0 & 0 & 0 \dots 2 & -1 & 0 \\
0 & 0 & 0 & 0 \dots -1 & 2 & -1 \\
0 & 0 & 0 & 0 \dots 0 & -1 & 2
\end{bmatrix}
\tag{14}
$$

The matrix $\underset{\sim}{B}$ is also diagonalized by a similarity transformation $\underset{\sim}{R}$,

$$
\underset{\sim}{R}^{-1}\underset{\sim}{B}\underset{\sim}{R} = \underset{\sim}{N}
\tag{15}
$$

where $\underset{\sim}{N}$ is a diagonal matrix with diagonal elements ν_k,

$$
\nu_k = 2(1-\cos\frac{k\pi}{N}) \qquad k=1, 2, \ldots, N-1
\tag{16}
$$

and where $\underset{\sim}{R}$ is an orthogonal matrix constructed from the (N-1) column eigenvectors $\underset{\sim}{\beta}_k$ of $\underset{\sim}{B}$,

$$
\beta_k(\ell) = \left(\frac{2}{N}\right)^{1/2} \sin\frac{k\pi\ell}{N} \qquad \ell=1, 2, \ldots, N-1
\tag{17}
$$

Here $\beta_k(\ell)$ designates the ℓth component of the kth eigenvector (5). We note that, with respect to our usual numbering system, we have replaced (N+1) everywhere by (N-1), and we have shifted our indices to run from 1 to (N-1) instead of from 0 to N. As before, we invert the matrix $[\underset{\sim}{I}+\zeta/\rho\underset{\sim}{B}]$,

$$[\underset{\sim}{I}+ \frac{\zeta}{\rho} \underset{\sim}{B}]^{-1} = \underset{\sim}{R}\underset{\sim}{K}\underset{\sim}{R}^T \tag{18}$$

where K is an (N-1) x (N-1)-dimensional diagonal matrix with diagonal elements κ_k,

$$\kappa_k = \frac{1}{1 + \frac{\zeta}{\rho} \nu_k} \qquad\qquad k-1, 2, \ldots, N-1 \tag{19}$$

We solve Equation 13 for the velocity v,

$$\underset{\sim}{v} = - \frac{E\mu}{\rho} e^{i\omega\tau} \underset{\sim}{R}\underset{\sim}{K}\underset{\sim}{R}^T \sin\theta - \frac{kT}{\rho} \underset{\sim}{R}\underset{\sim}{K}\underset{\sim}{R}^T \frac{\partial \ln\psi}{\partial\theta} \tag{20}$$

From here on, the remaining derivation is exactly parallel to the one in I. Execution of the indicated matrix multiplications yields an expression for the diagonal elements $(RKR^T)_{\ell\ell}$ of the matrix product $\underset{\sim}{R}\underset{\sim}{K}\underset{\sim}{R}^T$

$$(RKR^T)_{\ell\ell} = \sum_{j=1}^{N-1} \kappa_j \beta_j^2(\ell) \tag{21}$$

Also, we can show that

$$(RKR^T)_{N-\ell,N-\ell} = (RKR^T)_{\ell\ell}$$

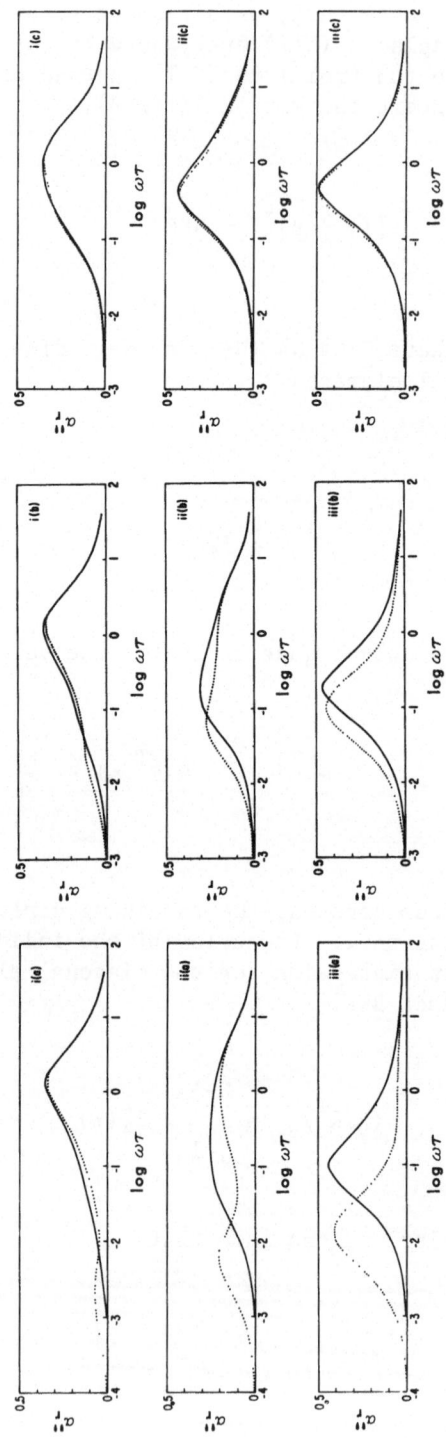

Figure 3.   Theoretical Loss Curves for Short Chains with Free Ends and with Fixed Ends.

The reduced loss factor α_r'' is plotted in solid lines for chains with free ends and in dotted lines for chains with fixed ends.   Calculations are for a 10-bead chain.

i(a). $c/kT = .24, \quad r = .01$
i(b). $c/kT = .24, \quad r = .1$
i(c). $c/kT = .24, \quad r = 1.0$
ii(a). $c/kT = 1.0, \quad r = .01$
ii(b). $c/kT = 1.0, \quad r = .1$
ii(c). $c/kT = 1.0, \quad r = 1.0$
iii(a). $c/kT = 10.0, \quad r = .01$
iii(b). $c/kT = 10.0, \quad r = .1$
iii(c). $c/kT = 10.0, \quad r = 1.0$

Finally, we obtain a result for the reduced polarizability α_r,

$$\alpha_r = \frac{2}{N-1} \begin{cases} \displaystyle\sum_{m=1}^{(N-3)/2} \frac{1}{1+i\omega\tau/(RKR^T)_{mm}} \, , & (N-1) - \text{even} \\[3em] \displaystyle\sum_{m=1}^{(N-4)/2} \frac{1}{1+i\omega\tau/(RKR^T)_{mm}} \\[3em] \quad + \frac{1}{2}\frac{1}{1+i\omega\tau/(RKR^T)_{\frac{N-2}{2},\frac{N-2}{2}}} \, , & (N-1) - \text{odd} \end{cases} \tag{22}$$

where we have computed the polarizability α_r for the (N-1) movable
dipoles. Again, we see that for very large N the end effects dis-
appear. The eigenvectors and eigenvalues of the $\underset{\sim}{A}$ and $\underset{\sim}{B}$ matrices
converge for infinite N.

Figure 3 presents results of these calculations for short
chains with ten beads. We show the dielectric loss curves α_r'' for
chains with free ends and with fixed ends on the same plot. The
solid lines designate free ends, and the dotted lines, fixed ends.
Figure 3i, 3ii, and 3iii indicate calculations for weak, moderate
and stiff springs, in that order. And, again, the subtitles (a),
(b), and (c) refer to computations for increasing values of r. We
shall commence with discussion of the results for free ends, since
we wish to understand the behavior of short chains within the context of our previous remarks for long chains. We shall summarize
our findings for both these cases in terms of the N-dependence of
the loss curve maximum $\alpha_r''(\text{max})$. Finally, we shall interpret the
effects of restricting the motion of the end beads.

For short chains with free ends, we again note that there is
essentially one dispersion. We also observe that our earlier general
comments, regarding the magnitude and relative location of the correlation length change and the effects on the shape of the dispersion curves, are still valid for short chains. The solid curves in
Figures 3i(a), 3i(b), and 3i(c), for weak springs, are very similar
to the corresponding long chain results shown in Figures 1i(a),
1i(b), and 1i(c). The interpretation of the previous section applies
here too. For springs of intermediate spring constant, we begin to
see some differences from the behavior of long chains. In Figures
3ii(a) and 3ii(b), the loss curves are somewhat higher and steeper,
especially on the low frequency side of the peak. It is a

straightforward matter to understand how this comes about. Recall
that, at low frequencies, the chain is relaxing against the ρ and
ρ' dashpots in segments of average length $2\lambda_{max}$ (compare Equation
7a). As the frequency increases and the springs begin to participate in the relaxation, the length of the rotating segments shifts
to λ, where λ now varies with frequency (compare Equation 9).
Between these two domains, we can imagine that the relaxations involve segments smaller than $2\lambda_{max}$ but greater than λ. All of this
discussion presupposes, of course, that the chain length is at
least $2\lambda_{max}$. Now for $r = .01$, $2\lambda_{max} = 20$; this is longer than our
short 10-bead chain. One can see that the average size of the
relaxing segments is curtailed at low frequencies for chains of
length comparable to the correlation length λ_{max}. Hence, the loss
curves for short chains, in the cases where the change in the
correlation length and the dielectric dispersion regions coincide,
will be narrower and higher than the corresponding long chain curves.

The height of the loss peak $\alpha_r''(max)$ is one of the characteristic parameters of dielectric dispersion experiments. In general,
it is a suitable barometer of deviations from Debye behavior. We
summarize our findings for the dependence of the polarizability
dispersion on the chain length, or molecular weight, in Figure 4.
Figure 4i gives the result for weak springs for various values of
r. The loss curve maximum achieves its long-chain value for chains
as short as 10 beads. For $r = .01$ and $r = .1$, we recall that this
behavior follows because the correlation length is quite small
throughout the dispersion region. Therefore, for weak springs, we
observe Debye peaks at $\omega\tau \simeq 1$, with low-frequency shoulders.
Figure 4ii shows the loss curves for moderate springs to be considerably depressed. We attribute this effect to the coincidence
of the change in the correlation length and the dispersion regions.
Figure 4iii indicates that the loss maxima for still springs have
come back up. We explain this result by noting that the correlation length remains close to its maximum value in the dispersion
region. As a consequence, we see low-frequency Debye-like peaks
at $\omega\tau \simeq 1/2\lambda_{max}$, for longer chains, with high-frequency shoulders.
The negative curvature of the curves in Figure 4iii derives from
the fact that, for chains of length $2\lambda_{max}$ or less, part of the
lower-frequency, larger-segment relaxation is excluded. For chains
longer than 100 beads, we do not anticipate further decreases in
$\alpha_r''(max)$.

We turn now to a consideration of the dielectric relaxation of
short chains with restricted ends. Perusal of the dotted curves
in Figure 3 reveals in some cases a second, low-frequency dispersion. Before discussing individual loss curves, we shall attempt
to elucidate the origin of the new low-frequency peak. Remember
that for small values of r, the chain length and the correlation
length may be of comparable magnitude. In the case of free ends, we
saw that, if the correlation length λ were equal to or greater than

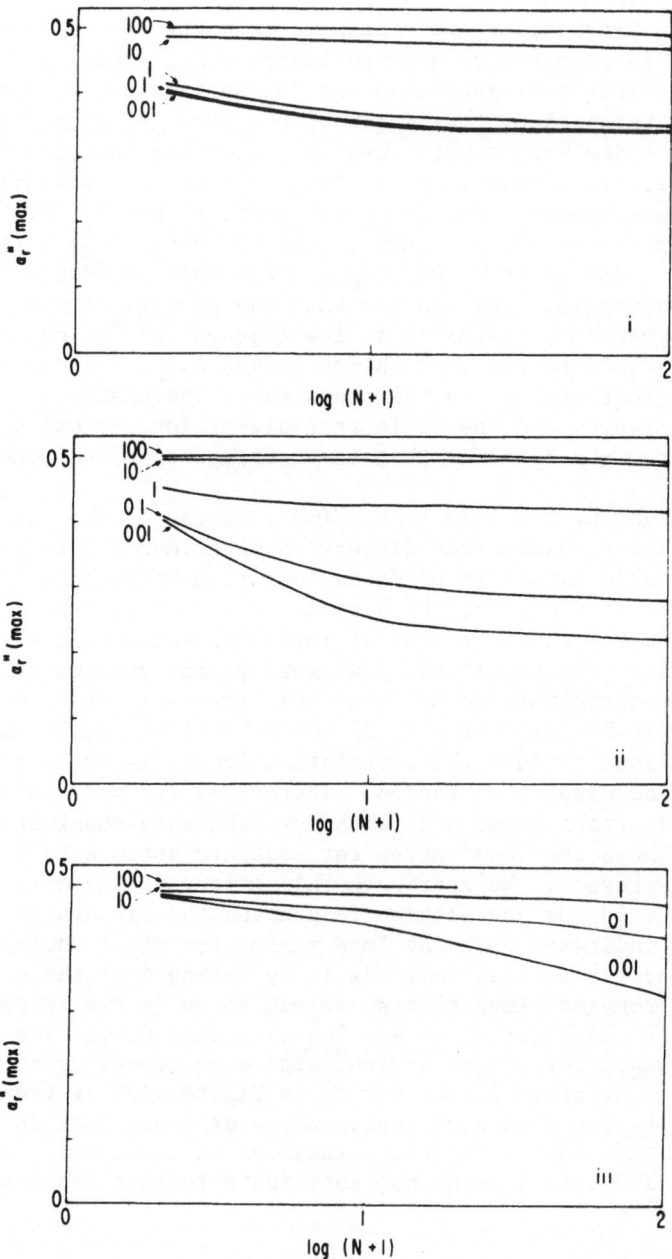

Figure 4.   Dependence of Peak Height at Maximum Loss on Chain Length.

The height of the loss peak $\alpha_r''(max)$ is plotted for
various values of r.

    i.   c/kT = .24
   ii.   c/kT = 1.0
  iii.   c/kT = 10.0

the chain length at frequencies within the dispersion region, a
Debye-like relaxation of the rigid chain against the ρ dashpots
was produced. For restricted ends, free rotation of the whole
chain is not possible. Instead, at low frequencies, in order for
the longer segments to relax, some of the ρ' dashpots must turn.
To further qualitative understanding, we may assume that rotation
of segments involving more than one or two beads is accomplished
solely by turning a few ρ' dashpots near the fixed ends. If we
must turn, for example, two ρ' dashpots in series at each end of
the chain, we encounter an effective resistance at the ends of
$\rho'/2$. The parallel resistances at the two ends then have a total
resistance of ρ'. On the other hand, for a 10-bead rigid chain
with free ends relaxing against the fluid alone, the frictional
resistance is 10ρ. The relative change in the effective frictional
coefficient for short chains with fixed or free ends is given by
the ratio of their respective resistances $(\rho'+10\rho)/10\rho$. For
$r = .01$, we expect an elevenfold increase in the frictional coefficient of a short chain with fixed ends giving rise to a low-
frequency peak near $\omega\tau \simeq .01$. We find such a dispersion in all
cases in Figures 3i(a), 3ii(a), and 3iii(a). The relative importance of the ends is determined by the extent of the contribution
of rigid chain motion to the dispersion in the case of free ends,
since it is primarily this motion which is shifted to lower frequencies when the ends are restricted. For correlation lengths
somewhat shorter than the chain length, one does not find a distinct second peak. Rather, in Figures 3i(b) and 3ii(b) for $r = .1$,
we see an enhanced shoulder at lower frequencies. These intermediate examples can be understood also as reflecting an increased
frictional resistance acting on the relaxing segments, due to the
presence of the fixed ends and to the consequent necessity of having
more motion in the stronger ρ' dashpots.

The remaining curves in Figure 3 can be understood by the application of similar principles.

DISCUSSION

We wish to discuss some of our theoretical results in relation
to experimental studies of dielectric loss in dilute solutions of
polymers having main-chain dipoles perpendicular to the polymer
chain axis. In this regard, we shall comment on the nature of the
rotational correlations as they appear in our model. We shall also
consider both the relaxation times, or frequencies of maximum loss,
and the shapes of the dispersion curves.

It has been established experimentally that the dielectric
relaxation time for a polymer with perpendicular dipoles is independent of molecular weight; it has long been suspected that local

relaxation processes are responsible for the dispersion (6).   Cor-
respondingly, a general conclusion of our calculations is that there
is no N-dependence of the relaxation time for long chains. That is
to say, the rotational correlations are short-range or local correlations. This result is in contrast to the treatments of Rouse (7)
and others (8), where long-range modes of distortion dominate the
relaxation phenomena. One can get a feeling for the origin of
this difference by the following simple argument. In the case of
the Rouse model, if one end of the chain molecule diffuses far
away from its equilibrium position, the only way that the polymer
can regain its equilibrium conformation is for the other end to
follow the first. Hence, the relaxation process occurs primarily
through large scale motions, i.e., the first few normal modes. For
rotational motion, a disturbance in the molecular conformation introduced by rotation about a bond has no such effect, since the
bond potential has an infinite number of equal minima, spaced at
the most 2π radians apart, perhaps closer. A rotational strain
can therefore be relieved by a relatively small rotation that need
not involve the whole molecule.

Two papers in recent years (9,10) have used the Ising lattice
as a model with which to represent the dielectric behavior of a
polymer chain with perpendicular dipoles. Their results show some
similarities to ours in that the relaxation times are affected by
the correlations between neighboring segments and become independent
of N when the chains are long enough. However, we must point out
that the models differ in a very important respect. The Ising model
introduces correlations between the rotational <u>positions</u> of neighboring segments; the Maxwell model tends to correlate the rotational
<u>velocities</u> of adjacent beads. (Position correlations could be intro-
duced into our model by another spring connected in parallel with
the Maxwell element between each bead.) Either model is only an
approximation to the real chain. We find our model attractive
because it reduces exactly to that of Debye in the degenerate case
of one element, although it must be granted that in this case the
equations of the dynamic Ising model are also formally similar to
those of Debye.

In the Introduction, we observed that the relaxation times,
calculated from Stokes' law for rotational diffusion, were larger
by at least an order of magnitude than the experimental relaxation
times for small molecules in dilute solution. This discrepancy is
slightly surprising, in view of the fact that the Stokes' expression for translational diffusion gives very good agreement with
experiment. We might understand this apparent anomaly by recalling
that, in the derivation of Stokes' equations, the velocity of the
fluid at the surface of the sphere is assumed equal to the velocity
of the sphere. In effect, the spherical molecule cannot slip through
the liquid. This stipulation may be quite reasonable for the case

of translational diffusion, where a molecule must actually displace
the surrounding solvent molecules in order to move. For rotational
diffusion, on the other hand, a compact molecule might rotate without necessarily disturbing all adjacent fluid molecules. Consequently, a small spherical molecule could encounter less frictional
resistance to rotatory motion than one would calculate from Stokes'
law. For polymer molecules, on the other hand, the Stokes' expression for the frictional coefficient of a monomer unit is probably
a more reasonable approximation. In the case of rotatory motion
in polymer chains with carbon-carbon backbones, the smallest unit
which can rotate freely is the so-called crankshaft.

This rotating polymer segment sweeps out a larger volume than an
equivalent small molecule spinning about an axis. Thus the polymer
unit must displace solvent molecules in order to rotate.

We shall summarize some experimental data for polymers with
perpendicular dipoles. To date, there is not much information published on dielectric relaxation of polymer molecules in dilute
solution. Dielectric measurements at -15°C on dilute solutions of
poly(vinylchloride) (11) in tetrahydrofuran and poly(vinylbromide)
(12) in dioxane yield relaxation times of $4x10^{-9}$ and $7x10^{-9}$ sec.,
respectively. For both polymers, the loss curves are quite Debye-
like in shape. The loss peak for poly(vinylchloride) is somewhat
depressed. For high-molecular weight samples of poly(p-chlorostyrene) (13,14) and poly(p-fluorostyrene) (15) in dilute solutions
of benzene or toluene at 25°C, the relaxation times are about
$5x10^{-9}$ and $4x10^{-9}$ sec., respectively. The loss curves for these
and other polysubstituted styrenes (14,15) are asymmetric, frequently with high-frequency shoulders. These loss peaks may be
considerably depressed and broadened. The relaxation time for
poly(ethylene oxide) (16) in benzene at 25°C is about 10^{-11} sec.
A remarkable concentration dependence has been observed for poly
(propylene oxide) in methyl cyclohexane solutions (17). In all of
these examples a single dispersion region from the relaxation of
the perpendicular dipoles is observed, and the frequency of maximum
loss is independent of molecular weight.

In relating our theoretical treatment to these experiments,
we feel that the model of a short chain of Maxwell elements with
fixed ends is the most appropriate caricature of a coiled polymer
molecule in solution. To our knowledge, though, there are no
dielectric dispersion experiments, which have been done on solutions

of polymer molecules having perpendicular dipoles only, and which show two loss peaks. It is evident that we must consider a distribution of sizes of these short linear sequences. A superposition of dispersion curves computed for various lengths of chains with fixed ends should yield a broadened single loss curve. For polymers, such as the polyvinyl halides and the polyhalo styrenes, with relaxation times several orders of magnitude longer than the calculated monomer unit relaxation times, we suggest that the motion of the short linear segments is highly correlated. For short chains with fixed ends, where the correlation length is greater than or equal to the chain length, and where the correlation length is large within the dispersion region, we would predict a longer relaxation time resulting from relaxation processes implicating the stiffer ρ' dashpots. We also expect a high-frequency shoulder on the loss curve arising from the decreasing correlation length and the concommitant relaxation of smaller segments. For example, the loss curves of polyvinyl halides are similar to the dotted curves in Figures 3iii(a) and 3iii(b). The dispersions observed for polyhalo styrenes could be approximated either by the dotted curve in Figure 3ii(b) or by an average over different chain lengths of Figures 3iii(b) or 3iii(c). In contrast, for a more flexible polymer molecule, such as poly(ethylene oxide), the relaxation time is comparable to that of a monomer. In this case, the motion is essentially uncorrelated and the predominant relaxation occurs against the external ρ dashpots.

ACKNOWLEDGMENTS

This research was funded from the following grants from the U.S. Public Health Service: GM11916 and GM 01045.

REFERENCES

1. M. B. Clark and B. H. Zimm, preceding paper.
2. P. Debye, POLAR MOLECULES, Dover Publications, New York, 1945.
3. C. P. Smyth, DIELECTRIC BEHAVIOR AND STRUCTURE, McGraw-Hill Book Company, New York, Toronto, and London, 1955.
4. R. E. D. McClung and D. Kivelson, J. Chem. Phys. 49, 3380 (1968).
5. The solution of the eigenvalue problem for the matrix B has been known for more than two centuries. For example, see L. Brillouin, WAVE PROPAGATION IN PERIODIC STRUCTURES, Dove Publications, New York, 1953.
6. W. H. Stockmayer, Pure Appl. Chem. 15, 539 (1967).
7. P. E. Rouse, Jr., J. Chem. Phys. 21, 1272 (1953).
8. B. H. Zimm, J. Chem. Phys. 24, 269 (1956).
9. R. N. Work and S. Fujita, J. Chem. Phys. 45, 3779 (1966).

10. J. E. Anderson, J. Chem. Phys. $\underline{52}$, 2821 (1970).

11. L. DeBrouckere and M. Mandel, Adv. Chem. Phys. $\underline{1}$, 77 (1958).

12. M. Kryszewski and J. Marchal, J. Polymer Sci. $\underline{29}$, 103 (1958).

13. W. H. Stockmayer, H. Yu and J. E. Davis. Am. Chem. Soc.,
     Div. Polymer Chem. Preprints $\underline{4}$, (2), 132 (1963).

14. J. H. Vreeland, Ph.D. Thesis, Massachusetts Institute of
     Technology, 1957.

15. B. A. Loury, B. Baysal and W. H. Stockmayer, unpublished
     results (1963-65).

16. M. Davies, G. Williams and G. D. Loveluck, Zeits. für. Electro-
     chem. $\underline{64}$, 575 (1960).

17. P. C. Lue, C. P. Smyth and A. V. Tobolsky, Macromolecules $\underline{2}$,
     446 (1969).

INTERNAL RELAXATION IN SHORT CHAINS BEARING TERMINAL POLAR GROUPS

R. L. Jernigan

Physical Sciences Laboratory, Division of Computer Re-

search and Technology, National Institutes of Health,

Bethesda, Maryland,[*] and Department of Chemistry, Univer-

sity of California at San Diego, La Jolla, California

ABSTRACT

Dielectric relaxation can be described in terms of correlations between an initial dipole vector, $\mu(0)$, and its value at later times, $\mu(t)$. Here, this dipolar correlation function is averaged over all short chain configurations with a rotational isomeric state model. The time dependent behavior of molecular configurations is developed from rates for passing over internal rotational energy barriers. Configurational perturbations caused by electric fields are treated; such effects are determined to be usually small. The theory developed is applied to calculate the high frequency dielectric dispersion for members of the family of α,ω-dibromo-n-alkanes, $Br-(CH_2)_{n-1}-Br$ for n = 4 to 6. Distributions of internal relaxation times are reported. For the longer chains these distributions lead to depressed and slightly skewed Cole-Cole dielectric constant diagrams.

INTRODUCTION

Molecular details have been ignored in previous treatments of relaxation processes in chain molecules. In the highly successful Rouse-Zimm (1,2) model, the polymer molecule is divided into an unspecified number of free jointed Gaussian subunits. The relaxation of each subunit is assumed to behave as if it were a Hookean

*Present Address.

99

spring. The molecule is represented by a chain of beads connected
with ideal springs. Frictional interactions with solvent occur only
at beads. This representation succeeds in predicting observed low
frequency relaxation behavior. Here a model is proposed to account
for high frequency relaxation which is characterized by short range
motions.

Resort to the devices of equivalent freely jointed chains and
Gaussian distributions results in a considerable loss (3,4) of
molecular information. In this paper a method for treating relaxation properties is set forth in which these artifices are avoided
by using structural features previously employed to calculate
equilibrium properties. These structural characteristics include
fixed bond lengths, rigid bond angles and rotational isomers (3,4).
Applications of these equilibrium theories have resulted in a fundamental understanding of a variety of electrical, geometric and
optical properties (4) of polymers.

In the present model, transitions between molecular configurations occur by rotations about backbone bonds. High energy barriers
to rotation permit a development of time dependent probabilities
for rotational isomeric states. The rates of transitions about a
given bond are assumed to be independent of states of neighboring
bonds. However, extension to an interdependent model is straightforward within the present framework. The important neighbor inter-
dependence of the equilibrium chain statistics (4) has been retained.
Because frictional and hydrodynamic interactions are ignored, the
treatment is appropriate only for isolated molecules. Applications
are limited to small chain molecules or short polymer segments.

This equilibrium and non-equilibrium rotational isomeric state
model has been applied to treat dielectric properties in several
short chains of the homologous series of α,ω-dibromo-n-alkanes.
These molecules were chosen because a previous treatment of their
equilibrium mean square dipole moments $\langle \mu^2 \rangle$ was available. After a
short review of the theory of this property, the effect on the dipole moments, at equilibrium, of an external electric field is considered. Also discussed is the approximate perturbation of probabilities of individual molecular configurations resulting from the
electric field. The development of time dependent configurational
probabilities makes possible a direct calculation of the internal
part of the dipolar auto-correlation function. It is also necessary
to include an external contribution which corresponds to rotational
diffusion of individual configurations. These external motions have
been assumed to be independent of all internal motions. The complex dielectric constant is obtained directly from the dipolar autocorrelation function. Frequency and temperature dependences of the
calculated dielectric dispersions are presented.

EQUILIBRIUM MEAN-SQUARE DIPOLE MOMENT (5)

Dipole moments of the class of molecules with polar groups at each terminus may be expressed as the sum of the two group moments, μ_1, and μ_n. For a fixed configuration, the molecular dipole moment is

$$\mu = \mu_1 + \mu_n \tag{1}$$

The values chosen for μ_1 and μ_n may be either the bond dipole moments or group moments so chosen as to also include induced effects in bonds adjacent to the polar bonds.

A right handed Cartesian coordinate system is assigned to every backbone bond. In the coordinate system based upon bond j, the x_j axis points along bond j. The y_j coordinate is chosen in the plane of bonds j-1 and j in a direction to form an acute angle with bond j-1. For such coordinate systems, the orthogonal transformation from the system j+1 to j is effected by

$$T_j = \begin{bmatrix} \cos\theta_j & \sin\theta_j & 0 \\ \sin\theta_j\cos\phi_j & -\cos\theta_j\cos\phi_j & \sin\phi_j \\ \sin\theta_j\sin\phi_j & -\cos\theta_j\sin\phi_j & -\cos\phi_j \end{bmatrix} \tag{2}$$

θ_j and ϕ_j are the bond angle supplement and rotational angle indicated for the alkane derivative shown in Figure 1. The angle ϕ_j is taken to be zero for the planar zigzag chain configuration.

The terminal bond dipole vectors of the hydrocarbon derivative shown in Figure 1 are given by

$$\mu_1 = \begin{bmatrix} \mu_1 \\ 0 \\ 0 \end{bmatrix} \quad \text{and} \quad \mu_n = \begin{bmatrix} -\mu_1 \\ 0 \\ 0 \end{bmatrix} \tag{3}$$

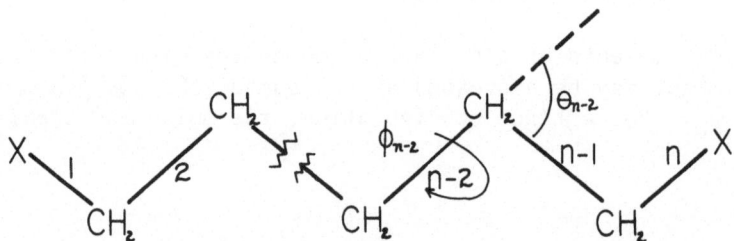

Figure 1.   α,ω-dihalo-n-alkane chain in the planar form.   An
            example of a rotational angle and bond angle supplement
            are shown.   For the applications in this paper, X =
            bromine.

For a given configuration, the dipole moment is expressed here in
the coordinate system of the first chain bond.   Combination of
Equations 1, 2 and 3 leads to

$$\mu = \mu_1 - \left(\prod_{j=1}^{n-1} T_j \right) \mu_1 \tag{4}$$

The required mean-square dipole moment is obtained by squaring and
averaging the last equation to give

$$\langle \mu^2 \rangle = 2\mu_1^2 - 2\mu_1^T \left\langle \prod_{j=1}^{n-1} T_j \right\rangle \mu_1 \tag{5}$$

Angle brackets denote averages over all configurations accessible
to the molecule and T indicates the transpose of the matrix to
which it is attached.

Further structural details are required in order to perform the average indicated in Equation 5. Near equilibrium, bond lengths and bond angles may be assumed to be fixed. The distribution of configurations arises solely because of variations in the set of rotational angles $\{\phi\}$. Enumeration of the n-2 backbone rotational angles serves to specify a molecular configuration.

For the n-alkanes and numerous other molecules (4), the configurational energy has been found to exhibit three minima which lie near $\phi = 0°$, 120° and 240°. The positions of these minima are designated trans(tr), gauche$^+$(g+) and gauche$^-$(g-). For the n-alkane chains, the minima are deep and nearly symmetric; hence, it is valid to assume that all molecular configurations may be generated by permitting each of the rotational angles to independently assume each of the three rotational positions. This yields a total of 3^{n-2} configurations for a chain of n bonds. In n-alkane chains (6) the g+ and g- states possess an energy approximately 0.5 kcal mole^{-1} above the trans energy. In addition, high energy configurations are encountered when pairs of bonds assume rotational states of g+g or g-g+. This effect may be accounted for by choosing the configurational weights to be neighbor dependent. It is convenient to account for this dependence in a matrix of statistical weights,

$$
\underset{\sim}{U}_j =
\begin{bmatrix}
1 & \sigma & \sigma \\
1 & \sigma & 0 \\
1 & 0 & \sigma
\end{bmatrix}
\quad 2 < j < n-1 \tag{6}
$$

where $\sigma = \exp(-0.5/RT)$. The rows of $\underset{\sim}{U}_j$ index rotational states for bonds j-1 and the column index states for bond j in the order tr, g+ and g-.

Rotational energies about carbon-carbon bonds 2 and n-1 are different because the bond lengths C-X and C-C are different and because of different van der Waals radii for methylene and for X. For α,ω-dibromo-n-alkane chains treated here, the form of these matrices is

$$
\underset{\sim}{U}_2 = \underset{\sim}{U}_{n-1} =
\begin{bmatrix}
1 & 1 & 1 \\
1 & 1 & 0 \\
1 & 0 & 1
\end{bmatrix}
\tag{7}
$$

The equilibrium partition function for these molecules is given by

$$Z = J^* U_2 \left(\prod_{j=3}^{n-2} U_j \right) U_{n-1} J \tag{8}$$

which is the sum of weights for all configurations. The row and column matrices in Equation 8 are defined as

$$J^* = [1 \ 0 \ 0] \text{ and}$$

$$J = \begin{bmatrix} 1 \\ 1 \\ 1 \end{bmatrix}$$

This partition function is utilized in a straightforward manner to average the product of matrices T in the r.h.s. of Equation 5

$$\left\langle \prod_{j=1}^{n-1} T_j \right\rangle = Z^{-1} (J^* \otimes I_3) \| T_1 \| \prod_{j=2}^{n-1} [(U_j \otimes I_3) \| T_j \|]$$

$$\cdot (J \otimes I_3) \tag{9}$$

I_3 is the unit matrix or order 3; $\otimes$ represents the direct matrix product. We define the pseudo-diagonal matrix $\| T \|$ by

$$\| T \| = \begin{bmatrix} T(\phi=0°) & & \\ & T(\phi=120°) & \\ & & T(\phi=240°) \end{bmatrix}$$

Substitution of this result into Equation 5 yields an expression for the mean-square equilibrium dipole moment

$$<\mu^2(t=\infty)> = 2\mu_1^2 - 2Z^{-1} \, \underset{\sim}{\mu}_1^T \, (J^* \underline{\otimes} I_3) \, \| T_1 \|$$

$$\prod_{j=2}^{n-1} [(U_j \underline{\otimes} I_3) \, \| T_j \|] \, (J \underline{\otimes} I_3) \, \underset{\sim}{\mu}_1 \qquad (10)$$

Generally $<\mu^2>$ will depend on the applied electric field or its residual influence after it has been removed. Here, t measures time after an initially applied field has been switched off; thus t = ∞ represents equilibrium.

Numerical calculations were carried out by Leonard, et al., (5) with this equation for the series of α,ω-dibromo-alkanes of different chain lengths at 25°. Effects of dipole-dipole inter-actions were also included. For chains longer than $Br-(CH_2)_5-Br$, agreement within about 5% of the experimental values was achieved by a choice of $\mu_1 = 1.90D$. A small dependence of $<\mu^2>$ on tempera-ture was computed which agrees with experiment. Inclusion of usually accepted small values for $u_{g+g-} = 0.02$ instead of the zeros in the 2,3 and 3,2 elements of U_j raised calculated moments by less than 1%. All following results were calculated using Equations 8 and 10 with $\theta_j = 68°$, $\mu_1 = 1.90D$, and $\sigma = \exp(0.5/RT)$. In addition, dipole-dipole interactions have been ignored; previous calculations indicated their effect to always be less than about 5%. Also g+g- and g-g+ pairs have been excluded throughout.

EQUILIBRIUM MEAN-SQUARE DIPOLE MOMENT IN
PRESENCE OF ELECTRIC FIELD

The molecular energy in an electric field which acts along the x axis is given by

$$E_F = E - \mu_x F - (\alpha_{xx}/2)F^2 \qquad (11)$$

where E is the internal energy in the absence of the field; F is the magnitude of the electric field. μ_x and α_{xx} represent the compo-nents of the dipole moment and the polarizability tensor in the direction of the field. This last term will be ignored because only polar molecules are treated here.

The partition function in the presence of an electric field is given by

$$Z_F = (8\pi^2)^{-1} \int \ldots \int \exp\,(-E_F/kT)\,\sin\chi\,d\chi\,d\psi\,d\rho\,d\{\phi\} \qquad (12)$$

where χ, ψ and ρ are the Euler angles and k is the Boltzmann constant. Substitution for E_F from Equation 11 and expansion of the exponential function $\exp\,(\mu_x F/kT)$ yields

$$Z_F = (8\pi^2)^{-1} \int \ldots \int \exp\,(-E/kT)\,[1 + \mu_x F/kT$$

$$+ \mu_x^2 F^2/(2k^2 T^2) + \mu_x^3 F^3/(6k^3 T^3)$$

$$+ \mu_x^4 F^4/(4!k^4 T^4) + \ldots]\,\sin\chi\,d\chi\,d\psi\,d\rho\,d\{\phi\} \qquad (13)$$

The first term is identical to the field free partition function. Terms in odd powers of μ_x vanish. The result is

$$Z_F = Z\,[1 + (1/2)\,<\mu_x^2>\,(F/kT)^2$$

$$+ (1/4!)\,<\mu_x^4>\,(F/kT)^4 + \ldots] \qquad (14)$$

Symmetry permits substitution of $<\mu_x^2> = (1/3)<\mu^2>$ and $<\mu_x^4> = (1/9)<\mu^4>$. Thus Equation 14 becomes

$$Z_F = Z\,[1 + (1/6)\,<\mu^2>\,(F/kT)^2$$

$$+ (1/216)\,<\mu^4>\,(F/kT)^4 + \ldots] \qquad (15)$$

This equation expresses the field dependent partition function in terms of the field free partition function and a series in even powers of (F/kT). Coefficients in the series are proportional to the even field free moments of the dipole vector.

Approximate expressions for any desired property in the presence of the electric field may be obtained by averaging with the
partition function in Equation 15. Only the mean-square dipole
moment is treated here. Direct application of Equation 13 to
obtain the requisite average yields

$$<\mu^2>_F = (Z/Z_F) \ [<\mu^2> + (1/2) <\mu^2\mu_x^2> \ (F/kT)^2$$

$$+ \ (1/4!) <\mu^2\mu_x^4> \ (F/kT)^4 + ...] \qquad (16)$$

Introduction of the symmetry of components of the even dipole
moments into Equation 16 gives

$$<\mu^2>_F = (Z/Z_F) \ [<\mu^2> + (1/6) <\mu^4> \ (F/kT)^2$$

$$+ \ (1/216) <\mu^6> \ (F/kT)^4 + ...] \qquad (17)$$

Exact calculation of terms in this series beyond $<\mu^2>$ is difficult
but can be accomplished by methods presented in References 4 and 7.
For short chains, these moments are obtained by explicit enumeration
of all configurations, their associated dipole vectors and statistical weights.

Maximum values of the even dipole moments are obtained for a
Gaussian distribution which corresponds to infinite chain length.
In this case, the higher even moments may be replaced by powers of
$<\mu^2>$ with appropriate numerical coefficients.

$$\text{Maximum}(<\mu^{2j}>) = [(2j+1)!/j!] \ (<\mu^2>/6)^j \qquad (18)$$

Substitution of these expressions into Equations 15 and 17 results
in the limiting expressions,

$$\langle\mu^2\rangle_F = (Z/Z_F) \, [\langle\mu^2\rangle + (5/18) \, \langle\mu^2\rangle^2 \, (F/kT)^2$$

$$+ (35/1944) \, \langle\mu^2\rangle^3 \, (F/kT)^4 + \ldots] \tag{19}$$

and

$$Z_F = Z \, [1 + (1/6) \, \langle\mu^2\rangle \, (F/kT)^2$$

$$+ (5/648) \, \langle\mu^2\rangle^2 \, (F/kT)^4 + \ldots] \tag{20}$$

Substitution of Equation 20 into Equation 19 yields

$$\langle\mu^2\rangle_F = \langle\mu^2\rangle + (1/9) \, \langle\mu^2\rangle^2 \, (F/kT)^2$$

$$- (2/243) \, \langle\mu^2\rangle^3 \, (F/kT)^4 + \ldots \tag{21}$$

This Gaussian expression for $\langle\mu^2\rangle_F$ represents the maximum value obtainable for the normal range of values of F, T and $\langle\mu^2\rangle$. Values of $\langle\mu^2\rangle_F$ calculated for 1,4-dibromo-butane at 25° are presented in Table I. Results of calculations of the maximum value in Equation 21 are included as well as those calculated by the more exact series in Equations 15 and 17. Normally $\mu F \ll kT$; therefore, we conclude that only large fields will have significant effects on $\langle\mu^2\rangle_F$. Within the range of small effects, deviations of $\langle\mu^2\rangle_F$ from $\langle\mu^2\rangle$ may adequately be described by means of Equation 21.

$\langle\mu^2\rangle_F$ differs from $\langle\mu^2\rangle$ because an external electric field changes the probabilities of individual rotational states. The interaction between the electric field and the dipole moment decreases the total energy; therefore, the probabilities of configurations with large dipole moments are favored. The treatment below of probabilities of bond rotational states in the presence of an electric field closely follows the derivation presented by Abe and Flory (8) for bond state probabilities with chain stretching.

TABLE I

Effect of Electric Field on Dipole Moment of
1,4-dibromo-n-butane (25°)

$$(<\mu^2> = 5.780D)$$

	$<\mu^2>_F$ (Units of Debye^2)		
F/kT (Units of Debye^{-1})	Eq. 17 with terms through $<\mu^4>$	Eq. 17 with terms through $<\mu^6>$	Gaussian Limit Eq. 21
0.03	5.784	5.784	5.784
0.06	5.793	5.793	5.794
0.09	5.809	5.810	5.812

Application of Equation 8 leads to an expression for the probability that bond i is in rotational state η,

$$P_{\eta;i} = Z_{\eta;i}/Z = Z^{-1} J^* \left[\prod_{j=2}^{i=1} U_j \right] U'_{\eta;i} \left[\prod_{k=i+1}^{n-1} U_k \right] J \quad (22)$$

where $U'_{\eta;i}$ is the matrix with two columns of zeros and the third column identical to the ηth column of U_i. By analogy with Equation 15, the statistical weight for bond i in rotational state η in the presence of the electric field is given by

$$Z_{\eta;i;F} \approx Z_{\eta;i} [1 + (1/6) <\mu^2_{\eta;i}> (F/kT)^2$$

$$+ (1/216) <\mu^4_{\eta;i}> (F/kT)^4 + ...] \quad (23)$$

The moments in this equation bearing subscripts $\eta;i$ are calculated subject to the constraint that bond i remains in rotational state η.

Likewise, the probability that bond i is in rotational state η in the presence of an electric field is defined by

$$P_{\eta;i;F} = Z_{\eta;i;F}/Z_F \tag{24}$$

Substitution of Equations 23 and 15 into Equation 24, followed by division of the two series, yields

$$P_{\eta;i;F} = (Z_{\eta;i}/Z) \; [1 + (1/6) \; (<\mu^2_{\eta;i}> - <\mu^2>) \; (F/kT)^2 + \ldots]$$

$$= P_{\eta;i} \; [1 + (1/6) \; (<\mu^2_{\eta;i}> - <\mu^2>) \; (F/kT)^2 + \ldots] \tag{25}$$

The factor $(<\mu^2_{\eta;i}> - <\mu^2>)$ in the correction term demonstrates the effect of η. If η produces a mean-square dipole moment $<\mu^2_{\eta;i}>$ which is larger than the total field free moment, then $P_{\eta;i;F}$ will be larger than $P_{\eta;i}$. If, however, $<\mu^2>$ is greater than $<\mu^2_{\eta;i}>$, then state η is less probable in the presence of the electric field than in its absence.

It is convenient to define a coefficient of the average effect of the electric field on the probabilities of bond i. For a chain of length n this coefficient is defined to be

$$C(i,n) = (1/3) \sum_{\eta=tr,g+,g-} |<\mu^2_{\eta;i}> - <\mu^2>|/m_1^2$$

In Figure 2 the dependence of $C(i,n)$ on n is shown for i=2 and i=3. Except for the very shortest α,ω-dibromoalkanes, $C(i,n)$ is smallest for values of i nearest the middle of the chain. The zigzag character of the curves is a consequence of the bond geometry. The results depicted in Figure 2 indicate that the effect of the electric field upon bond rotational probabilities is usually largest for bonds nearest the bond dipoles.

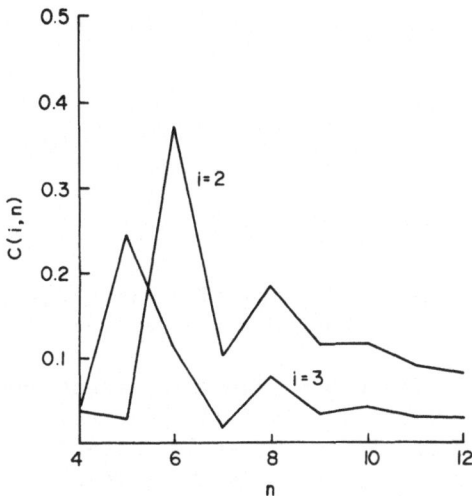

Figure 2.   Relative average influence of an electric field on the
            rotational probabilities for the ith bond in n bond
            α,ω-dibromo-n-alkane chains.   Results are for 25°.

DIPOLAR TIME CORRELATION FUNCTION

The frequency dependence of the dielectric constant is
directly related (9-13) to the dipolar time correlation function,

$$\Phi(\tau+t) \equiv <\underset{\sim}{\mu}(\tau) \cdot \underset{\sim}{\mu}(\tau+t)>/<\mu^2(\tau)> \tag{27}$$

For a stationary process the origin in time is arbitrary so that

$$\Phi(\tau+t) = \Phi(t) \equiv <\underset{\sim}{\mu}(0) \cdot \underset{\sim}{\mu}(t)>/<\mu^2(0)> \tag{28}$$

This function is properly normalized so that $\Phi(t = 0) = 1$ and
$\Phi(t = \infty) = 0$.   Decay of $\Phi(t)$ is assumed to occur by two independent
processes, external rotational diffusion of the whole molecule and

internal relaxation by transitions between rotational isomeric
configurations. Thus the total dipolar time correlation function
is expressed as

$$\Phi(t) = \Phi_{int}(t) \exp(-t/\tau_{rot}) \tag{29}$$

where τ_{rot} is the rotational relaxation time of a rigid body with
dimensions equivalent to the average molecular dimensions.

Hoffman (14) proposed a mechanism of rotational barriers to
explain dielectric relaxation in solid polymers. In his model
barriers arise because of intermolecular interactions. In the
treatment for isolated molecules which follows, intramolecular
interactions are the source of rotational barriers.

Calculation of Φ_{int} requires a complete description of the
time dependent probabilities of rotational states. A formulation
of these statistics is developed below in which rotation rates
about individual bonds are assumed to be independent.

Consider rotation about a single carbon-carbon bond located
within a sequence of identical bonds. The rotational energy about
this bond possesses a lowest minimum at $\phi = 0°$ for the planar form
(tr). Two additional somewhat higher minima appear near $120°$ (g+)
and $240°$ (g-). They lie at an energy approximately 0.5 kcal mole^{-1}
above the trans minimum. Two equal maxima appear near $60°$ and
$300°$. The magnitude of these barriers is 3 to 4 kcal mole^{-1} above
the trans zero energy position. Additionally, there is a high
barrier near $180°$. The height of this barrier, 9 to 10 kcal mole^{-1},
is so great as to be experimentally inaccessible. Such a barrier
is sufficiently high to preclude the direct transition g+$\rightleftharpoons$g-.
Transitions between rotational states therefore occur by passage
over the two smaller barriers between trans and the two gauche
states. Permitted transitions and their associated rates are
described by

$$g- \underset{r_1}{\overset{r_2}{\rightleftharpoons}} tr \underset{r_2}{\overset{r_1}{\rightleftharpoons}} g+$$

The rates of transitions are assumed to be proportional to Boltzmann factors of the barrier heights. Values used here for calculations are $r_1 = (kT/h) \exp(-3/RT)$ and $r_2 = (kT/h) \exp(-2.5/RT)$,
where h is Planck's constant.

If $P_{tr;j}(t)$ is the probability of rotational state trans for the jth bond at time t, then a vector formed from such probabilities is defined by

$$\underset{\sim}{p}_j(t) = \begin{bmatrix} P_{tr;j}(t) \\ P_{g+;j}(t) \\ P_{g-;j}(t) \end{bmatrix} \tag{30}$$

The first order differential equation to describe the rates of transitions for this vector is

$$d\underset{\sim}{p}_j(t)/dt = \underset{\sim}{A}_j \underset{\sim}{p}_j(t) \tag{31}$$

The matrix $\underset{\sim}{A}$, expressed in terms of the transition rates r_1 and r_2, is

$$\underset{\sim}{A}_j = \begin{bmatrix} -2r_1 & r_2 & r_2 \\ r_1 & -r_2 & 0 \\ r_1 & 0 & -r_2 \end{bmatrix}, \qquad 2 < j < n-1 \tag{32}$$

The $-2r_1$ in the first element represents the rate and two directions in which the bond is permitted to turn away from trans. The gauche rotational states are only permitted to turn in the one direction toward trans; hence, the other diagonal terms are $-r_2$.

The solution to differential Equation 31 is

$$\underset{\sim}{p}_j(t) = \exp(\underset{\sim}{A}_j t) \; \underset{\sim}{p}_j(t=0) \tag{33}$$

Matrix A_j can be diagonalized by a similarity transformation; hence

$$B_j^{-1} A_j B_j = \Lambda_j \tag{34}$$

B is a matrix formed from eigenvectors of A and Λ is the diagonal array of eigenvalues of A. Substitution of Equation 34 into Equation 33 yields

$$p_j(t) = B_j \, \exp(\Lambda_j t) \, B_j^{-1} \, p_j \ (t=0) \tag{35}$$

For the vector and matrix in Equations 30 and 32, the solution in Equation 35 is given by the matrix depicted on page 115.

$$
\underset{\sim}{p}_j(t) =
\begin{bmatrix}
p_{tr;j}(\infty) + [p_{g+;j}(\infty) + p_{g-;j}(\infty)] s_1 & p_{tr;j}(\infty)\,[1 - s_1] & p_{tr;j}(\infty)\,[1 - s_1] \\[2mm]
p_{g+;j}(\infty)\,[1 - s_1] & p_{g+;j}(\infty) + (1/2)\,s_1 + [p_{tr;j}(\infty)/2]\,s_1 & p_{g+;j}(\infty) - (1/2)\,s_2 + [p_{tr;j}(\infty)/2]\,s_1 \\[2mm]
p_{g-;j}(\infty)\,[1 - s_1] & p_{g-;j}(\infty) - (1/2)\,s_2 + [p_{tr;j}(\infty)/2]\,s_1 & p_{g-;j}(\infty) + (1/2)\,s_2 + [p_{tr;j}(\infty)/2]\,s_1
\end{bmatrix}
\cdot \underset{\sim}{p}_j(0) \qquad (36)
$$

The eigenrates are defined as $s_1 = \exp(-2r_1t-r_2t)$ and $s_2 = \exp(-r_2t)$.

If bond rotations are assumed to be independent of one another then this method may be extended to treat longer chains. The appropriate probability vector for a general n-bond chain will contain 3^{n-2} elements,

$$\underset{\sim}{p}^{(n)}(t) = \underset{\sim}{p}_2(t) \otimes \underset{\sim}{p}_3(t) \otimes \ldots \otimes \underset{\sim}{p}_j(t) \otimes \ldots \otimes \underset{\sim}{p}_{n-2}(t) \otimes \underset{\sim}{p}_{n-1}(t) \qquad (37)$$

The time derivative of Equation 37 is

$$d\underset{\sim}{p}^{(n)}/dt = (d\underset{\sim}{p}_2/dt) \otimes \underset{\sim}{p}_3 \otimes \ldots \otimes \underset{\sim}{p}_j \otimes \ldots \otimes \underset{\sim}{p}_{n-2} \otimes \underset{\sim}{p}_{n-1}$$

$$+ \underset{\sim}{p}_2 \otimes (d\underset{\sim}{p}_3/dt) \otimes \ldots \otimes \underset{\sim}{p}_j \otimes \ldots \otimes \underset{\sim}{p}_{n-2} \otimes \underset{\sim}{p}_{n-1}$$

$$\vdots \qquad\qquad \vdots$$

$$+ \underset{\sim}{p}_2 \otimes \underset{\sim}{p}_3 \otimes \ldots \otimes (d\underset{\sim}{p}_j/dt) \otimes \ldots \otimes \underset{\sim}{p}_{n-2} \otimes \underset{\sim}{p}_{n-1}$$

$$\vdots \qquad\qquad \vdots$$

$$+ \underset{\sim}{p}_2 \otimes \underset{\sim}{p}_3 \otimes \ldots \otimes \underset{\sim}{p}_j \otimes \ldots \otimes (d\underset{\sim}{p}_{n-2}/dt) \otimes \underset{\sim}{p}_{n-1}$$

$$+ \underset{\sim}{p}_2 \otimes \underset{\sim}{p}_3 \otimes \ldots \otimes \underset{\sim}{p}_j \otimes \ldots \otimes \underset{\sim}{p}_{n-2} \otimes (d\underset{\sim}{p}_{n-1}/dt) \qquad (38)$$

The explicit functional time dependence of the probabilities has been removed from Equation 38 and later equations. Unless otherwise stated, all probabilities to follow are for time t. Time derivatives on the right side of Equation 38 can be removed by substitution of Equation 31.

$$d\underset{\sim}{p}^{(n)}/dt = \underset{\sim 2}{A_2}\underset{\sim 2}{p_2}\otimes\underset{\sim 3}{p_3}\otimes\cdots\otimes\underset{\sim j}{p_j}\otimes\cdots\otimes\underset{\sim n-2}{p_{n-2}}\otimes\underset{\sim n-1}{p_{n-1}}$$

$$+ \underset{\sim 2}{p_2}\otimes\underset{\sim 3}{A_3}\underset{\sim 3}{p_3}\otimes\cdots\otimes\underset{\sim j}{p_j}\otimes\cdots\otimes\underset{\sim n-2}{p_{n-2}}\otimes\underset{\sim n-1}{p_{n-1}}$$

$$\vdots$$

$$+ \underset{\sim 2}{p_2}\otimes\underset{\sim 3}{p_3}\otimes\cdots\otimes\underset{\sim j}{A_j}\underset{\sim j}{p_j}\otimes\cdots\otimes\underset{\sim n-2}{p_{n-2}}\otimes\underset{\sim n-1}{p_{n-1}}$$

$$\vdots$$

$$+ \underset{\sim 2}{p_2}\otimes\underset{\sim 3}{p_3}\otimes\cdots\otimes\underset{\sim j}{p_j}\otimes\cdots\otimes\underset{\sim n-2}{A_{n-2}}\underset{\sim n-2}{p_{n-2}}\otimes\underset{\sim n-1}{p_{n-1}}$$

$$+ \underset{\sim 2}{p_2}\otimes\underset{\sim 3}{p_3}\otimes\cdots\otimes\underset{\sim j}{p_j}\otimes\cdots\otimes\underset{\sim n-2}{p_{n-2}}\otimes\underset{\sim n-1}{A_{n-1}}\underset{\sim n-1}{p_{n-1}} \qquad (39)$$

Matrices $\underset{\sim j}{A_j}$ are given by Equation 32 except for $\underset{\sim 2}{A_2}$ and $\underset{\sim n-1}{A_{n-1}}$ which are

$$\underset{\sim 2}{A_2} = \underset{\sim n-1}{A_{n-1}} = \begin{bmatrix} -2r_1 & r_1 & r_1 \\ r_1 & -r_1 & 0 \\ r_1 & 0 & -r_1 \end{bmatrix}$$

The direct product theorem for any arbitrary conformable matrices A, B, C and D permits the rearrangement $(AB)\otimes(CD) = (A\otimes C)(B\otimes D)$. Successive applications of this theorem to Equation 39 yields

$$dp^{(n)}/dt = [(A_2 \otimes I_3 \otimes \ldots \otimes I_3 \otimes \ldots \otimes I_3 \otimes I_3)$$

$$+ (I_3 \otimes A_3 \otimes \ldots \otimes I_3 \otimes \ldots \otimes I_3 \otimes I_3)$$

$$\vdots \qquad \vdots$$

$$+ (I_3 \otimes I_3 \otimes \ldots \otimes A_j \otimes \ldots \otimes I_3 \otimes I_3)$$

$$\vdots \qquad \vdots$$

$$+ (I_3 \otimes I_3 \otimes \ldots \otimes I_3 \otimes \ldots \otimes A_{n-2} \otimes I_3)$$

$$+ (I_3 \otimes I_3 \otimes \ldots \otimes I_3 \otimes \ldots \otimes I_3 \otimes A_{n-1})] \, p^{(n)} \qquad (40)$$

Combination of terms on the right side of Equation 40 gives a result similar in form to Equation 31

$$dp^{(n)}/dt = \mathcal{a} \, p^{(n)} \qquad (41)$$

Likewise, matrix $\mathcal{a}$ may be diagonalized and the solution expressed in a form similar to Equation 35,

$$p^{(n)} = \mathcal{B} \exp(\mathcal{L} t) \, \mathcal{B}^{-1} \, p^{(n)}(t=0)$$

$$= \mathcal{C} \, p^{(n)}(t=0) \qquad (42)$$

$\mathcal{B}$ is the matrix formed from eigenvectors of $\mathcal{a}$ and $\mathcal{L}$ is the diagonal array of eigenvalues of $\mathcal{a}$. Inspection of Equation 41 reveals that the matrix $\mathcal{a}$ has dimensions $3^{n-2} \times 3^{n-2}$. Practical computations with this method are therefore limited to $n \le 7$. Another method (15) is more appropriate for treating longer chains.

Results of calculations above indicate that the effect of normal electric fields on rotational state probabilities is usually negligible. Hence, $p^{(n)}(t=0)$ in Equation 42 may be replaced with $p^{(n)}(t=\infty)$, the equilibrium value. In order to obtain the same

initial time values of $\mu^2(0)$ from both the partition function in
Equation 8 and the statistics in Equation 42, the probability
vector in Equation 37 is replaced for computational purposes with

$$
\underset{\sim}{p}^{(n)}(t=0) = \begin{bmatrix}
P_{tr,tr,\ldots,tr,tr} \\
P_{tr,tr,\ldots,tr,g+} \\
P_{tr,tr,\ldots,tr,g-} \\
P_{tr,tr,\ldots,g+,tr} \\
\cdot \\
\cdot \\
\cdot \\
P_{g-,g-,\ldots,g-,g+} \\
P_{g-,g-,\ldots,g-,g-}
\end{bmatrix}
\tag{43}
$$

Each element of this vector represents the total equilibrium
probability for an individual configuration. These probabilities
may be calculated with the partition function in Equation 8.
Neighbor dependence of the equilibrium statistics is thus reintro-
duced into Equation 42.

In order to reduce the sizes of matrices $\mathcal{Q}$, the rows and
columns corresponding to occurrence of g+g- and g-g+ pairs of states
have been eliminated. Diagonal terms of $\mathcal{Q}$ are adjusted appropriately
to retain zero sums of each column. By this method, the dimensions
of matrix $\mathcal{Q}$ for n=6 are reduced from 81 x 81 to 41 x 41.

Matrix $\mathcal{C}$ in Equation 42 is composed of time conditional probabilities. The row indexes the configuration at time t and the
column indexes the initial configuration. The total time dependent
joint probability array is therefore given by

$$
\underset{\sim}{\mathcal{P}}^{(n)} = \underset{\sim}{\mathcal{C}} \mathrm{diag}[\underset{\sim}{p}^{(n)}(t=0)]
\tag{44}
$$

This matrix contains a complete description of the time dependence
of the rotational state probabilities. The internal portion of the
dipolar time correlation function is obtained by direct application
of this matrix.

$$\Phi_{int} = \underset{\sim}{M}^T (\underset{\sim}{\mathcal{A}}^{(n)} \otimes \underset{\sim 3}{I}) \underset{\sim}{M} \tag{45}$$

Vector M is formed from the dipole moment vectors for each chain configuration, calculated by Equation 4.

$$\underset{\sim}{M} = \begin{bmatrix} \mu_{tr,tr,\ldots,tr,tr} \\ \mu_{tr,tr,\ldots,tr,g+} \\ \cdot \\ \cdot \\ \cdot \\ \mu_{g-,g-,\ldots,g-,g-} \end{bmatrix} \tag{46}$$

Inspection of Equations 42 and 44 indicates that the internal dipolar time correlation function in Equation 45 may be expressed in the form

$$\Phi_{int} = \sum_j k_j \exp(\lambda_j t) \tag{47}$$

where λ_j is the jth eigenrate from the diagonal array $\underset{\sim}{\mathcal{A}}$. The relative importance of eigenrates is shown in Figures 3 to 5 for chains of 4, 5 and 6 backbone bonds. The eigenrate is the inverse of the relaxation time; therefore, these diagrams are directly related to the distributions of relaxation times. With increasing chain length, additional relaxation times contribute to the correlation function. The mean relaxation time is shifted to shorter times for the longer chains. This is a consequence of the mechanism of independent relaxation about all chain bonds.

The total correlation function is expressed by

$$\Phi = \sum_j k_j \exp(\lambda_j' t) \tag{48}$$

where $\lambda_j' = \lambda_j - 1/\tau_{rot}$. The effect of including this independent
external rotation is to shift all internal rates by the same
amount. Approximate values of τ_{rot} were chosen from the results
of Rayleigh scattering experiments for the similar liquid alkyl
bromides (16). Use of empirical equations given in Reference 16
for the chain length and temperature dependences of the orientational relaxation times yields the values presented in Table II.
These values for the monobromo compounds are substituted for the
values required for the dibromo compounds with the same number of
carbon atoms. It is assumed that introduction of the additional
bromine atom does not significantly change the rotational diffusion rate. The magnitude of the error introduced by assuming
that the two experimental methods yield the same relaxation time
depends upon the chain length. Estimates (16) of this error indicate that it is probably less than 50%.

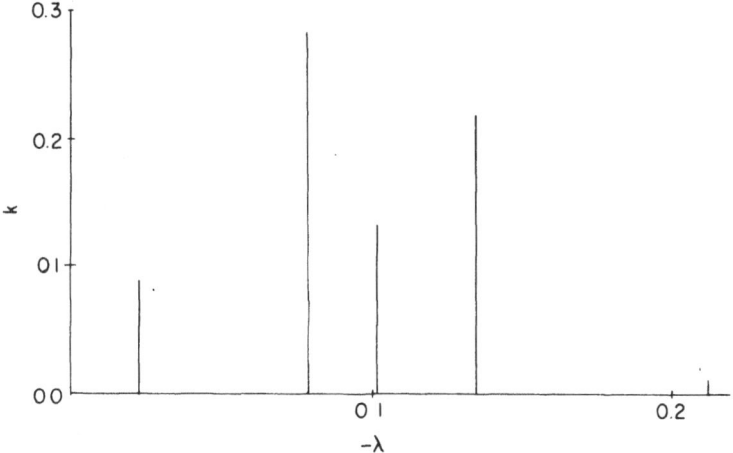

Figure 3.   Contributions k of eigenrates λ to Φ_{int}, the normalized
internal dipolar correlation function, for 1,3-dibromo-n-
propane at 25°. Units of λ are 10^{12} sec^{-1}.

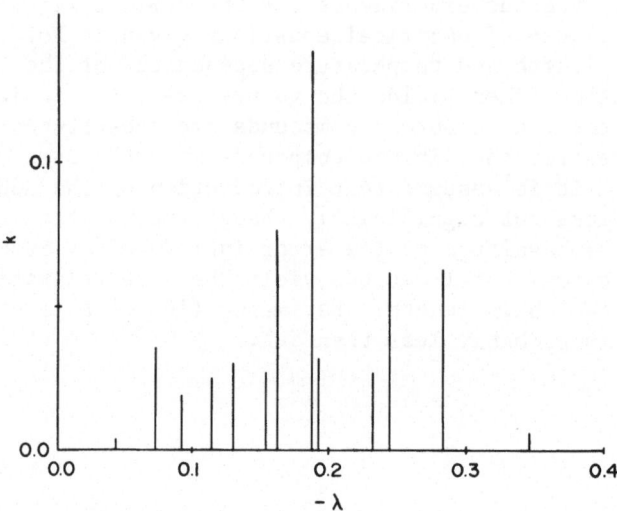

Figure 4.   Contributions k of eigenrates λ to Φ_{int}, the normalized
             internal dipolar correlation function, for 1,4-dibromo-n-
             butane at 25°.   Units of λ are 10^{12} sec^{-1}.

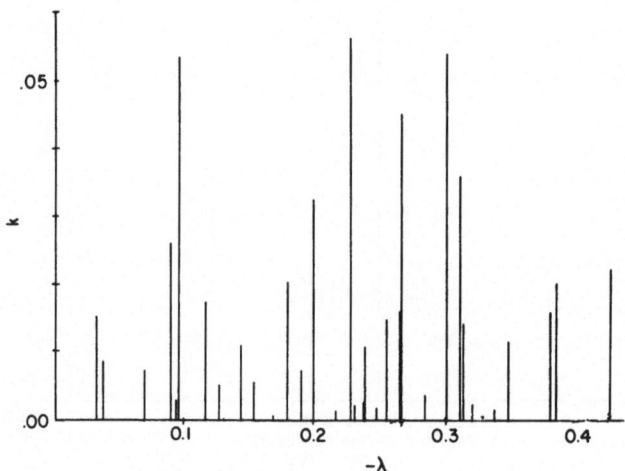

Figure 5.   Contributions k of eigenrates λ to Φ_{int}, the normalized
             internal dipolar correlation function, for 1,5-dibromo-n-
             pentane at 25°.   Units of λ are 10^{12} sec^{-1}.

TABLE II

Orientational Relaxation Times from
Rayleigh Scattering Experiments (16)

Liquid	Temperature	τ_{rot} x 10^{12} (sec)
1-propyl bromide	25°	3.4
1-butyl bromide	0°	8.0
1-butyl bromide	25°	5.8
1-butyl bromide	50°	4.7
1-pentyl bromide	25°	9.9

A correlation time for the normalized total correlation function in Equation 48 is defined by

$$\tau_c = \int_0^\infty \phi \, dt$$

$$= \sum_j \int_0^\infty k_j \exp(\lambda_j' t) \, dt \tag{49}$$

Integration of this equation yields

$$\tau_c = - \sum_j k_j / \lambda_j'$$

Values of τ_c calculated at 25° for the α,ω-dibromo-n-alkanes
are 2.84 x 10^{-12} sec for n = 4, 4.22 x 10^{-12} sec for n = 5 and
6.41 x 10^{-12} sec for n = 6.

Figure 6 displays the decay of the total correlation function
at 25° for n = 4 to 6. The smallest molecule relaxes fastest only
because the smallest values of τ_{rot} were used. If the same value
of external relaxation time had been used for all chain lengths,
then the curves would appear in the opposite order. The reason for
this behavior is that longer chains in this model manifest shorter
mean internal relaxation times.

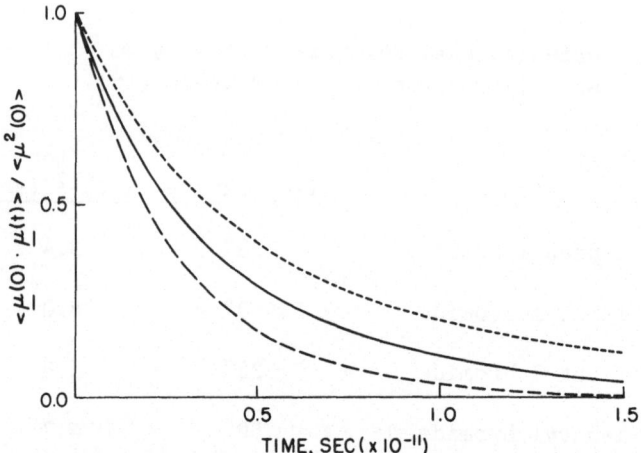

Figure 6.   Time decay of total normalized dipolar time correlation
             function at 25°. Upper curve represents behavior for a
             chain length n=6. Middle curve is for n=5 and lowest
             curve is for n=4.

DIELECTRIC CONSTANT

The complex dielectric constant ε^* is directly proportional to
the Laplace transform L of the time derivative of the total dipolar
correlation function.

$$(\varepsilon^*-\varepsilon_\infty)/(\varepsilon_o-\varepsilon_\infty) = L(-d\Phi/dt) \tag{51}$$

ε_o and ε_∞ are respectively the static and infinite frequency
dielectric constants. Substitution for the argument of the Laplace
transform from Equation 48 yields

$$(\varepsilon^*-\varepsilon_\infty)/(\varepsilon_o-\varepsilon_\infty) = -\sum_j \lambda_j' k_j L[\exp(\lambda_j' t)] \tag{52}$$

Evaluation of the Laplace transform gives

$$(\varepsilon^* - \varepsilon_\infty)/(\varepsilon_0 - \varepsilon_\infty) = - \sum_j \lambda'_j k_j / (i\omega - \lambda'_j) \tag{53}$$

where ω is identified with frequency. If $-\lambda_j$ is expressed as the inverse of relaxation time τ_j, then the real part of the dielectric constant is given by

$$(\varepsilon' - \varepsilon_\infty)/(\varepsilon_0 - \varepsilon_\infty) = \sum_j k_j / (1 + \omega^2 \tau_j^2) \tag{54}$$

The imaginary part of the dielectric constant is

$$\varepsilon'' / (\varepsilon_0 - \varepsilon_\infty) = \sum_j k_j \omega \tau_j / (1 + \omega^2 \tau_j^2) \tag{55}$$

Each term of the series in Equations 54 and 55 is identical to the result for the Debye one relaxation time model. The dielectric constant for the present model is given by a sum of these Debye-like terms. If the real part for the Debye model is plotted on one axis and the imaginary part on the other axis, the familiar Cole-Cole semicircle is obtained. The results of such plots are displayed in Figure 7 for the chains with n = 4, 5 and 6. The uppermost curve for 1,3-dibromo-n-propane is indistinguishable from the one relaxation semicircle. Curves for longer chains are depressed and slightly skewed from this limit. Such depressions and high frequency broadening is commonly observed for systems with multiple relaxation times. These deviations in Figure 7 are direct results of the distributions of relaxation rates depicted in Figures 3 - 5.

The temperature dependence of the real and imaginary dielectric constant for $Br-(CH_2)_4-Br$ is presented in Figures 8 and 9. Temperature affects the value of τ_{rot} chosen from Table II as well as the Boltzmann factors for rates of passage over barriers and for equilibrium statistical weights of rotational isomers. At higher temperatures the dispersion region is shifted to higher frequencies.

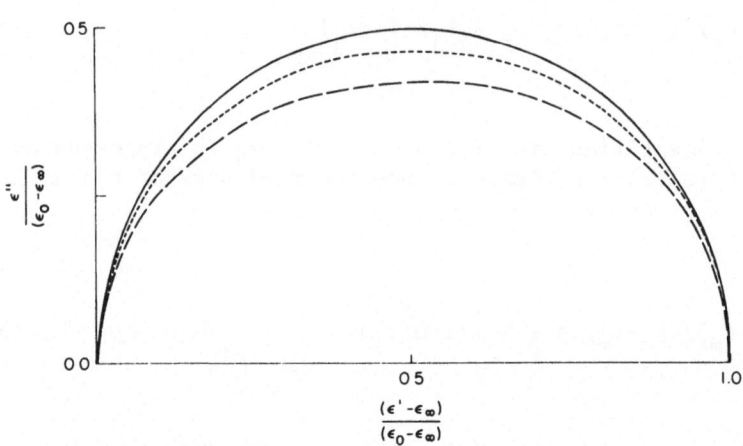

Figure 7.   Cole-Cole diagram of dielectric behavior. Ordinate represents the imaginary part of the dielectric constant,
and the abscissa is the real part.   Upper curve is the
Debye one relaxation time semicircle.   The depressed
curve of short dashes is for 1,4-dibromo-n-butane.
Curve of long dashes represents behavior of 1,5-dibromo-
pentane.   All calculated curves are for 25°.

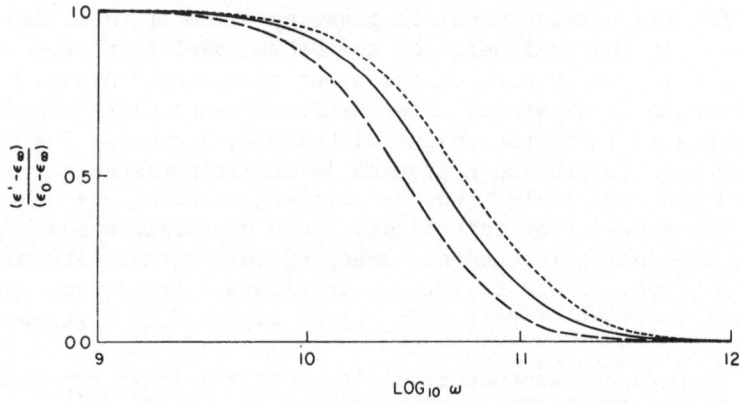

Figure 8.   Frequency dependence of real part of dielectric constant
for 1,4-dibromo-n-butane at three temperatures.   Curve of
small dashes is for 50°; solid curve is for 25°; long dash
curve is for 0°.

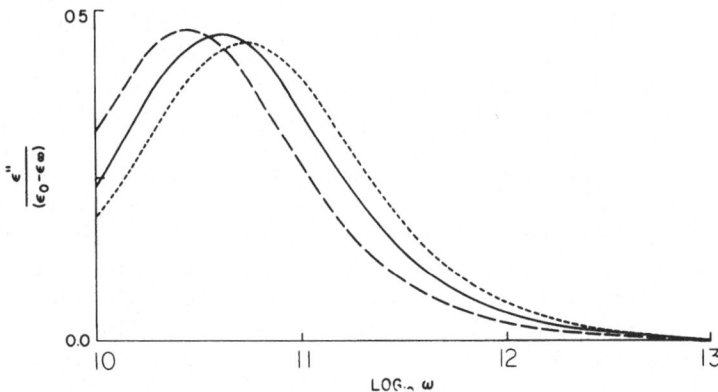

Figure 9.   Frequency dependence of imaginary part of dielectric
            constant for 1,4-dibromo-n-butane at three temperatures.
            Curves are designated as in Figure 8.

        The present theory has been developed for isolated molecules.
Any comparison of these calculated results with experiment would
require consideration of interactions with the surrounding medium.
Numerous approximate theories (9-13) for this purpose exist. However, unavailability of experiments in the very high frequency
range of simple configurational transitions does not encourage
further pursuit. A preliminary estimate of the effect of intermolecular interactions in the pure liquids was performed by the
method of References 12 and 13. The deviations from the Debye
model manifested in Figure 7 were considerably reduced.

        The present method for short chains admits of extension to more
complex configurational transition schemes. Simultaneous transitions
about several bonds may be accommodated by modifying matrix Q in
Equation 41 to include additional non-zero terms. In this manner
any desired motions, such as the so-called crankshaft motion, may
be included. The range of bonds encompassed by such motions is
limited only because this method is limited to short chains. Also,
it is possible to treat relaxation properties of other molecules
(4,17) using this model, if their structural features are known.

The model presented is a first attempt to account for the effect of detailed structural features on non-equilibrium properties. It may be appropriately applied only in the case of small deviations from equilibrium. Account has not been taken of large non-equilibrium effects such as bond length distortions and significant distortions of bond angles. However, such effects are not expected to be important for the application of small external fields. In its present application, transitions about all bonds have been assumed to occur independently and at the same rate. Frictional resistance to motions will modify transition rates. These modifications will show dependence on both chain length and configuration. For example, in a viscous medium it is more difficult to rotate about a bond in the middle of a long chain than about a terminal bond. The present isolated molecule theory accounts for neither hydrodynamic nor electrical intermolecular interactions.

ACKNOWLEDGMENT

The author wishes to thank B. H. Zimm and G. H. Weiss for helpful discussions

REFERENCES

1. P. E. Rouse, Jr., J. Chem. Phys. $\underline{21}$, 1272 (1953).
2. B. H. Zimm, J. Chem. Phys. $\underline{24}$, 269 (1956).
3. M. V. Volkenstein, CONFIGURATIONAL STATISTICS OF POLYMERIC CHAINS, Interscience, New York, 1963.
4. P. J. Flory, STATISTICAL MECHANICS OF CHAIN MOLECULES, Interscience, New York, 1969.
5. W. J. Leonard, Jr., R. L. Jernigan and P. J. Flory, J. Chem. Phys. $\underline{43}$, 2256 (1965).
6. A. Abe, R. L. Jernigan and P. J. Flory, J. Am. Chem. Soc. $\underline{88}$, 631 (1966).
7. P. J. Flory and R. L. Jernigan, J. Chem. Phys. $\underline{42}$, 3509 (1965).
8. Y. Abe and P. J. Flory, J. Chem. Phys. $\underline{52}$, 2814 (1970).
9. R. H. Cole, J. Chem. Phys. $\underline{42}$, 637 (1965).
10. S. H. Glarum, J. Chem. Phys. $\underline{33}$, 1371 (1960).
11. E. Fatuzzo and P. R. Mason, Proc. Phys. Soc. $\underline{90}$, 741 (1967).
12. D. D. Klug, D. E. Kranbuehl and W. E. Vaughan, J. Chem. Phys. $\underline{50}$, 3904 (1969).
13. J.-L. Rivail, J. Chim. Phys. $\underline{66}$, 981 (1969).
14. See, for example, J. D. Hoffman and H. G. Pfeiffer, J. Chem. Phys. $\underline{22}$, 132 (1954).
15. R. L. Jernigan, to be published.
16. D. A. Pinnow, S. J. Candau and T. A. Litovitz, J. Chem. Phys. $\underline{49}$, 347 (1968).
17. T. W. Bates and W. H. Stockmayer, Macromolecules $\underline{1}$, 12 (1968).

DIELECTRIC PROPERTIES OF METHYL STEARATE-A MODEL CRYSTALLINE

POLYMER*

M. G. Broadhurst

Institute for Materials Research, National Bureau of

Standards, Washington, D.C.

INTRODUCTION

This paper gives the procedures and initial results from a
study of the effect of pressure on dielectric relaxation parameters for crystalline polymers. Knowledge of pressure dependence
is important since it makes possible the separation of volume and
temperature effects. This separation cannot be done by varying
temperature alone. The choice of crystalline polymers for the
study was made because the structure of the crystal is better known
that that of the liquid or glass states and because considerable
dielectric pressure data already exist for amorphous polymers.
Finally, a dynamical description of the relaxation process is given
and the relaxation behavior is described in terms of bulk thermodynamic properties.

MEASUREMENT SYSTEM

A simple system has been developed for measuring the dielectric
properties of solid materials at temperatures down to $-60°C$ and at
pressures up to 2 kbar (1 kbar = 1013 atmospheres = 10^5 N/m^2). The
electrodes shown in Figure 1 consist of rectangular brass plates
4 cm x 8 cm x 6 mm thick. One plate serves as the high potential
electrode and the other as the guard. In the guard electrode are
located three circular low potential electrodes, a central 1.9 cm
diameter "sample" and two 1.25 cm diameter "spacing" electrodes
wired in parallel. The electrodes are cemented into and insulated

*Contribution of the National Bureau of Standards, not subject to
 copyright.

from the guard plate by epoxy cement. The electrodes must be re-
gularly lapped flat to keep them planar. The sample, in the form
of a flat circular disk, 2.5 cm in diameter, is placed between the
plates so that it covers the "sample" electrode but not the spacing
electrodes. The plates are held together against the sample by
spring loaded nylon screws as shown in Figure 2. The measured
capacitance and conductance of the sample electrodes yield the
dielectric data for the sample and the measured capacitance of the
spacing electrodes yields data on the sample thickness. The cell
is used like the one described by McCammon and Work (1).

Figure 1.   Unassembled electrodes and sample.

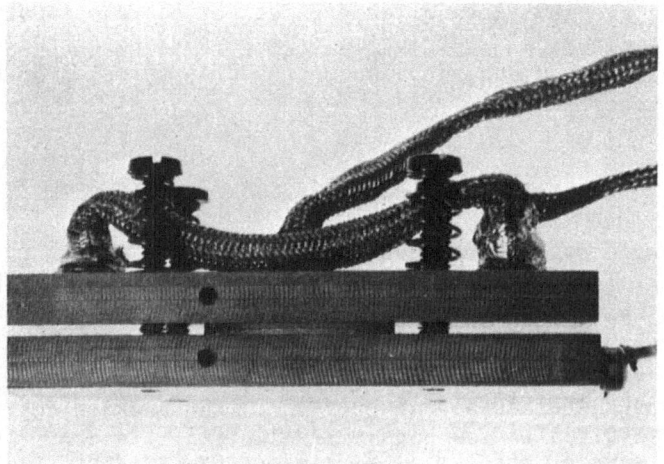

Figure 2.   Electrode assembly with sample in place.

Hydrostatic pressure is applied to the sample through a pressure fluid. We chose hexane because it has a fairly low freezing point, still remains liquid at room temperature, and has reasonably low polarity and high dielectric stability. Many polymers are dissolved or swollen by hydrocarbons so to protect the sample from hexane we sealed the sample in a plastic pouch. The plastic pouch was made from a commercial heat sealable polyester - polyolefin sandwich film (2). To check possible leaks, the sealed sample was placed in hexane. It was found that air diffused out of a well sealed pouch when submerged in hexane, leaving the sample in close contact with the film. The measurements were corrected for the series impedance of the air (or hexane) gap and the film without serious loss of accuracy or sensitivity. The dielectric constant of the hexane was measured as a function of temperature and pressure (3) so that the capacitance of the hexane-immersed spacing electrodes could be used to calculate the electrode spacing and hence the sample thickness.

The electrodes containing the sealed sample were suspended in a pressure bomb in a thermostated bath. The pressure and bath systems have been described elsewhere (3). The electrical measurements were made using a Harris ultra low frequency bridge (4) from 10^{-2} to 10^4 Hz, a GR 1615A decade transformer bridge from 10^2 to 10^4 Hz and a Cole-Gross type transformer bridge (3) from 10^2 to 4×10^5 Hz. The overlap in frequency range between instruments was used to check accuracy.

SAMPLE

Methyl stearate (MS), $CH_3-(CH_2)_{16}-\overset{\overset{\textstyle O}{\textstyle \|}}{C}-O-CH_3$, is a member of a large class of paraffin-like molecules which exhibit crystal α and γ relaxations (5). These relaxations are typical of those found in crystalline polymers (6). Because the paraffin derivatives can be better purified and characterized than polymers, they are useful as model polymer systems.

The MS samples used for this data were either used as received from a commercial source or were melted, degassed in vacuum and zone refined several times in a 7 mm glass tube until the central portion of the sample formed nearly transparent well ordered crystals. This central material was then melted and quenched into a 1 mm thick disc. Even extensive zone refining was unable to remove neighboring homologs and gas chromatography showed about 1% of an undetermined impurity. The disk was sealed in a pouch which was then degassed in hexane.

RESULTS

Figure 3 shows a typical frequency plot of the imaginary part of the dielectric constant, ε'', for the different samples of methyl stearate (MS) at –115 and –118°C. The higher frequency γ relaxation is centered at 10^{+5} Hz and the lower frequency α relaxation at 10^{-1} Hz. The γ peak at room temperature is larger than the α peak, according to Dryden and Welsh (7), but decreases rapidly in amplitude with decreasing temperature. The γ peak does not seem as sensitive to thermal treatment as does the α peak, but more extensive measurements must be made to better characterize its behavior.

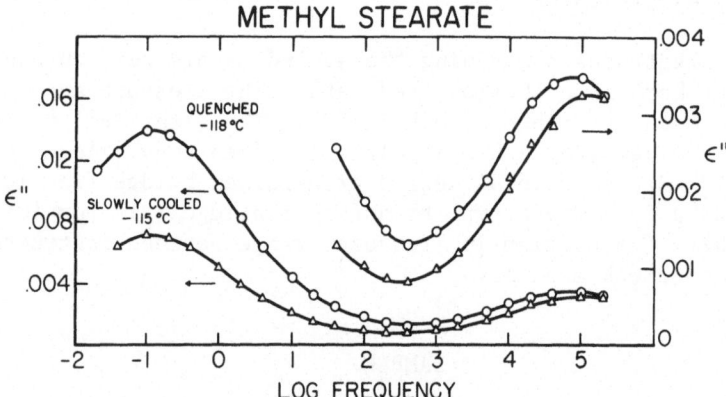

Figure 3. ε'' vs frequency for two samples of methyl stearate with different thermal history.

The decreases in α peak amplitude with purification and annealing are typical of the long chain solids (5,8-10). Figure 4 shows the effect of purification and more gradual solidification on the amplitude of the α relaxation (note the effect of purification on the low frequency wing). However, there is no detectable shift in the frequency of maximum absorption, f_{max}. As shown in Figure 5, there is no effect on the dispersion width as measured by the Cole-Cole α parameter. Thus to a good approximation, only the number of rotating molecules are affected and not the details of the molecular process.

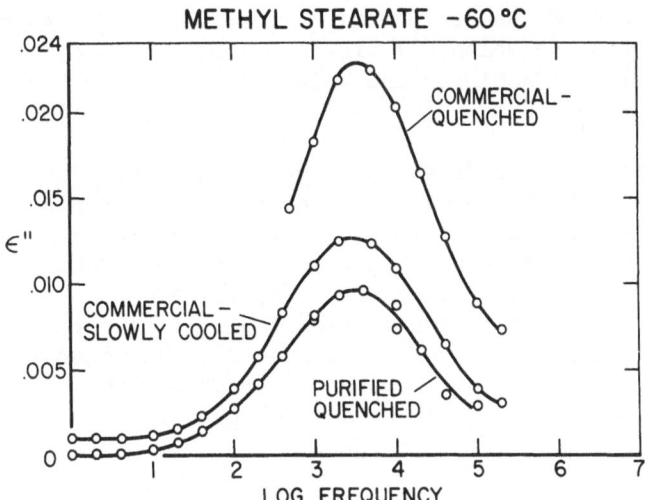

Figure 4. ϵ'' as a function of frequency at T-60°C and P=1 bar
showing the effects of purity and crystallization.

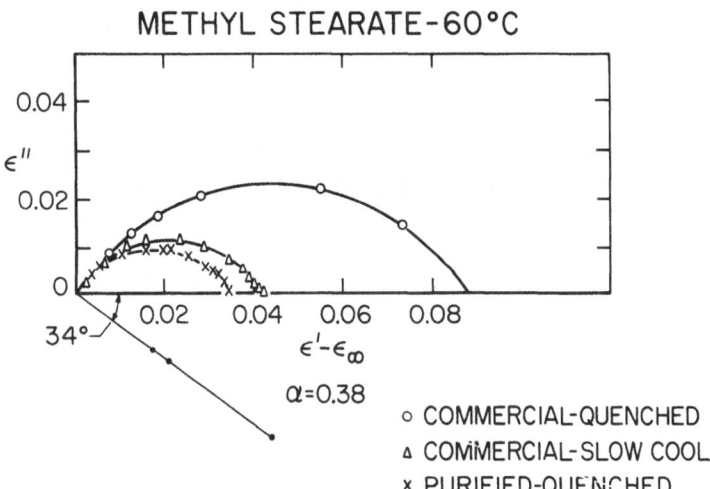

Figure 5. Cole-Cole plot showing constancy of relaxation width
(α parameter) with changes in amplitude for data shown
in Figure 4.

In Figure 6 is shown the Cole-Cole plots for the α relaxation at zero pressure, P, for a commercial sample as a function of temperature. A similar sample measured at various temperatures and pressures is shown in a plot of ε" versus frequency in Figure 7. Figures 8 and 9 show the linear dependence of the log f_{max} for the α relaxation in MS on reciprocal temperature and pressure. The data of Dryden and Welsh (7) are also included in Figure 8 and more recent results of Robinson and Welsh (11) give a curve very similar to that in Figure 8.

A summary of the data from Figure 7 is given in Table I. Remember that the dispersion amplitude, Δε, is sample dependent but at these temperatures is stable with time for a given sample. Hence, the behavior of Δε with temperature and pressure is significant only for a single sample and comparison should not be made between Δε values from Figure 6 and Table I. A careful analysis of uncertainties has not yet been done and the uncertainties shown in Table I are estimated from the apparent lattitude in derived values and are meant merely as a rough guide.

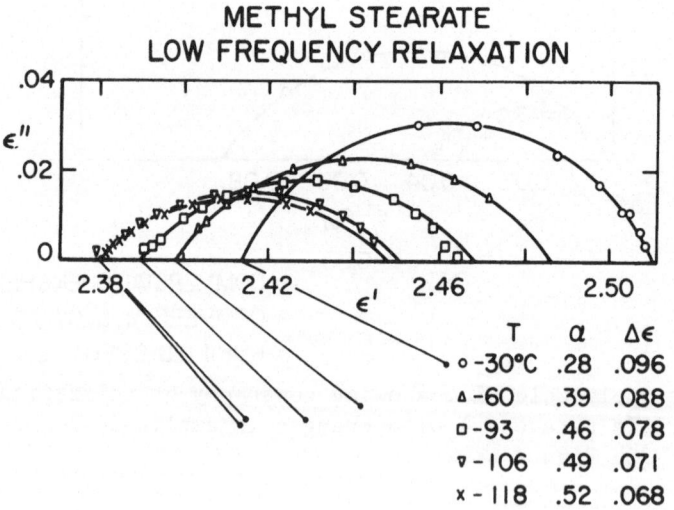

Figure 6.   Cole-Cole plots as a function of temperature for a
            commercial sample.

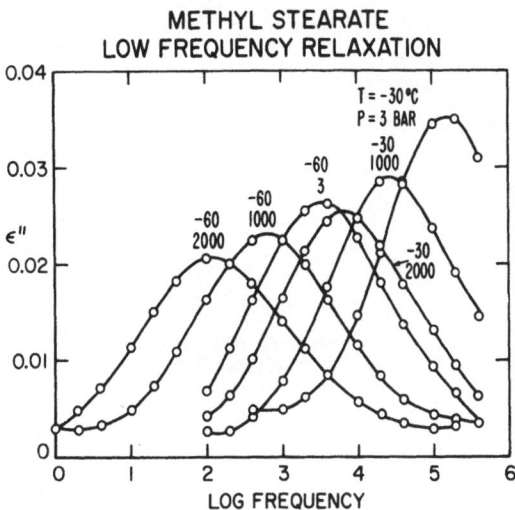

Figure 7.   ε" as a function of frequency at different temperatures
and pressures.

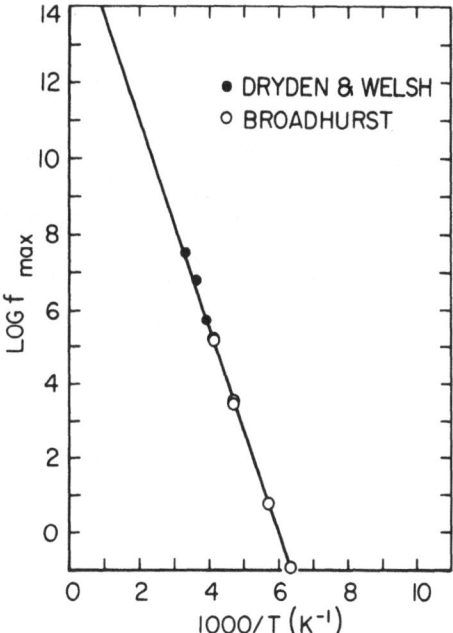

Figure 8.   Log of the frequency of maximum ε" as a function of
reciprocal temperature for methyl stearate.

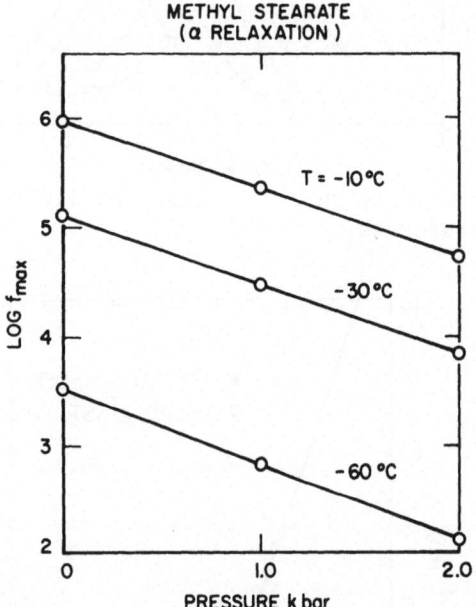

Figure 9.   Log of the frequency of maximum ε'' as a function of
pressure at different temperatures for methyl stearate.

TABLE I

Parameter	T(K)	P(kbar)		
		0	1	2
Log f_{max} (Hz) ± 0.1	213	3.52	2.85	2.11
	243	5.11	4.48	3.83
	263	5.98	5.36	4.78
Cole-Cole α ± 0.05	213	0.38	0.42	0.44
	243	0.32	0.35	0.37
	263		0.27	0.32
$\Delta\varepsilon$ ± 0.05	213	0.100	0.983	0.088
	243	0.124	0.106	0.095
	255		0.102	0.099
ε_∞ ± 0.01	213	2.520	2.520	2.528
	243	2.464	2.500	2.514
	263		2.492	2.506
$\Delta H^{\neq}$ (kcal/mol) ± 0.2		12.6	12.9	13.6
Log A (A in sec^{-1}) ± 0.5		16.4	16.0	16.1
$\Delta S^{\neq}$ (e.u.) $\pm 1.$		20	18.	19.
$\Delta V^{\neq}$ (cm^3/mol) $\pm 2.$		29.0	30.2	30.7

(1 kcal/mol = 4.184 kJ/mol, 1 e.u. = 4.184 J/mol K)

DISCUSSION OF RESULTS

The dependence of log f_{max} on temperature and pressure is similar to that reported by Williams (12) for the glass transition relaxation in amorphous polymers well above T_g. The α relaxation width increases with increasing pressure and decreasing temperature. The Cole-Cole α values shown in Figure 6 fall on the same linear α-temperature curve as those for ethyl stearate as reported by Davidson (8). The ε_∞ value changes in the proper way to reflect the change in sample density with T and P. One expects the quantity $(\partial T/\partial P)_{\varepsilon_\infty}$ to equal $(\partial T/\partial P)_V = (1/\alpha_V B_T)$ where $\alpha_V \approx 2 \times 10^{-4}/°C$ is the volume coefficient of expansion (13) and $B_T \approx 50$ kbar is the isothermal bulk modulus (14). The values in Table I show this to be true within experimental error with $(\partial T/\partial P)_{\varepsilon_\infty}$ 100°C/kbar. The ratio $[\varepsilon_0(solid)-\varepsilon_\infty(solid)]/[\varepsilon_0(liquid)-\varepsilon_\infty(solid)]$ gives a rough estimate of the fraction of molecules participating in the relaxation. For ES this fraction was shown (10) to vary from nearly 0 to about 0.5. Methyl stearate gives similar magnitudes for $\varepsilon_0-\varepsilon_\infty$.

It thus seems that the α relaxation results from the existence of a metastable phase rather than crystal defects. This phase should be characterized mostly by rotational disorder of the type discussed by Brot and Darmon for chlorobenzenes (15). Such a model is compatible with the proposal of Davidson that the molecules are first in a metastable rotational state which has alternative available angular positions, but that upon annealing the molecules progressively rotate into a state with only a single "deep" position. It is reasonable that the cooperation needed for the molecules to "decide" on which will be the final crystal orientation (as far as the C-C-C zigzag planes are concerned) might require considerable time for completion.

The quantity $(\partial T/\partial P)_{f_{max}} \approx 15°C/kbar$ is of the same magnitude and has the same sort of f_{max} dependence as the amorphous β relaxation (12) [using the notation of Hoffman, Williams and Passaglia (6)]. The ratio of the constant volume to constant pressure activation energies is $1-(\partial P/\partial T)_V(\partial T/\partial P)_{f_{max}}$ (12) which equals 0.85 for MS. This is quite close to Williams' values for polymethyl acrylate (12) extrapolated to similar temperatures.

The results found here for methyl stearate are consistent with generally observed behavior of crystalline long chain compounds. Surprisingly, they also have similarities with the rubbery β relaxation in polymers well above T_g.

A DYNAMICAL DESCRIPTION OF THE RELAXATION PROCESS

The crystalline α relaxation in polymers arises from rotational motions of large polymer segments about their extended chain axes. We will ignore the effects of chain twisting (6,16) and consider only the "rigid-rod" rotations of lower molecular weight polymers. This rotational diffusion occurs by occasional rotational "jumps", often through large angles, from one equilibrium orientation to another. The equilibrium orientations or "sites" are minima (with respect to angular position, θ) in the potential energy $U(\theta)$ of a segment. Since a segment and its neighbors are in continual motion, the potential energy will naturally vary with time and it is important to recognize that $U(\theta)$ is a time average function.

The orientational sites are separated by energy barriers, W [maxima in $U(\theta)$], which in cases of present interest are considerably greater than the mean thermal energy (W ≈ 10 kT). Most of the time the segments will merely librate in their sites. But occasionally, fluctuations will occur in the local vibrational motions and produce a combination of circumstances allowing an angular "jump" from one site to another over the intervening barrier. For instance, the rotational kinetic energy of a segment can increase to a value sufficient to carry it over the barrier (1/2 $I\omega^2$ ≈ W >> kT where I is the moment of inertia and ω the angular velocity at the minimum in $U(\theta)$. Or, a severe distortion in $U(\theta)$ can reduce W to some value (W ≈ kT) such that the segment will cross the barrier during a normal librational cycle. Thermodynamically, the kinetic energy fluctuation corresponds to a temperature fluctuation and the barrier fluctuation to a volume fluctuation. Clearly both can be expected to occur and contribute to the jumps.

The above dynamical description of the rotational jump diffusion process has been applied to ordinary translational diffusion for many years (17,18) and we can directly apply the results of that work. A general jump frequency f_{max} can be calculated in terms of a critical value R of a reaction coordinate and a mean vibrational frequency, ω_o. The result (18) is the familiar form

$$f_{max} = (\omega_o/2\pi)e^{-W/kT} \tag{1}$$

where

$$W = \frac{1}{2} m\omega_o^2 |R|^2 \tag{2}$$

and m is the mass of the diffusing particle. A similar result was obtained directly from the normal mode solution of a mass-spring model where R was taken as the critical radial expansion of nearest neighbors away from a central mass (19).

In related problems using elasticity theory to calculate the work to expand a local volume v to a critical value $v+\Delta v_c$ (20) and using thermodynamic fluctuation theory to calculate the probability of a spontaneous local volume fluctuation from v to $v+\Delta v_c$ (21), it was found that

$$W = \frac{1}{2} B_T \Delta v_c^2 / v \tag{3}$$

where $B_T = - (\partial P / \partial \ell n V)_T$ is the isothermal bulk modulus. One can easily see the connection between Equations 2 and 3 by remembering the relationship (approximate) between the frequency, $(\omega_o / 2\pi)$, and wavelength, a, of intermolecular vibrations and the bulk modulus of the solid,

$$a \; \omega_o \; \infty \; (B_T)^{1/2} \tag{4}$$

A common form of Equation 2 is given (22) in terms of the Debye θ_D for the solid and it should be clear that $\theta_D \approx h \, \omega_o / k$ where h and k are the Planck and Boltzmann constants.

Thus, the proportionality between W and the lattice stiffness (B_T or ω_o^2 or θ_D^2) is quite general. What is not clear is how to calculate the critical values R or Δv_c. Attempts to do so have frequently been made (23,24) without satisfying results. Let us consider a simple model and use that result as a guide in assuming a more general result.

For example, take a simple rigid diatom of length 2r. Each atom, 1 and 2, interacts with its surroundings with a one-dimensional potential U_1 and U_2. The center of the diatom is at a distance X from its surroundings and ϕ is the angle between the diatom axis and the X axis. If the potential minimum is at X_o and the diatom is short compared to its distance from its neighbors so that $X_o \gg r$, then the total potential can be expanded in the region of X_o to give,

$$U(X,\phi) = U(X) + \frac{1}{2} U''(X) \, r^2 \cos^2\phi + \dots \tag{5}$$

That is, we have an equivalent monatomic problem with super-imposed but largely uncoupled rotational motions. In this case the potential energy barrier will be $U(X,\pi/2) - U(X,0)$ or

$$W = r^2 U''(R) \tag{6}$$

For this simple model (which should be a good first approximation for nearly-round molecules) the barrier is proportional to the lattice force constant U'' and the essentially volume independent length of the molecule. Thus, remembering the connection between B_T and $U''(R)$ (14), there is considerable precedent and simple-model support for assuming that the result

$$W(v) = \text{const. } B_T(v) \tag{7}$$

has sufficient generality to cover the case of the crystalline α relaxation in polymers.

Having assumed Equation 7, it is easy to describe the volume dependence of W by expanding about some reference volume v_0, the equilibrium volume at pressure P_0 and temperature T_0. The result is

$$W(v) = W(v_0) \, [1+(d\ell nW/d\ell nv)(v-v_0)/v] \tag{8}$$

Writing Equation 8 explicitly as a function of temperature and pressure,

$$W(v) = W(v_0) \, \{1+(d\ell nW/d\ell nv)[\alpha(T-T_0)-(P-P_0)/B_T]\} \tag{9}$$

where $\alpha = (\partial \ell nV/\partial T)_P$ and $B_T = -(\partial P/\partial \ell nV)_T$ are the volume coefficient of expansion and bulk modulus. Equation 9 was used by Frenkel (17) for ordinary diffusion. From Equation 7 and the quasi-harmonic approximation for solids (14,25), one can evaluate the coefficient

$$(d\ell nW/d\ell nv) = (d\ell nB_T/d\ell nV) = -2\gamma \tag{10}$$

where γ is the Grüneisen constant. Since γ is well known and reasonably constant for linear crystalline polymers (14) Equation 9 completely determines the pressure and temperature dependence of the relaxation frequency. In addition, we can take into account the fact that each chain unit contributes equally to the barrier so that W is also proportional to the number of units n. For convenience, the const. in Equation 9 can be expressed in terms of f_{mo}, the empirical value of f_{max} for reference conditions $n=n_o$, $T=T_o$ and $P=P_o$,

$$f_{max} = f_o \exp\{-(n/n_o)(T_o/T)\ell n(f_o/f_{mo})[1-2\gamma\alpha(T-T_o)$$

$$+ 2\gamma(P-P_o)/B_T]\} \tag{11}$$

Three derivatives of Equation 11 are of interest:

$$(\partial logf_{max}/\partial P)_{n,T} = -(n/n_o)(T_o/T)(2\gamma/B_T)log(f_o/f_{mo}) \tag{12a}$$

$$[\partial logf_{max}/\partial(1000/T)]_{n,P} = -(n/n_o)(T_o/1000)log(f_o/f_{mo})$$

$$[1+2\gamma\alpha T_o+2\gamma(P-P_o)/B_T] \tag{12b}$$

$$(\partial logf_{max}/\partial n)_{T,P} = -(1/n_o)(T_o/T)log(f_o/f_{mo})$$

$$[1-2\gamma\alpha(T-T_o)+2\gamma(P-P_o)/B] \tag{12c}$$

We can evaluate these derivatives using typical values: $\gamma = 6$ (14), $\alpha = 3 \times 10^{-4}$/K (25), $B_T = 30$ kbar (25) and $f_o = 10^{12}$/sec (26). The reference conditions are chosen to be n = 20 (to correspond to MS), $T_o = 300$ K and $P_o = 0$, and for these conditions log f_{mo} was experimentally found to be 7.3 (from Figure 8). These data yield the predicted values in Table II for Equations 12.

TABLE II

Predicted (Equation 12) and Experimental Chain Length, Temperature and Pressure Dependence of the Rigid-Rod α Relaxation in Polymers

Quantity	Predicted	Experimental
$(\partial \log f_{max}/\partial n)_{T=300K, P=0}$	$- 0.24$/unit	$- 0.25$[a]
$(\partial \log f_{max}/\partial(1000/T))_{n=20, P=0}$	$- 2.9$ K	$- 2.7$ K[b]
$(\partial \log f_{max}/\partial P)_{n=20, T=263}$	$- 2.1$/kbar	$- 0.61$/kbar[b]

[a]Reference 5

[b]This work

The dynamical theory applied to the crystalline α relaxation gives results which are quite encouraging except for the discrepancy with the pressure data. Even though there is considerable simplification in the development given here, the reason for so large a discrepancy is yet to be explained.

Finally, because polymer relaxation data is frequently summarized in terms of reaction rate parameters, it is useful to explicitly mention the correspondence between the dynamical theory (11) and reaction rate theory,

$$f_{max} = (kT/h) \exp-(\Delta H^{\neq} - T\Delta S^{\neq})$$

$$= (kT/h) \exp-(\Delta U^{\neq} - T\Delta S^{\neq} + P\Delta V^{\neq}) \tag{13}$$

According to dynamical theory, the reaction rate parameters are not independent but are connected through the bulk properties α, B_T and γ. Thus,

$$\Delta S^{\neq}/\Delta H^{\neq} = 2\gamma\alpha/(1+\gamma\alpha T_o) \tag{14a}$$

$$\Delta S^{\neq}/\Delta U^{\neq} = \alpha B_T \tag{14b}$$

Empirical relationships like Equations 14 have been reported for a variety of relaxation processes (27,28).

In summary, it seems that the dynamical theory of diffusion in the quasi-harmonic approximation provides a simple and largely correct description of the relaxation behavior of the type reported here for (MS).

REFERENCES

1. R. D. McCammon and R. N. Work, Rev. Sci. Instr. 36, 1169 (1965).
2. Scotchpak No. EX-31224 heat sealable polyester film made by the 3M Company. We have also tried FEP film for this purpose.
3. F. I. Mopsik, J. Res. NBS 71A, 287 (1967).
4. W. P. Harris, Annual Report, Conference on Electrical Insulation and Dielectric Phenomena NAS-NRC (1966) p. 72.
5. R. J. Meakins, PROGRESS IN DIELECTRICS, 3, 180, John Wiley, New York, 1961.
6. J. D. Hoffman, G. Williams and E. Passaglia, J. Polymer Sci. C, 14, 173 (1966).
7. J. S. Dryden and H. K. Welsh, Aust. J. Scient. Res. A4, 616 (1951).
8. D. W. Davidson, Can. J. Chem. 39, 1321 (1961).
9. J. Crissman of NBS grew a single crystal of $n-C_{20}$ which had no mechanical α peak (private communication).
10. M. G. Broadhurst, J. Chem. Phys. 33, 210 (1960). Ethyl stearate single crystals showed essentially no α relaxation.

11. W. E. Robinson and H. K. Welsh, Aust. J. Chem. 22, 1299
    (1969).
12. G. Williams, Trans. Faraday Soc. 60, 1556 (1964); 61, 1564
    (1965); 62, 1321, 1329, 2091 (1966).
13. M. Craig, J. Am. Oil. Chem. Soc. 34, 30 (1956).
14. M. G. Broadhurst and F. I. Mopsik, J. Chem. Phys. 52, 3634
    (1970).
15. C. Brot and I. Darmon, J. Chem. Phys. 53, 2271 (1970).
16. G. Williams, J. I. Lauritzen, Jr. and J. D. Hoffman, J. Appl.
    Phys. 38, 4203 (1967).
17. J. Frenkel, KINETIC THEORY OF LIQUIDS, Oxford University Press,
    Oxford, 1946.
18. M. D. Feit, Phys. Rev. B3, 1223 (1971).
19. F. Bueche, J. Chem. Phys. 30, 748 (1959).
20. J. S. Dryden and R. J. Meakins, Disc. Faraday Soc. 23, 39
    (1957).
21. M. G. Broadhurst, Conference on Electrical Insulation and
    Dielectric Phenomena - 1969 Annual Report, National Academy
    of Sciences, Washington, D.C. (1970).
22. H. R. Glyde, J. Phys. Chem. Solids 28, 2061 (1967); B. N.
    Oshcherin, Phys. Status Solidi (Germany) 43, K59 (1971).
23. J. Chichocki, J. de Phys. 9, 129 (1938).
24. R. W. Keyes, J. Chem. Phys. 29, 467 (1958).
25. M. G. Broadhurst and F. I. Mopsik, J. Chem. Phys. 54, 4239
    (1971).
26. M. Tasumi and S. Krimm, J. Chem. Phys. 46, 755 (1967).
27. A. W. Lawson, J. Chem. Phys. 32, 131 (1960).
28. R. K. Eby, J. Chem. Phys. 37, 2786 (1962).

THE MEASUREMENT OF DIELECTRIC CONSTANT AND LOSS OF POLYMERIC SOLIDS

AND MELTS OVER A WIDE FREQUENCY RANGE

Christopher H. Porter and Richard H. Boyd

Department of Chemical Engineering and Division of Materials

Science and Engineering, University of Utah, Salt Lake City,

Utah

INTRODUCTION

Dielectric measurements have been widely used to characterize the relaxation behavior of polymers, and a considerable literature of such work exists. The basic techniques for making measurements over a very wide frequency range, from considerably less than 1 Hz to greater than 10 GHz, have existed for some time. However, the bulk of the work on polymers has been done either on materials that behave as solids or on solutions and in the frequency region readily accessible to impedance-bridge techniques (~20 Hz to ~1 MHz). In the latter range, accurate data is readily obtainable (often using guarded cells) with commercial bridges. Above this frequency range, the techniques are less straightforward and the results subject to greater error. In the course of determining the effects of melting on dielectric relaxation in crystalline polymers (1), it was necessary to develop methods that would yield data as accurate as possible up to the microwave region and that would be as consistent as possible between differing measuring techniques. Further, the following capabilities were demanded:

(a) the samples could be viscous melts (including some rather thermally unstable ones), as well as solids or solutions and could be very lossy;

(b) a wide temperature range could be studied, up to at least 250°C.

A coaxial sample-cell configuration was found best to meet the above requirements in all frequency regions. Standard bridge techniques sufficed in the appropriate frequency region (up to ~1 MHz). Above this range (1-65 MHz), a resonant circuit technique was developed that yields, under the above conditions, data whose accuracy is equal to or better than that traditionally associated with this method. In the region, 300 MHz to 8 GHz, refinements and modifications to the slotted-line method were made that permitted accurate data to be obtained under the above circumstances. Figures 1 and 2 illustrate typical overall results obtained. In the belief that a detailed unified description of these techniques would be of general use, we present the following account.

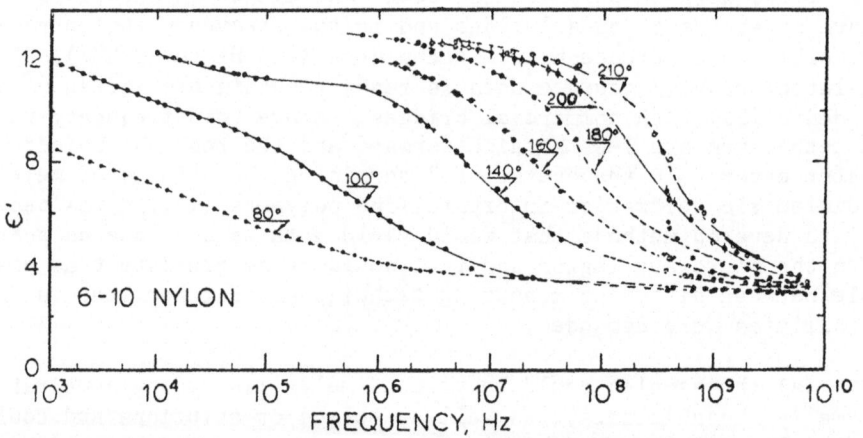

Figure 1. Dielectric constant of 6-10 Nylon as a function of frequency. (From R. H. Boyd and C. H. Porter, Reference 1; full paper to be published.)

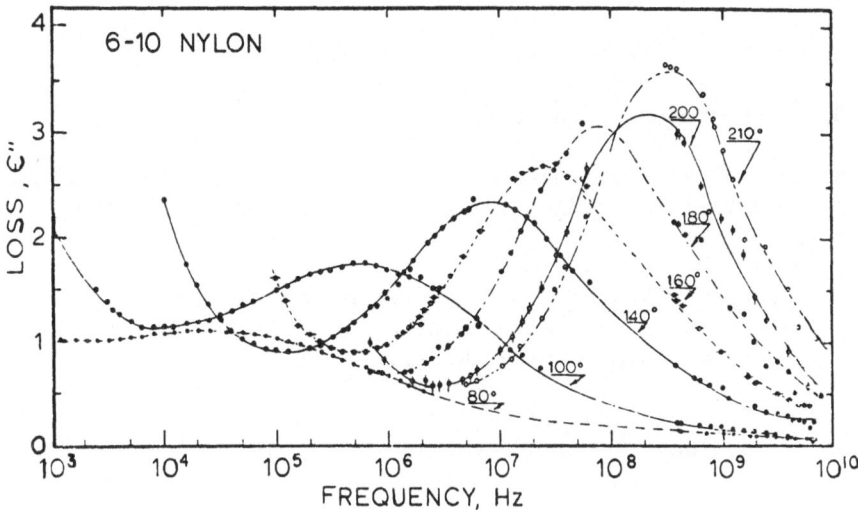

Figure 2. Loss factor of 6-10 Nylon as a function of frequency.
(From R. H. Boyd and C. H. Porter, Reference 1; full
paper to be published.)

EXPERIMENTAL PROCEDURE

There are basically two approaches used in measuring the
dielectric properties of materials. These methods are dictated by
the frequency of interest. At low frequencies, where all sample
lengths, electrical paths, and leads are short as compared with
1/4 wavelength, a lumped circuit approach is used. These techniques
have been used successfully from 0 to 10^8 Hz. At higher frequencies,
where the dimensions of the circuit approach 1/4 wavelength, a
distributed circuit approach is used. This approach can be applied
from 10^8 through the frequencies of visible light.

Lumped Circuits

The frequencies used in lumped circuits fall into three basic
categories, determined by the method required for measurements.
These are:

1. low-frequency range (0-10 Hz),

2. medium-frequency range (1-10^7 Hz),

3. high-frequency range (10^5-10^8 Hz).

Reviews of standard methods of measuring dielectric constants in the aforementioned ranges are found in Von Hippel (2) and McCrum (3).

Low-Frequency Range: The low-frequency range was not measured during the course of this experiment with the exception of measurements of the D.C. conductivity (0 Hz).

Medium-Frequency Range: The measurements in the medium-frequency range were obtained by using two separate impedance bridges. The basic circuit of both these bridges is illustrated in Figure 3.

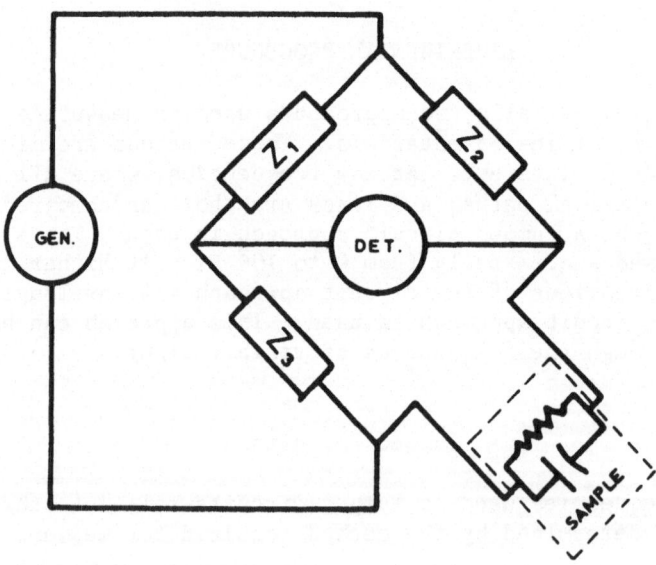

Figure 3.   Schematic diagram of a typical impedance-bridge circuit.

The principle of operation of these bridges is similar to that of a Wheatstone bridge with the exception that impedances are measured instead of pure resistances. The voltage across the detector reads a minimum when the bridge is balanced. Balancing involves adjusting one or a combination of the bridge impedances (Z_1, Z_2 and Z_3) in order that the magnitude and the phase of the voltage drops across Z_1 and Z_2 are equal.

A General Radio Type 1615-A transformer-arm bridge was used for frequencies between 20 and 10^5 Hz. Referring to Figure 3, Z_2 is an arm of a transformer. Z_1 is a winding which is tapped in ten places and Z_3 is a combination of precision capacitors and resistors. The oscillator, General Radio model 1310, is inductively coupled to the circuit. A General Radio model 1232-A null detector was used. This detector can be tuned continuously from 20 to 20,000 Hz and has fixed filters for 50 to 100 KHz. To fill in the gaps between 20 and 100 KHz, a parallel-filter network was constructed to pass frequencies of 25, 32, 40, 63 and 80 KHz while the detector was operating in the "flat" or untuned mode.

The bridge was operated in the three-terminal (guarded) position. This bridge is constructed in such a manner that the guard point is at ground potential. The high-voltage (low-impedance side) lead was connected to the shell of the sample cell and the low voltage (high impedance side), which is susceptible to pickup from external sources, was connected to the inner conductor of the coaxial cell. Both leads were shielded relative to ground from the bridge to a terminal box, which contained the connector for attaching the sample cell.

The described arrangement provided a circuit in which the capacitance of the leads were eliminated. There was some residual capacitance between the inner and outer conductor of the sample cell "lead in," the portion of the sample cell between the connector and the actual measurement area above the gasket. As a result the actual circuit measured by the bridge is illustrated in Figure 4. After converting the bridge readings from the series to parallel configuration model (note: dissipation factor is same for both models), the total values of the capacitance and dissipation factor are,

$$C_p = C_{res} + C_{px} \tag{1}$$

$$D = 1/(\omega C_p R_{px}) \tag{2}$$

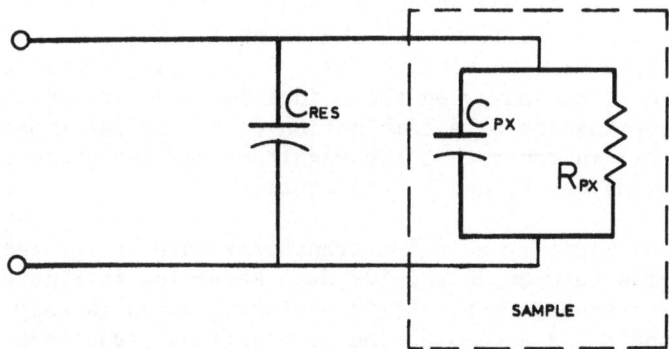

Figure 4.   Equivalent circuit assumed for capacitance measurement.

Noting that the capacitance of the sample is, by definition,

$$C_{px} = \varepsilon' \, C_o \tag{3}$$

and that the capacitance reading of the bridge with an empty sample holder (ε'_{air} taken to be equal to 1.000) is

$$C_{mt} = C_{res} + C_o \tag{4}$$

the capacitance reading at the bridge with a sample in the cell then becomes

$$C_p = C_{mt} + (\varepsilon' - 1) \, C_o \tag{5}$$

The dissipation factor of the sample is given by the expression,

$$D_x = 1/(\omega C_{px} R_{px}) = DC_p/C_{px} \tag{6}$$

Rearrangement of the above equations leads to the following equations for ε' and ε'',

$$\varepsilon' = (C_p - C_{mt})/C_o + 1 \tag{7}$$

and

$$\varepsilon'' = \varepsilon' D_x = DC_p/C_o \tag{8}$$

Frequently, when making measurements at low frequency, the dissipation factor of the sample exceeded the range of the bridge. When this occurred, mica capacitors were connected in parallel with the sample cell. This increased the circuit capacitance without substantially increasing the circuit resistance. Inspection of Equation 2 reveals that an increase in C_p results in a decrease in D at constant frequency and R_{px}.

In general, the mica capacitors placed in parallel had a loss that was small but not negligible in comparison with that of the sample. The equivalent circuit of the sample cell and shunt capacitor is shown in Figure 5. Analysis of this circuit reveals that the expressions for ε' and ε'' are

$$\varepsilon'' = DC_p/C_o - D_{sh}C_{sh}/C_o \tag{9}$$

and

$$\varepsilon' = (C_p - C_{sh} - C_{mt})/C_o + 1 \tag{10}$$

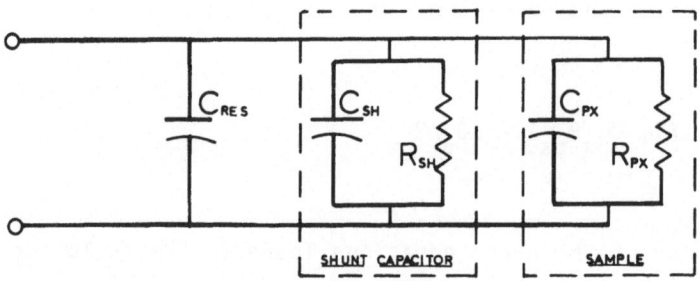

Figure 5.   Equivalent circuit assumed for capacitance measurements
when using a shunt capacitor.

Although in principle ε' could be calculated from Equation 10, in
practice this leads to highly inaccurate answers. The values of
C_p and C_{sh} were in all cases greater than 7,000 pf, and in most
cases greater than 100,000 pf, but only differed from one another
by about 100 pf. As a result, ε' calculated from Equation 10
involved finding a small difference between two large numbers
leading to unreliable and inaccurate results. For this reason, ε'
data are sometimes not reported at low frequencies, where loss
data is satisfactory.

The frequency range between 100 KHz and 3 MHz was studied
using a General Radio Type 716 CSI high-frequency bridge. Referring
to Figure 3, the bridge impedances are as follows:

1. Z_1 is a variable capacitor in parallel with a resistor.
This capacitor is calibrated to measure .01 D/f(MHz) directly.

2. Z_2 is a fixed capacitor and resistor in parallel.

3. Z_3 is a precision variable capacitor marked off in 0.2 pfd
divisions and is used to balance the capacitance of the cell.

The bridge is constructed so that both series and substitution methods
can be used: the substitution being more accurate but less flexible.

The null detector available, GR 1232 A, had a maximum frequency capability of 100 KHz and could not be used directly to detect the null on the bridge. As a result, a R.F. mixer was constructed which had a tuned 16 KHz output. The detector output signal from the bridge was mixed with a reference signal from a Knight KG-686 R.F. generator which operated at 16 KHz greater than the bridge frequency, the bridge being driven at 30 volts with a Hewlett-Packard 606A signal generator. The output of the R.F. mixer was then fed to the GR 1232 A null detector, tuned to 16 KHz, which now detected the "beat" between the bridge output and the reference signal. This procedure is commonly referred to as "heterodyning."

The measured values of capacitance of this bridge can be read to the closest 0.1 pf which leads to typical three-place accuracy in ε' provided that C_o is larger than 10 pf. In some cases, such as that of Nylon 6-10, C_o had to be made small in order that the dissipation factor did not exceed the range of the bridge at low frequencies. Also, this bridge is not "guarded," and the value of C_{res} in Equation 1 included the lead capacitance which was about 100 pf. In this case the calculated value of ε' involved a small difference in large numbers with the expected loss of accuracy.

The bridge was connected to the sample cell through the same terminal box as the 1615 bridge and the cell circuit transferred by means of a rotary switch. As a result, the value for C_{mt} in Equation 4 could be determined by comparing values of capacitance between the two bridges at 100 KHz, or in other words, the values for ε' at 100 KHz were forced to agree. This was not true for the value of ε'', however, and in all cases good agreement between both bridges was observed.

High-Frequency Range: The high-frequency range (0.6 MHz-63 MHz) was studied using a parallel-resonance circuit. Ideally, the circuit used in this method is shown in Figure 6. An analysis of this circuit shows that at resonance (voltage reads a maximum, admittance a minimum), the sum of the imaginary parts of the admittance between points a and b is equal to zero,

$$Y_{i\ a,b} = j\ \omega(C_d + C_v + C_{px} + C_{leads}) - j/\omega L = 0 \qquad (11)$$

A sample is placed in the cell and is perturbed by frequency $\omega/2\pi$. The circuit is tuned to resonance by adjusting capacitor C_d to value C_{d1}. The sample is then removed and the circuit is returned to resonance with C_d to value C_{d2}. The difference between the readings of C_{d2} and C_{d1} is the difference in the capacitance of the sample and empty cell,

$$C_{d2} - C_{d1} = \Delta C_d = C_{px} - C_o \qquad (12)$$

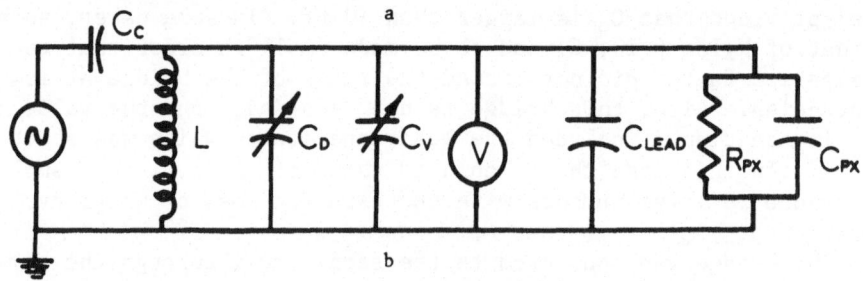

Figure 6.   Idealized resonant circuit.

Substitution of Equation 3 for C_{px} results in

$$\frac{\Delta C_d}{C_o} = (\varepsilon' - 1) \qquad (13)$$

The value of R_{px} is found by finding the width of the resonant peak. The circuit is tuned to resonance and the maximum voltage is noted. The circuit is then detuned on both sides of the resonant peak to voltage V using a vernier capacitor, C_v. The difference in capacitance between V through the maximum and back to V on the other side of the peak is then ΔC_v. The loss of the sample is then

$$\epsilon'' = \frac{\Delta C_{v-s} - \Delta C_{v-mt}}{2 C_o \sqrt{m^2 - 1}} \tag{14}$$

where $m = V_{max}/V$. The derivation of the above equation can be found in the Appendix.

In reality, the circuit illustrated in Figure 6 is over-simplified and a more realistic model is illustrated in Figure 7. The essential difference in the two models is that the impedance of the leads cannot be represented by merely a capacitance but rather as a distributed reactance. To permit analysis of this reactance, the leads to the cell were constructed in a 50 ohm characteristic impedance configuration. This allows the use of transmission-line equations to find the impedance of the sample.

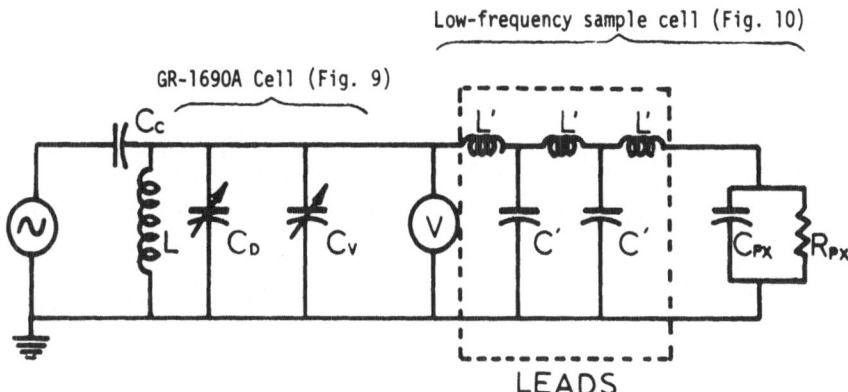

Figure 7. Resonant circuit containing lead parameters.

The equations for finding the sample parameters using trans-mission-line equations are found in the Appendix. In summary, the resulting equations are:

$$\frac{\Delta C_{v-s} - \Delta C_{v-mt}}{2 \, C_o \, \sqrt{m^2 - 1}} = \varepsilon'' \left(\frac{1 + x^2}{DEN} \right) \tag{15}$$

$$\frac{\Delta C_d}{C_o} = \frac{\varepsilon' - A\varepsilon'^2 x + x/A - \varepsilon' x^2 - A\varepsilon''^2 x}{DEN} - CRF \tag{16}$$

where

$$DEN = (1 - Ax\varepsilon')^2 + (Ax\varepsilon'')^2 \tag{17}$$

$$CRF = (1 - Ax + x/A - x^2)/(1 - Ax)^2 \tag{18}$$

$$x = \tan(2\pi \, fS/c) \tag{19}$$

$$A = 2\pi \, fZ_o C_o \tag{20}$$

In the actual circuit used, $Z_o = 50$ ohm, $C_o = 1.699$ pf and $S = 11.9$ cm.

ε' and ε'' were found by using Equation 15 and rearranging Equation 16 into an expression for ε' and then using an interative technique. The interative equations used were,

$$\varepsilon''_{j+1} = \frac{\Delta C_{v-s} - \Delta C_{v-mt}}{2 \, C_o \, \sqrt{m^2 - 1}} \cdot \frac{DEN_j}{(1 + x^2)} \tag{21}$$

$$\varepsilon'_{j+1} = \frac{DEN_j}{DE_j} \left[\Delta C_d/C_o + (1 - Ax + x/A - x^2)/(1 - Ax)^2 \right] \tag{22}$$

$$DEN_j = (1 - Ax\varepsilon'_j)^2 + (Ax\varepsilon''_j)^2 \tag{23}$$

$$DE_j = 1 - Ax\varepsilon'_j + x/(A\varepsilon'_j) - x^2 - \varepsilon''^2_j A/\varepsilon'_j \tag{24}$$

These equations were programmed and solved on a digital computer. This program can be found in the Appendix.

The magnitude of the correction depends on the values of ε' and ε'' but typically the corrected values of loss and dielectric constant were 10 to 15% lower than those predicted by Equations 13 and 14 at frequencies around 50 MHz, ε' of 5 and ε'' of 1.2 (typical nylon data). This is an important correction especially when the loss peak falls in the 50 MHz range as it did in many 6-10 nylon samples. Also, the use of these equations will allow measurements at higher frequencies with good accuracy. In the experiment, the upper frequency of 63 MHz was dictated by the fact it was the maximum frequency of the oscillator used, the HP 606 A.

The circuit shown in Figure 7 was constructed by modifying a General Radio 1690-A dielectric sample holder. These modifications included:

1. Construction of a coupling capacitor, C_c,

2. Attaching a GR-900 series connector to the cell and leading out to it with 50 ohm coaxial line,

3. Attaching terminals for the voltmeter and inductors.

A schematic drawing of the modified unit is included as Figure 8. The separate GR-1690-A cell and low-frequency sample cell are shown in Figures 9 and 10.

The main capacitance changes, those involving C_d, were accomplished by moving the plates in and out. The capacitance change of these plates as a function of separation was calibrated at 10 KHz using the GR type 1615 A bridge. The plate separation was taken from the drum knob and could be read to the nearest 1/1000 inch. This leads to four-place accuracy for capacitance changes around 10 pf and three-place accuracy for changes above 40 pf. The width of the resonant peak was measured using the vernier capacitor, C_v, which had a 6 pf range. This capacitor was also calibrated in the same manner as C_d resulting in capacitance readings accurate to .01 pf.

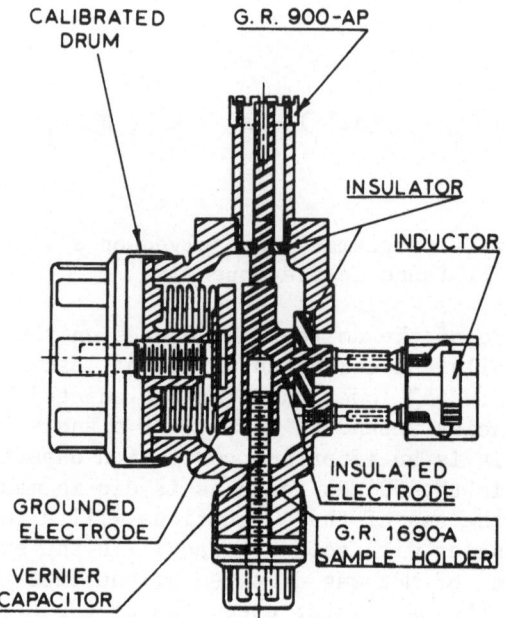

Figure 8.   Combined sample and tuning cells for resonant circuit
            technique.   See Figure 7 for equivalent circuit and
            Figures 9 and 10 for details of separate tuning and
            sample cells.

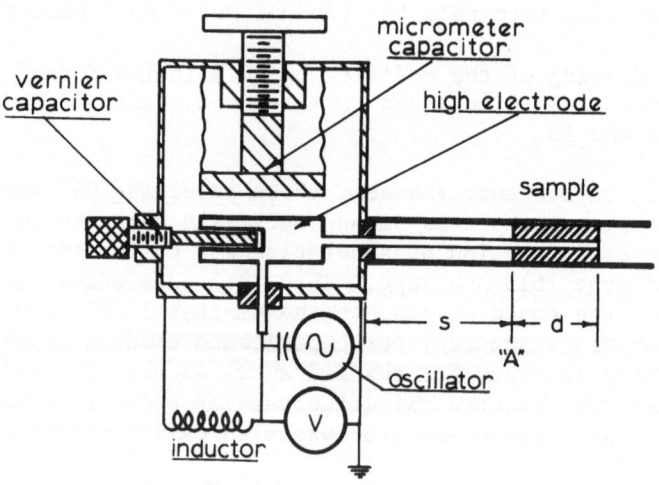

Figure 9.   Modified G.R. 1690-A sample holder used in resonant
            circuit.

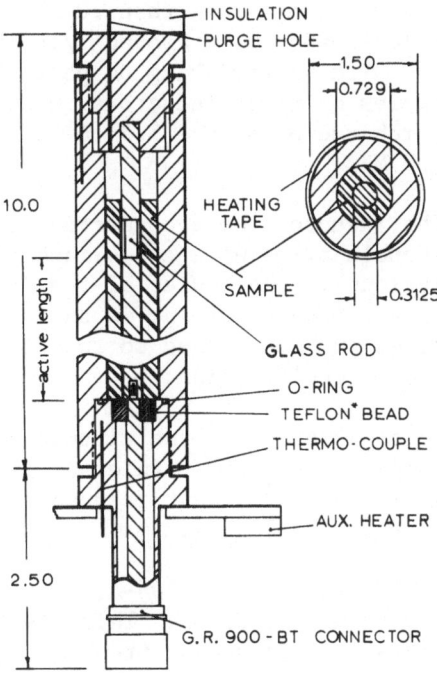

Figure 10. Low-frequency sample cell.

The inductors were (22 in all) either purchased or made. The purchased inductors were all Miller encapsulated R.F. chokes with powdered-iron cores. These are listed below together with typical resonant frequencies obtained with a sample in the cell (sample capacitance about 10 pf):

Miller No.	Rated Inductance (μH)	$F_{res.}$ (MHz)
9350-32	1000	0.75
9350-26	560	0.95
9350-18	270	1.4
9350-14	180	1.7
4632	100	2.2
4631	82	2.6
4630	62	3.0
4629	55	3.1
4628	39	3.7
4626	24	4.9

Inductors with inductances less than 20 µH were made by wrapping 18 gage enamel-copper wire around 3/16 inch diameter glass rod and anchoring the wire by shrinking on polyethylene tubing. The resonant frequencies of these inductors, under the same operation conditions as the purchased ones were 5.3, 6.8, 10.9, 13.7, 16.5, 20, 25, 33, 43 and 60 MHz.

The leads of the inductors were connected to the jacks of standard-insulated double plugs with a 3/4 inch jack spacing (GR 274-MB and equivalent). The bodies of the inductors were then taped to the body of the plug resulting in an inductor that could be reproducibly removed and reattached to the circuit. The plugs were then numbered for identification.

The approximate capacitance of the circuit with the drum at 0.300 inches (max. separation), the vernier at 5.00 pf, and no cell attached was 16 pf. The self-resonant frequency with an empty sample cell was in excess of 250 MHz.

The oscillator, Hewlett Packard 606-A, was loosely coupled to the circuit by a very small capacitor. This was constructed by looping a wire around a rod protruding from the bottom plate of C_d. The coupling capacitor was kept at a very small value to insure constant current flow in the circuit. The frequency of the oscillator was determined by calibrating it against a 100 KHz or a 1 MHz internal crystal oscillator. This provided a frequency accuracy of better than 1% and a resetability better than 0.1% (ability to return to the same frequency).

The voltage drop across the capacitors was monitored using a HP 410C voltmeter outfitted with a HP 11036-A A.C. probe. The insertion impedance of this unit was negligible. The ground lead of the probe was connected to the case while the probe tip was connected to the fixed plate of the GR 1690 A. The voltmeter was operated in the 1.5 vac max and the 0.5 vac max ranges.

The actual procedure used in taking measurements is outlined below.

1. The cell with sample in it was heated to the required temperature and allowed to equilibrate for 30 minutes.

2. An inductor was plugged into the circuit.

3. The drum was placed at 300 and the vernier at 2.50.

4. The oscillator was then set (using the crystal calibrator) at a frequency at which the circuit could be made to resonate by turning in the vernier capacitor (resonates at C_v settings between 2.50 and -0.50).

5. The circuit was then tuned to resonance using the vernier capacitor and the reading on the vernier recorded. Note: In general the resonant peak was broad due to the sample loss and the maximum was difficult to locate accurately. As a result, the circuit was tuned to some voltage, V', on either side of the loss peak using the vernier capacitor. The vernier setting for resonance, V_{max}, was then the average reading to the two aforementioned settings.

6. The vernier was then returned to 2.50 pf and the circuit approximately tuned back to resonance using the drum (main) capacitor.

7. The output of the oscillator was adjusted for full scale deflection of the voltmeter at resonance.

8. The circuit was detuned with the vernier to some voltage V and the setting on the vernier and V recorded. The circuit is then tuned back through resonance, V_{max}, and back to V on the other side of the peak and that value of C_v recorded. The difference in the two vernier readings was ΔC_{v-s} and the ratio of V_{max}/V was the voltage ratio, m. The value of V was determined as that value which would require around a 5 pf change in the vernier readings, ΔC_{v-s}, to maintain maximum accuracy.

9. Steps 3 through 8 were repeated using the other inductors.

10. The sample was allowed to cool and then removed from the cell. The cell was then reheated to the same temperature as before.

11. An inductor was placed in the circuit and the oscillator returned to that frequency (using the crystal calibrator) used with this inductor in step 4.

12. The vernier capacitor was placed at the value recorded in step 5 and the circuit tuned to resonance using the drum capacitor, C_d. (Note: The note in step 5 applied here also except that the drum capacitor was not linear as was the vernier and the value sought was the average value of capacitance rather than average drum reading.) The resulting reading was C_{d2}.

13. Step 8 was repeated for the empty cell using the same values for V and V_{max} as were used before. This results in ΔC_{v-mt}.

14. Steps 11 through 13 were repeated for the other inductors.

Sample Cell: The same sample cell, illustrated by Figure 10, was used for the bridge and resonant circuits. The empty cell was basically an air-filled coaxial cable with the inner electrode supported by Teflon* washers. It was constructed in three parts, machined from brass, which screwed together. These parts included:

1. The "lead-in" section, located at the bottom of the cell.

2. The sample section, comprising the center portion of the cell,

3. The cap, located at the top of the cell.

The "lead-in" section consisted of a G.R. 900-BT connector supporting the end of a center conductor, the other end being supported by a Teflon gasket, and an outer conductor. The conductors' diameter ratio, b/a, was chosen to conform to that of 50 ohm characteristic impedance. This corresponds to a (b/a) ratio of 2.303 for air-filled sections and a ratio of 3.24 for Teflon-filled sections. The outer conductor I.D. was 0.5625 inch, and the inner conductor O.D. was 0.2442 inch. A thermocouple was imbedded in this section to control the power input to the auxiliary heater. The heater assembly consisted of a Chromalox A-20-3 ring heater, 500 watt/115 vac, attached to a quarter-inch thick, 4 inch diameter aluminum disc which, in turn, was bolted to the face of the section. The heater power input was controlled using a 115 vac auto-transformer.

The center section of the cell had an outer conductor I.D. of 0.729 inches. The inner conductor could be interchanged resulting in different values of C_o ranging from 10.18 pf for a 6 inch long, 0.3125 inch diameter rod to 1.698 pf for a 1 inch long, 0.2242 inch diameter rod. C_o, the empty capacitance of the active length, was found by calibrating the cell with benzene, carbon tetrachloride and toluene using the values for dielectric constant given in the HANDBOOK OF CHEMISTRY AND PHYSICS (4). A rubber "0-ring" was necessary to prevent leakage of the calibration fluid.

Two thermocouples were imbedded in the wall; one was used as a control thermocouple and the other as a monitoring thermocouple. The thermocouples penetrated the cell to a depth of 3 inches from the top and were situated 90° apart.

The cap of the cell was drilled to accept a 0.3125 rod which acted as a centering support for the inner conductor. A hole was drilled and tapped through the cap to provide an inlet for purge gases. The top was insulated to minimize axial temperature gradients. Radial gradients were so small that they could not be detected by the two thermocouples placed 90° apart.

The cell was heated electrically at temperatures in excess of 80°C and by a circulating fluid system at lower temperatures. The electrical system consisted of Van Waters and Rodgers Briskeat

33737-143 576 watt/115 vac fiberglass heating tape wrapped around
the cell. The heat input to the tape was controlled by a potentiometric controller, described elsewhere, connected to the control
thermocouple. The rate of heat loss down the neck of the "lead-in"
section caused an axial temperature gradient but this could be
eliminated by either using the auxiliary heater or by insulating
the top of the "lead-in" section and the neck. The auxiliary heater
proved easier to use because the amount of insulation necessary was
dependent on the cell temperature.

The fluid circulating system is also described elsewhere. To
adapt the cell to this system it was jacketed with a 2 inch I.D.
pipe which was connected to the cell in the same location as the
auxiliary heater assembly. Due to the high temperature range
encountered during the experiments, 0° to 250°C, thermal expansion
effects affected the cell readings in three ways. These included
cell capacitance changes due to thermal expansion of the brass cell
itself, the Teflon* gaskets supporting the center electrode, and
the sample contained within the cell.

The first effect, change in cell dimensions, was minimized by
constructing both inner and outer conductors from brass. This
eliminates capacitance changes due to radial expansion since the
capacitance of the cell is dependent on the ratio of diameter,
(b/a). An axial effect was experienced due to the change in the
lengths of both the active length and the "lead-in" length of the
cell. Empty cell capacitances (which are the sum of the capacitance
of the "lead-in" section and the active length) were taken at
several temperatures, ranging from room temperature to 250°C. It
was found that the specific change in capacitance, dC'_{mt}/dT, could
be well represented by the value, 15×10^{-6} pf/pf-°C, which is in
fairly good agreement with the value given by Perry's HANDBOOK (5)
for the linear coefficient of brass ($20.4 \times 10^{-6}/°C$). C_0 at room
temperature in this run was 10.18 pf and C_{mt} was 17.67, leading to
a capacitance of the "lead-in" section and residual junction box
capacitance of 7.59 pf. If it can now be assumed that the specific
change in C_0 is the same as the linear coefficient of brass, and
this change is subtracted from the total change, the resulting
specific capacitance for the "lead-in" section and junction box
becomes $dC'_{res}/dT = 7.5 \times 10^{-6}$ pf/pf-°C. The change of capacitance
of C_{mt} is then the sum of the changes in C_0 and C_{res}. It should be
pointed out that typically the uncorrected values of capacitance
differed from the corrected values by about 0.2%/100°C, and for most
work it is negligible. The only time one is really justified in
using it is when C_0 is very small, such as 1.698 pf, which can lead
to corrections as high as 1.2%/100°C.

Gasket expansion was important due to the high thermal expansion of Teflon*. This expansion was especially evident when a new

gasket had been placed in the cell, which was pressed into the gasket seat to make the cell liquid tight. When the cell was heated, this gasket would expand and extrude into the "lead-in" section. The gasket was not only deformed but on cooling it would shrink resulting in an air space between the Teflon* and the wall. The difference in C_{mt}, when measured before and after the heating-cooling cycle, varied about 0.1 pf which could lead to capacitance errors with a C_0 of 1.7 pf of 6%.

It will be shown later than experimentation with Teflon* revealed that the product of its dielectric constant and density as a function of temperature was essentially constant. As a result, after the initial deformation of the gasket, the capacitance due to gasket changes should be negligible because the increased expansion was offset by a decreased dielectric constant. To account for these effects, C_0 was measured using a new gasket (which was necessary to hold the calibration fluid) and C_{mt} was determined after the cell had been heated to the highest temperature for that particular experiment and then cooled to room temperature. In other words, a cell with new gaskets had to be "broken-in" before using it with samples.

Sample thermal expansion was both radial and axial. The effect of axial expansion could be eliminated by cutting the sample longer than the center conductor, whose end was left free when using solid samples or supported by a glass-brass rod combination when using liquid samples. The radial expansion was beneficial as it forced the sample into the walls of the respective electrodes thereby eliminating any air spaces.

Temperature Control System: All sample cells had thermocouples imbedded in the walls to monitor and control temperature. The monitoring was accomplished by using a Leeds and Northrup 8686 millivolt potentiometer with the thermocouples referenced against similar thermocouples in an ice bath. Two separate temperature control systems were used, an electrical system and a circulating-fluid system.

The electrical system consisted of a potentiometric controller regulating a heating tape wrapped around the sample cell being monitored by a control thermocouple. The potentiometric controller consisted of a 0-1 mv full scale recorder, a modified Leeds and Northrup Speedomax G, outfitted with a 20% transmitting control sidewire connected to a Leeds and Northrup Series 60 three-mode controller. The output of the controller was connected to a Leeds and Northrup 11905-25 power package which in turn was connected to the inputs of the heating tape.

The output of the control thermocouple was bucked against a controlled 0-20 mv signal from a voltage dividing network. This signal, the voltage from the thermocouple minus the bucking voltage, was then fed to the recorder controller. This allowed the 0-1 mv recorder to be used with T.C. outputs ranging from 0-21 mv with the recorder responding only to the "top" 0-1 mv of the T.C. signal.

The thermocouples used were iron constantan type constructed by spot welding Leeds and Northrup T.C. wire together and insulating the junction with 0.060" O.D. glass capillary tubing. The glass tube was then inserted into a 3/32" O.D. brass tube for structural strength. Electrical insulation of the tip was necessary to isolate the thermocouple circuit from the bridge circuits when measurements were taken at low frequencies. The thermocouples were checked against a calibrated thermometer between 0° and 100°C. It was found that the output in all cases never deviated from those published in the HANDBOOK OF CHEMISTRY AND PHYSICS (4) by more than .01 mv (about 0.2°C).

The aforementioned system, with the heating tapes around the cell, could heat a sample cell from room temperature to 200°C in about ten minutes. The system was in control typically within two more minutes, never deviating from the set point by more than 0.005 mv (0.1°C) for temperatures in excess of 100°C.

The electrical system could not be used for temperature control below 80°C due to excessive cycling and cycle times. A liquid circulation system, consisting of a constant temperature bath and circulating pump, was constructed for lower temperatures.

This system consisted of a well-stirred bath heated with 500 watt heaters controlled by a Philadelphia Roto-Stat mercury thermo-regulator in series with a Fisher Model 32 transistorized relay. Temperature control better than 0.05°C was typical in this system.

The liquid, ethylene glycol, was circulated through a jacket surrounding the cell at approximately 5 gallon/minute. Ethylene glycol was used to insulate the bath electrically from the cell and because it had a relatively low vapor pressure at high temperatures.

Some runs were conducted, using the microwave cell, at 0°C using ice water as the circulating fluid. It was not necessary to electrically isolate the cell from ground when working at microwave frequencies.

Distributed Circuits

When the frequency of interest approaches 1/4 wavelength of the electrical paths, a distributed circuit analysis must be employed.

Practically, this involves frequencies in excess of 10^8 Hz (microwave and higher) with the upper frequency limit usually determined
by equipment.

There are many methods that are used to determine the impedance
of samples in the microwave region. These methods have been reviewed
by McCrum (21), von Hippel (38), Sucher (33,34), Harvey (43)
and others. As only transmission line techniques were used in this
work, the interested reader is referred to the aforementioned
references for other methods.

Roberts-von Hippel Method: The basis of the equations used in
analyzing microwave circuits is commonly referred to as the Roberts-
von Hippel method. This work is reviewed in von Hippel (38). The
method basically involves finding the impedance of electromagnetic
waves in a lossless transmission line terminated with a sample and
then a short circuit. The impedance in the lossless section is
then transformed to the sample interface and the dielectric properties of the sample calculated.

It has been shown (2) that a standing wave pattern is set up in
a transmission line due to the interaction of incident and reflected
electromagnetic waves, such as is illustrated in Figure 11. At the
voltage maximum, the incident and reflected voltages are in phase,

$$E_{max} = E'_i + E'_r \tag{25}$$

and, at the voltage minimum, they are 180° out of phase,

$$E_{min} = E'_i - E'_r \tag{26}$$

At the voltage minimum, the current is a maximum, due to the fact
that the incident current is in phase with the incident voltage; and
the reflected current is 180° out of phase with the reflected voltage,

$$I_{E_{min}} = I'_i + I'_r \tag{27}$$

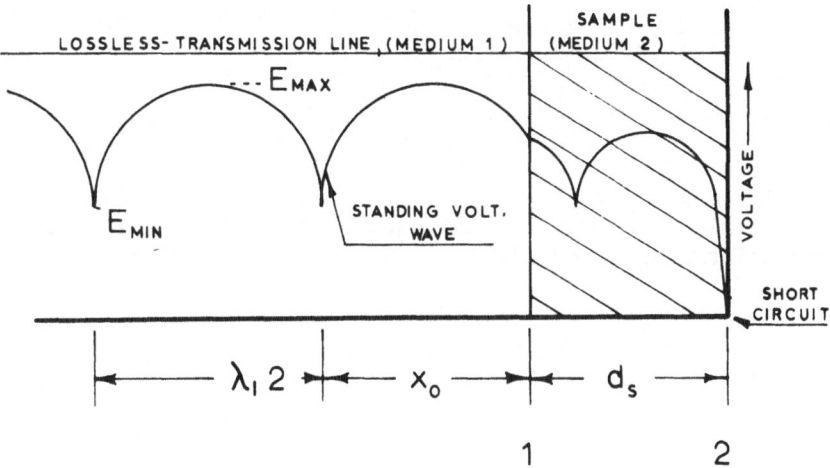

Figure 11.   Schematic diagram of standing waves in a sample-filled, short-circuited transmission line.

Recalling that the respective voltages and currents are related by the characteristic impedance and that this impedance is purely resistive for lossless lines, Equation 27 can be written as

$$I_{E_{min}} = (E'_i + E'_r)/Z_o = E_{max}/Z_o \qquad (28)$$

The impedance at the voltage minimum, the ratio of the voltage and the current at that point, is then

$$Z_{E_{min}} = E_{min}/I_{E_{min}} = Z_o/S \qquad (29)$$

where S = voltage standing wave ratio, E_{max}/E_{min}. Seeing the impedance is known at the minimum, the impedance at the sample interface can be found by using an impedance transformation equation for lossless lines (Reference 2). The appropriate equation for this case is

$$Z(x) = Z_o \left(\frac{Z_p - jZ_o \tan(2\pi\, d/\lambda)}{Z_o - jZ_p \tan(2\pi\, d/\lambda)} \right) \tag{30}$$

where $Z(x)$ = impedance at point x, Z_p = impedance (known) at point p, d = electrical length between p and x, Z_o = characteristic impedance of transmission line.

Substitution of Equation 29 for Z_p and referring to Figure 11 results in the expression for the impedance at point (1), the sample interface,

$$Z(1)_1 = Z_{o1} \left(\frac{1 - jS\tan(2\pi\, x_o/\lambda_1)}{S - j\tan(2\pi\, x_o/\lambda_1)} \right) \tag{31}$$

Seeing there can be no voltage at a short circuit, point (2) the impedance at that point is 0. The impedance at point (1) can be found by transforming the impedance at point (1) to point (2) using a "lossy line" impedance-transformation equation (2).

$$Z_p = Z_{o2} \left(\frac{Z_r + Z_{o2}\tanh(\gamma_2 d_s)}{Z_{o2} + Z_r\tanh(\gamma_2 d_s)} \right) \tag{32}$$

where Z_r = load impedance, Z_{o2} = characteristic impedance of the section of line containing the sample, and d_s = distance from load (short) to sample interface.

Substituting in 0 for Z_r results in the expression for the impedance at the interface of the sample in medium 2,

$$Z(1)_2 = Z_{o2}\tanh(\gamma_2 d_s) \tag{33}$$

Snell's laws of reflection and refraction, see von Hippel (2), state that the tangential components of the electromagnetic waves must be continuous in traversing the interface between the two media. In

other words, the voltage and current are of equal magnitude in
both media at the interface which leads to the result that the
impedance at the interface is the same in both media, namely,

$$E(1)_1 = E(1)_2 \tag{34}$$

and

$$I(1)_1 = I(1)_2 \tag{35}$$

from which follows,

$$Z(1)_1 = Z(1)_2 \tag{36}$$

As a result, Equations 31 and 33 can be equated. Doing this and
using the relationship,

$$Z_{o2}/Z_{o1} = j2\pi/\gamma_2\lambda_1 \tag{37}$$

and then rearranging results in

$$\frac{\tanh(\gamma_2 d_s)}{(\gamma_2 d_s)} = -\frac{j\lambda_1}{2\pi d_s}\left(\frac{1 - jS\tan(2\pi x_o/\lambda_1)}{S - j\tan(2\pi x_o/\lambda_1)}\right) \tag{38}$$

The former equation can now be solved for $\gamma_2 d_s$ by using a Newton
method in the complex plane of the form,

$$x_n = x_{n-1} - \frac{f(x_{n-1})}{f'(x_{n-1})} \tag{39}$$

with $x = \gamma_2 d_s$, $f(x) = \tanh(\gamma_2 d_s) - \gamma_2 d_s(CC)$, $f'(x) = 1 -$
$\tanh^2(\gamma_2 d_s) - (CC)$, $CC = -\dfrac{j\lambda_1}{2\pi d_s} \dfrac{1 - jStan(2\pi x_o/\lambda_1)}{S - jtan(2\pi x_o/\lambda_1)}$,

resulting in the iterative equation,

$$(\gamma_2 d_s)_n = (\gamma_2 d_s)_{n-1} - \frac{\tanh(\gamma_2 d_s)_{n-1} - CC(\gamma_2 d_s)_{n-1}}{1 - \tanh^2(\gamma_2 d_s)_{n-1} - CC} \tag{40}$$

The above equation was programmed and solved using a digital computer. The "FORTRAN" language permits the Newton method to be
employed directly in the complex plane. This program is included
in the Appendix.

The values of ε'_2 and ε''_2 are then found by noting that

$$\gamma_2/\gamma_1 \doteq (\varepsilon'_1/\varepsilon^*_2)^{1/2} \tag{41}$$

This equation can be expanded to the final result,

$$\varepsilon^*_2 = \varepsilon'_2 - j\varepsilon''_2 = -\varepsilon'_1 (\lambda_1\gamma_2/2\pi)^2 \tag{42}$$

Gasket Interface Method: When studying melts it was necessary
to place a gasket at the air sample interface for reasons which will
be discussed later in this section. Due to the high temperatures
encountered and electrical requirements, Teflon* (polytetrafluoroethylene) was used. Teflon* has the advantage that it is a very
low-loss material (and can be considered lossless) with a dissipation factor of less than .0004 at 1 GHz and 25°C. ε' of this
material is not a function of frequency in the range used and has
a relatively low value of 2.1. This material has the disadvantages
that it has a very high coefficient of expansion (20 x 10^{-5} in/in °C)
and an accompanying decrease in dielectric constant as a function of

temperature. Also, due to the manufacturing processes involved in
the production of Teflon*, the electrical properties will vary from
sample to sample.

Figure 12 is a diagram of a short-circuited transmission line
containing a sample (medium 2). The sample is separated from the
air-filled section of coaxial-line (medium 2) by a lossless gasket
(medium 3) of width d_t and relative dielectric constant of ε'_3.
The impedance at the air-gasket interface (point a) is the same as
that found in Equation (31); namely

$$Z(A)_1 = Z_{o1}\left(\frac{1 - jS\tan(2\pi x_o/\lambda_1)}{S - j\tan(2\pi x_o/\lambda_1)}\right)$$

$$\text{(31)}$$
$$\text{(Repeated)}$$

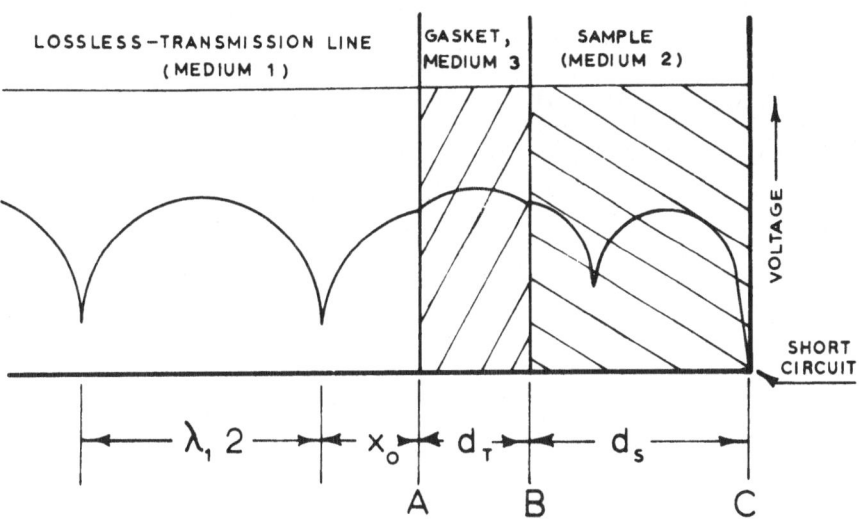

Figure 12.   Schematic diagram of standing waves in a sample-filled
              short-circuited transmission line with a lossless
              gasket preceding the sample.

The impedance at point B, the gasket sample interface, can be found by noting:

1. $Z(A)_1 = Z(A)_3$,

2. Medium 3 is lossless and Equation 30 can be used to transform the impedance from A to B.

3. $\varepsilon'_1 = 1.000$,

4. $Z_{o3}/Z_{o1} = 1/(\varepsilon'_3)^{1/2}$,

5. $\lambda_3 = \lambda_1/(\varepsilon'_3)^{1/2}$.

These result in the expression for the impedance at B,

$$Z(B)_3 = \frac{Z_{o1}}{(\varepsilon'_3)^{1/2}} \left(\frac{S(1+x_1^2)/G - j[(S^2-1)x_1/G + x_3/(\varepsilon'_3)^{1/2}]}{[1/(\varepsilon'_3)^{1/2}-(S^2-1)x_1x_3/G) - jS(1+x_1^2)x_3/G} \right) \quad (43)$$

where $G = S^2 + x_1^2$, $x_1 = \tan(2\pi x_o/\lambda_1)$, and $x_3 = \tan[2\pi(\varepsilon'_3)^{1/2}d_t/\lambda_1]$.

Noting now that $Z(B)_3 = Z(B)_2$, and that $Z(B)_2$ is still given by Equation 33, rearrangement results in

$$\frac{\tanh(\gamma_2 d_s)}{\gamma_2 d_s} = -\frac{j\lambda_1 Z(B)}{2\pi d_s Z_{o1}} \quad (44)$$

This can be solved in exactly the same manner as Equation 40, but now with

$$CC = -\frac{-j\lambda_1}{2\pi d_2\sqrt{\varepsilon'}_3} \left(\frac{S(1+x_1^2)/G - j[(S^2-1)x_1/G + x_3/(\varepsilon'_3)^{1/2}]}{[1/(\varepsilon'_3)^{1/2}-(S^2-1)x_1x_3/G] - jS(1+x_1^2)x_3/G} \right) \quad (45)$$

The above equations were programmed in FORTRAN, and this problem is included as part of TANPLT in the Appendix.

Numerical Solution: Equation 40 has several properties that cause some difficulty in obtaining a solution for $\gamma_2 d_s$. Some of these are:

1. The solution is Newton method in the complex plane which makes it a Newton-Raphson method with two functions; one real and the other complex.

2. The complex portion of γ_2, the variable β_2, is periodic which leads to an infinite number of solutions for ε' and ε''.

3. ε' and ε'' are interdependent at high loss.

Property 1 can lead to either of the following:

a. The iteration will never converge on the correct solution even if the starting point is extremely close to the correct one.

b. The iterative equations will not converge unless the value for $\gamma_2 d_s$ (initial) is close to the solution.

The criteria of convergence of a two-equation, iterative method is (6),

$$\left|\frac{\partial F}{\partial x}\right| + \left|\frac{\partial F}{\partial y}\right| < 1, \quad \left|\frac{\partial G}{\partial x}\right| + \left|\frac{\partial G}{\partial y}\right| < 1, \tag{46}$$

where F = complex part of Equation 40, G = real part of Equation 40, x = real part of $\gamma_2 d_s$ ($\alpha_2 d_s$), and y = imaginary part of $\gamma_2 d_s$ ($\beta_2 d_s$).

When the criteria in Equation 46 are not met, the iterative equations are divergent, no matter how close the initial values are to the correct solution. As a result there are some values of F and G that cannot be solved by using a Newton method, although they have solutions. Typically, 5% of the solutions would not converge.

The value $\gamma_2 d_s$ (initial) was found by using the "low-loss" approximations (2). If $\tan\delta_2$ was sufficiently low, the real and imaginary parts of Equation 40 could be separated and approximate values for α_2 and β_2 could be found, which were then substituted in Equation 23 to find γ_2 (initial). This was a satisfactory method, except for some cases where high loss was encountered.

The solution for $\gamma_2 d_s$ was actually a number of solutions due to the fact that β_2 was periodic. Equation 40 was derived by comparison of impedances at the sample interface, but there was no

way of telling without additional experimentation how many 1/2
wavelengths were contained in the sample. This gave rise to an
infinite number of possible answers for ε' and ε''; each dependent
on the number of 1/2 wavelengths between the interface and the
short circuit, of which only one pair was correct. There were a
number of methods that could have been used to determine the correct set of values. In most cases one has a good idea of what ε'
should have been at the lowest frequencies, and by using that value
as a starting point, and taking all the other values in context,
the dielectric constant at the next highest frequency could be
deduced. This method was satisfactory where the dielectric constant either did not change or only changed slightly as a function
of frequency. The principal objection to this method was that at
high frequencies the possible values of ε' and ε'' became closely
spaced and it was really a question of judgement as to which was
the correct solution, especially in the case where the dielectric
constant was a strong function of frequency. There were other
complications such as sometimes the computer missed the correct
solution, the solutions were in error, or there was a relatively
large interval where no data existed in regions where the solutions
did not converge.

A method that tied down exactly the correct solution set was
that of making two separate runs with samples of different lengths.
The two samples, for each frequency, then had two families of solutions; the solution in common being the correct one (see Figure 13).
The extra work involved in doubling the number of runs was justified by the assurance that the answers obtained were the correct
ones. This method also had the advantage of providing two data
points instead of one for each frequency which was statistically
better. Also, the fact that the second sample was of a different
length frequently allowed the solution to converge at frequencies
it did not converge using the first sample, thereby filling in some
of the gaps.

In the case where $\tan \delta_2$ approached 1, the solution of Equation
40 becomes complicated. The major factor was that the procedure
for starting off the solution, the low-loss approximations, caused
the solution to skip values. The "low-loss" approximation for the
complex part of Equation 40 was

$$\frac{\tan \beta_2 d_s}{\beta_2 d_s} = \frac{\lambda_1}{2\pi d_s} \tan \frac{2\pi x_o}{\lambda_1} \tag{47}$$

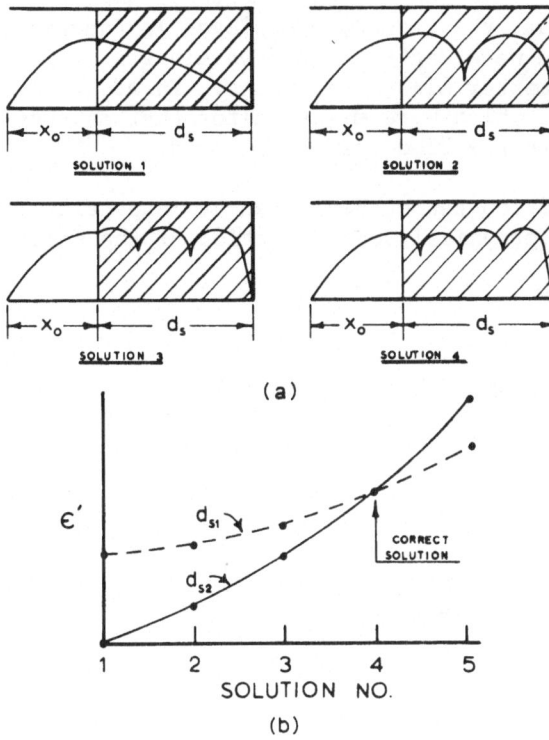

Figure 13. (a) Representation of the possible number of solutions for a given x_0 and d_s as a function of the number of 1/2 wavelengths contained in the sample. (b) Correct solution is that one in common of two solution sets; using two sample lengths, d_{s1} and d_{s2}.

The value of $\gamma_2 d_s$ was found by a Newton method starting this solution off by letting,

$$\beta_2 d_s = \frac{2n+1}{2}\pi \tag{48}$$

which is equivalent to setting λ_2 at 1/4 for the first solution and increasing it by $\lambda_2/2$ for each successive solution. This procedure should take care of the fact that there could be any number of 1/2 wavelengths in the sample. At high loss, however, the "low-loss" approximation for wavelength becomes poor; and there was no assurance that the procedure employed in Equation 48 would increase the true wavelength by half-wavelength intervals. As a result, the computer missed solutions. In addition, there existed other solutions which were missed altogether as they do not relate in any manner to the "low-loss" solutions.

This situation can be best seen by inspecting Figure 14. Illustrated is a graphical representation of the solution to the function,

$$F(\gamma_2 d_s) = \tanh (\gamma_2 d_s) - CC (\gamma_2 d_s) \tag{49}$$

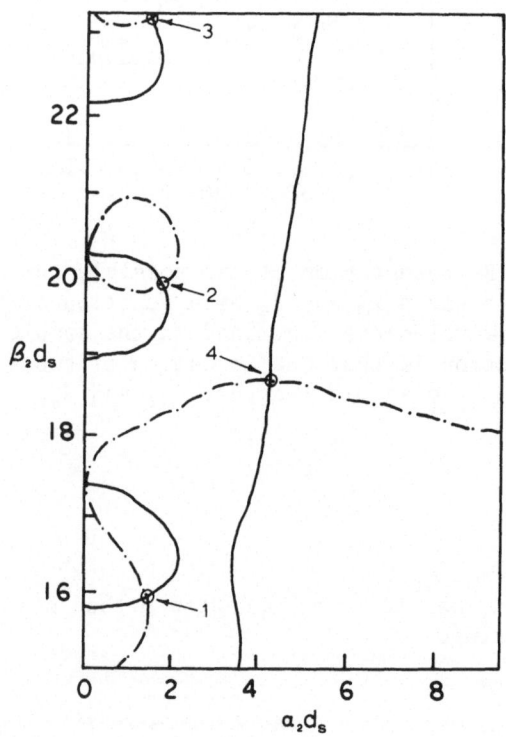

Figure 14.   Solution map of the equation, $F(\gamma_2 d_s) = \tanh (\alpha_2 d_s + j\beta_2 d_s) - CC (\alpha_2 d_s + j\beta_2 d_s)$. The dashed line represents the locus of points where imaginary parts of $F(\gamma_2 d_s)$ go from positive to negative. The same boundary for the real part is represented by the solid line. Points 1, 2, 3 and 4 represent solutions to $F(\gamma_2 d_s)$.

The sought values of $(\gamma_2 d_s)$ are those values which cause both the real and imaginary parts of $F(\gamma_2 d_s)$ to be 0. The solid line in Figure 14 represents the boundary between positive and negative values of the real part of $F(\gamma_2 d_s)$ and the dashed line is the boundary of the imaginary part. The imaginary part can also change sign when it goes from − to + infinity, commonly called a pole, which can cause a problem in interpretation. The intersection of the real and imaginary boundary lines (neglecting poles) represent the values of $\beta_2 d_s$ and $\alpha_2 d_s$ which represents a solution to Equation 49, $F = 0$.

Figure 14 was generated from data obtained on POM at 5.910 GHz with $\beta_1 d_s = 9.797$. This figure shows four acceptable solutions within the range of ε' of 5.00 max and 1.50 min and ε'' of 2.00 max and 0 min. The intersections of the boundaries at $\alpha_2 d_s = 0$ do not represent solutions. The procedure outlined previously for starting off the Newton interation , the low loss approximation, would find solutions 1, 2 and 3, but would skip solution 4, which in this case happened to be the correct solution.

The problem of the imaginary part of the solution of Equation 49 changing sign at both the points where it crosses the 0 value and at poles could be circumvented by defining a new function,

$$G(\gamma_2 d_s) = 1 + F(\gamma_2 d_s) \tag{50}$$

This could then be converted to the polar form,

$$G(\gamma_2 d_s) = R\exp(j\ \psi) \tag{51}$$

where

$$R = \sqrt{(\text{real } G)^2 + (\text{imag} G)^2} \tag{52}$$

and

$$\psi = \arctan\ (\text{imag} G/\text{real } G) \tag{53}$$

Solutions exist at the boundary of R being greater and less than 1 and ψ/m = changing from 2π to 0, for integer m's. This form of the solution contained no boundaries due to poles.

Both methods were programmed for the computer but the polar-coordinate method was finally discarded as it resulted in excessive computation time. In this method both the real and imaginary parts had to be evaluated for every point, while in the other method it was possible to just test the imaginary part for sign change at a real part boundary. The polar method typically required four times the computation time as the cartesian-coordinate method. The cartesian-coordinate method is included in TANPLT in the Appendix. The solutions obtained from this method were then fed to the Newton iteration, previously mentioned, to find improved values for ε' and ε''.

Description of Apparatus: The equation developed for finding dielectric properties from standing waves involved two sections of coaxial waveguide; a lossless section, where the measurements are taken, and a "lossy" section, where the sample is situated. It should be pointed out that the equations, with some modification, are valid for waveguides in general. Coaxial waveguide was used because it allows measurements to be taken at relatively low frequencies (there is no cut-off frequency) and the mode of propagation of the electromagnetic waves (TEM) makes analysis easier. It did have the disadvantage that there was an upper-frequency limitation on coaxial line. Above this frequency limit other modes of propagation predominate.

The standing-wave patterns were measured by a traveling detector which probed the electric field in a coaxial tube. The whole assembly consisted of a slotted section of coaxial tube, an electric probe, and a detector.

The slotted section consisted of a 50 ohm configuration 0.5625 inch O.D. silver-plated coaxial line. This configuration allows frequencies up to 8.5 GHz to be studied. The probe detector assembly was mounted on a movable carriage whose position was determined by a scale-vernier combination accurate to 0.01 cm. The probe was a stiff wire that penetrated the voltage field acting as an antenna. The depth of the probe was positioned by a screw mechanism which was calibrated to read the distance between the inner conductor O.D. and the probe tip. The R.F. output of the probe was rectified by a crystal diode and that signal sent to a standing wave indicator. The probe and detector were "matched" for maximum power transfer by a variable short circuit in a coaxial tube above the diode. This matching also tuned out unwanted signals introduced at the oscillator. This whole assembly, the slotted coaxial line, the tuner-detector, and movable carriage is commonly called a slotted line.

In this experiment a General Radio type 900-LB precision slotted line was used in conjunction with a Hewlett Packard 415E SWR meter and one of the following oscillators:

1. H.P. model 3200-B (10-500 MHz)

2. H.P. model 612-A (450-1230 MHz)

3. H.P. 8690-B sweep oscillator with one of the following plug-in units:

    a. 8692-B R.F. Unit (2-4 GHz)

    b. 8693-A R.F. Unit (4-8 GHz)

    c. 8699-B R.F. Unit (0.1-4 GHz)

Oscillator 3c was not available for the polyoxymethylene studies; consequently, oscillators 1, 2, 3a and 3b were used. All the rest of the work was done using 3c and 3b. For frequencies below 500 MHz, a G.R. 874-500-L low-pass filter was used. A G.R. 874-1000-L low-pass filter was used for frequencies between 500 and 1000 MHz. Harmonics in the 8690 B became negligible above 1 GHz except for very low loss measurements.

The slotted line was mounted vertically, pointed downwards. This configuration made it necessary to counter-weight the carriage probe assembly. The counter-weight chord was attached to the carriage-drive chord support which eliminated torque on the carriage. It had been found previously that torque caused the outer conductor to deflect relative to the inner conductor with a resulting loss of performance. The equipment is pictured schematically in Figure 15.

The lower limiting frequency of the slotted line was determined by one of the following:

1. The length of a wavelength,

2. Harmonics and spurious noise in the oscillator,

3. Tuning the probe.

The first limitation was that at least one minima must appear in the slotted line for measurements and two are required for an accurate determination of wavelength. The length of the measurement section of the slotted line was 50 cm and a minima as far as 82 cm from the sample interface could be detected. Practically, this restricted measurements to a lower frequency limit of about

0.2 GHz, provided the wavelength could be determined by some method other than measuring the distance between two minima. The minimum frequency could have been lowered if precision air line was placed between the sample cell and the slotted line to increase the sample cell length, L_{cell}.

Spurious noise and harmonics were important at low frequencies because they are frequently of the same order of magnitude as E_{min}. This leads to serious errors in the measure value of S, the standing wave ratio (SWR), resulting in unreliable values of ε''.

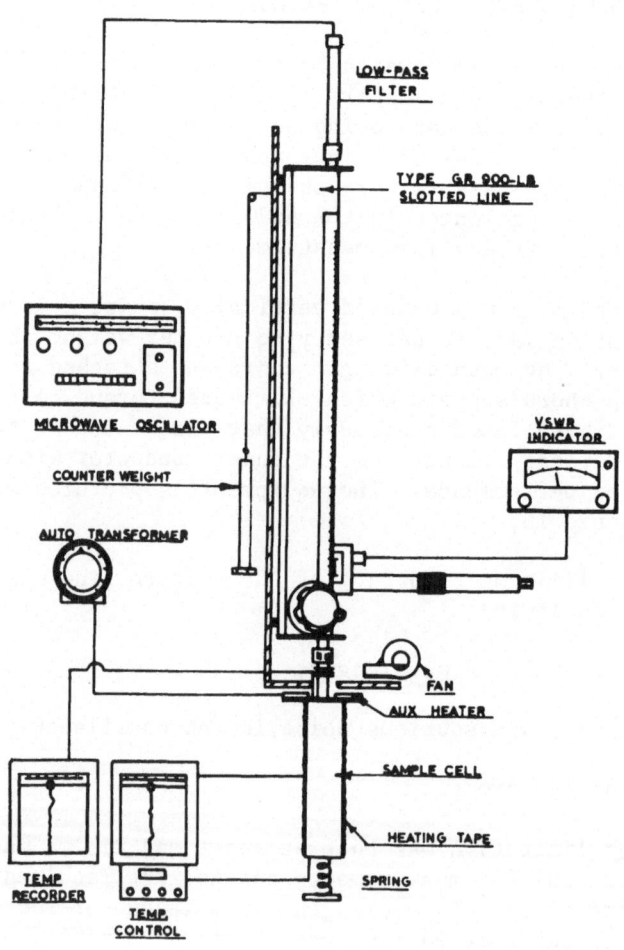

Figure 15.   Schematic diagram of microwave equipment.

The tuned probe supplied with the slotted line, G.R. 900-3070 adjustable probe tuner assembly, could be used to 0.32 GHz minimum frequency. Lower frequencies were studied by replacing the GR 900-3070 with a G.R. 874-4670 R.F. probe connected to a G.R. 874-VQL voltmeter detector. A G.R. 874-D20L adjustable stub (variable length, short-circuited coaxial line) was attached to the R.F. output plug of the detector. The rectified voltage was then sent to the SWR meter as before. The above arrangement resulted in a tuned detector that could be used directly down to 0.18 GHz and down to 0.10 GHz with a section of coaxial air line between the voltmeter and the stub.

Slotted Line Errors: Errors possible from the slotted line included:

1. Errors due to the manufacturing and design of the unit,

2. Reflections from the probe in the slotted section,

3. Deviation from square-law behavior of the detector,

4. Line losses.

The first source, that of the fundamental design of the slotted line, the experimentor had no control over. The G.R. 900-LB has specifications that make these errors negligible for the type of work done in this experiment. The accuracy of the machine can be improved by calibration, but unless four-place accuracy is mandatory, it is unnecessary. Periodic checks revealed that the machine was always operating well within General Radio's published specifications.

The depth of penetration of the probe is important for two reasons:

1. Excessive penetration causes reflections,

2. Too shallow a penetration causes errors due to variation in inner and outer conductor spacing.

The presence of the probe caused a field distortion which could be minimized by using a matched oscillator, a tuned detector and reducing the depth of the probe. Even with all these there was a depth of penetration that caused excessive field coupling and thereby measurement errors. The analysis of this effect is complicated and is done by considering the probe as a shunting admittance comprised of a conductance and suseptance component. This model leads to the relative effect the probe has on the measurements, principally that of the SWR, by comparing phase shifts in open and short circuited lines. As this procedure is complicated, though intellectually satisfying, a pragmatic approach was taken to the problem. Standing wave ratios of lossy samples were taken as a function of probe depth at various frequencies. In all cases no shift in the minima or SWR was detected for probe distances 0.100 inch or greater from the inner conductor, while some differences were detected at a distance of 0.080 inch from the inner conductor. As a result, 0.100 inch was considered to be the maximum allowable depth of penetration. Another factor related to probe penetration was that the inner and outer conductors were not perfectly parallel to each other and that the carriage track was not perfectly parallel to the outer conductor. The combination of these factors can lead to probe depth variation as the carriage is moved along the trans- mission line. The voltage detected was proportional to the depth of penetration which resulted in a relative error in the output proportional to the depth of penetration. The combination of probe effects led to the conclusion that the optimum depth of penetration was the maximum depth that did not cause measurable reflections in the line, namely 0.100 inches from the center conductor.

At low power levels, the output of a crystal diode was propor- tional to the square of the voltage averaged over a period of time. Above this power level the detector began to deviate from square- law behavior and approached linear-law behavior. As a result it was important to always operate at power levels which insured square-law behavior. Another factor that had to be considered was that low power levels were difficult to detect accurately using the SWR meter, due to noise in the meter circuit. The optimum power level was then the maximum power level in the square law region. This power level was determined by shorting the end of the slotted line and measuring the power level as a function of distance from a minima. The detected voltage was then

$$V_d = V_{max} \sin \frac{2\pi d}{\lambda_1} \tag{54}$$

where V_d = voltage at point d, V_{max} = voltage at the maximum, and λ_1 = wavelength in coaxial line.

In terms of the meter reading (V_{max} being set at a SWR of 1) the equation becomes

$$SWR_d = \frac{1}{\sin \frac{2\pi d}{\lambda_1}} \tag{55}$$

The SWR meter responded directly to the output of the diode; consequently, it was possible to express the maximum power in terms of the meter amplifier settings. In terms of these settings, deviation from square law behavior was detected at that power which caused a full scale meter deflection with the amplifier at 30 db and the gain at 1/2 full power. Accordingly, the apparatus was operated at maximum power levels sufficient to cause a full-scale deflection at 30 db with the gain set at full power. In terms of the output of the diode, the maximum square law power was about 1 mv peak-to-peak square wave voltage.

Line losses were neglected in the derivation of the equations used in this work. This is justified in all cases except where very low losses are being studied or at low frequencies. In these cases the line losses can be subtracted out. A detailed discussion of this procedure can be found in the Appendix.

Experimental Measurements: Inspection of Equations 38 and 45 reveals that the following parameters had to be obtained to effect a solution:

1. x_0, the distance from the first minima to the air-gasket interface (to the air-sample interface for runs with no gasket used).

2. λ_1, the wavelength in the slotted line,

3. S, the ratio of E_{max} and E_{min} in the slotted line,

4. d_s, the sample length,

5. d_t, the thickness of the Teflon* gasket (if a gasket was used),

6. ε'_3, the relative dielectric constant of Teflon*.

Figures 16 and 17 are presented to illustrate the quantities measured.

It was not possible to measure x_0 directly because it generally occurred within the sample cell. As a result, the position of the first minima found in the slotted line, relative to an arbitrary zero, x_0', was measured. The location of this arbitrary zero was not needed since all measurements were taken relative to it and it cancelled out of the equations. Inspection of Figure 17 shows that x_0 and x_0' are related by

$$x_0 = x_0' + L_{cell} - (d_s + d_t) - n'\lambda_1/2 \qquad (56)$$

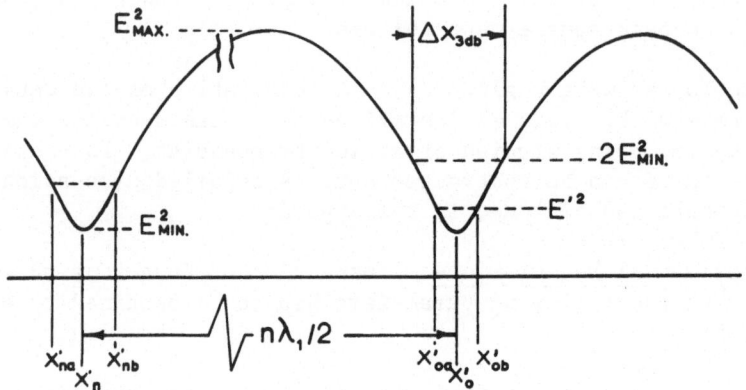

Figure 16.   Diagram defining nomenclature.

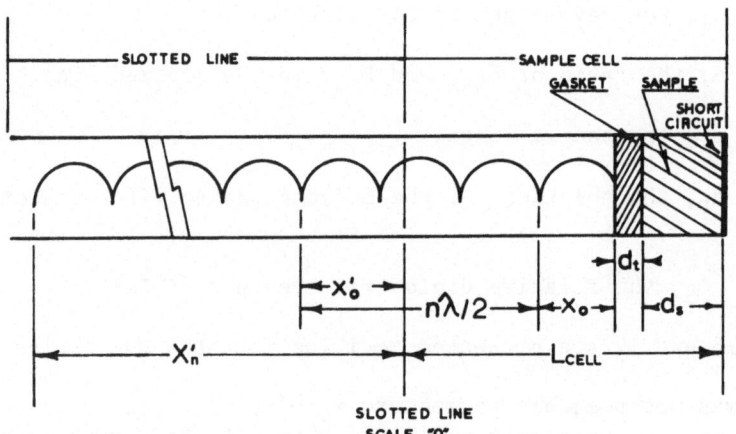

Figure 17.   Diagram defining nomenclature.

The last term of the right-hand side of the above equation was not necessary, due to the fact that every time x_o appeared in an equation it was in the expression, $\tan(2\pi x_o/\lambda_1)$. This function repeats every 1/2 wavelength and has the same value if the expression

$$x_o = x_o' + L_{cell} - (d_s + d_t) \tag{57}$$

is used instead of Equation 56.

The shape of the minima, for samples with a fairly low standing wave ratio, was rounded and thereby could not be located accurately by a single measurement. The minima were found by recording the position of some voltage E' on either side of the minima, the respective values being x_{oa}' and x_{ob}'. The minima is then the average of these values, specifically,

$$x_o' = (x_{oa}' + x_{ob}')/2 \tag{58}$$

The wavelength, λ_1, was determined by finding the distance between two minima. A second minima, x_n', was located as many half wavelengths away from x_o' as the 50 cm length of the slotted line would allow. The number of half wavelengths between x_o' and x_n' was estimated using the oscillator frequency reading.

$$n = 2f_{osc} (x_n' - x_o')/c \tag{59}$$

The value of λ_1 was then determined by using the equation

$$\lambda_1 = 2(x_n' - x_o')/n \tag{60}$$

The frequency settings of the oscillator were not accurate enough to determine wavelength directly but were close enough to detect an error in x_n' or x_o' by comparing the frequency calculated from λ_1 and the oscillator settings.

The standing wave ratio, S, was found by one of three methods:

1. Direct reading of the SWR meter,

2. Measuring ΔDb between E_{max} and E_{min},

3. Width-of-minimum method.

Direct reading of the meter was accomplished by setting the power output of the oscillator in order that E_{max} would read SWR = 1 at 30 db and full gain. The meter scale was calibrated in such a manner that S could be read directly by moving the probe to E_{min}. SWR's up to about 10 can be obtained with this method, the accuracy depending on the precision of the needle reading.

The second method, also used primarily between SWR of 1 and 10, was that of measuring ΔDb between the maximum and minimum voltages. This could be done on an expanded scale on the meter which allowed a five-fold increase in meter reading precision over the direct method. Seeing ΔDb is a power index, the voltage standing-wave ratio was related to this quantity by the equation,

$$VSWR = \log^{-1} \frac{\Delta Db}{20} \tag{61}$$

The third method, the width of minimum, was used for SWR's in excess of 10. The voltage, as a function of distance, in a coaxial tube as measured by a square law detector was

$$E_x'^2 = E_{max}'^2 \sin^2\theta + E_{min}'^2 \cos^2\theta \tag{62}$$

where $\theta = \pi \Delta x / \lambda_1$ and $\Delta x / 2 = $ distance from x to the minimum.

If $E_x'^2$ was chosen as $2E_{min}'^2$, which is equivalent to $E_x'^2$ being 3Db greater than $E_{min}'^2$, then Equation 62 could be rearranged into

$$S = E_{max}/E_{min} = \frac{(2 - \cos^2\theta)^{1/2}}{\sin\theta} \tag{63}$$

where $\theta = \dfrac{\pi \Delta x_{3Db}}{\lambda_1}$ and Δx_{3Db} = distance separating $E'^2 = 2E'^2_{min}$ through the minima (see Figure 26).

At high values of S, Equation 63 could be well approximated by simply

$$S = \frac{\lambda_1}{\pi \Delta x_{3Db}} \tag{64}$$

The value Δx_{3Db} was measured using a micrometer accurate to 10^{-3} cm which was attached to the slotted line.

Use of the width-of-minimum method was mandatory for VSWR's in excess of 40. The noise in the amplifier circuit made direct methods impossible for VSWR's taken on the 60 Db scale. S was determined by increasing the oscillator power to a level that E_{min} was measured on the 40 Db range of the meter. Note here that E_{max} was now at least in the 20 Db range which was not a square law region. This did not matter as the width of minimum on the 40 Db scale was square law and the E_{max} value was not needed in the calculations. This procedure, in principle, could be used for VSWR's in excess of 1000 providing that the oscillator had sufficient power and that harmonics and spurious noise were absent. The harmonics and noise proved to be the limitation. These, at best, were about 60 Db down from the desired signal. These unwanted signals also set up standing waves superimposed on the main pattern, but of different wavelengths. The superposition of these patterns on E_{min} limited the precision to which this could be determined. In practice, SWR's in excess of 100 led to unreliable results.

The sample length, d_s, was found by measuring the original sample at room temperature using a micrometer accurate to .001 inch. These room temperature lengths varied between 1 and 3 inches. The sample expansion as a function of temperature was found in one of two piston-cylinder devices similar to that illustrated in Figure 18. Polyoxymethylene was studied using a 1 inch diameter piston cylinder. The linear expansion of the apparatus and sample was measured using a micrometer which in turn was connected to an ohm-meter to indicate contact. The expansion of the apparatus was measured using an aluminum rod, of approximately the same length as the sample, as a standard.

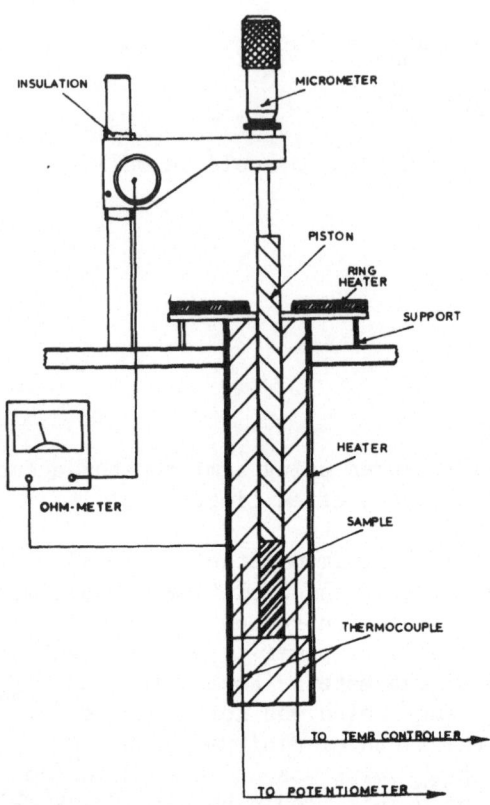

Figure 18. Representation of density measuring apparatus.

The sample diameters were usually machined to 0.001 inch less than that of the cylinder. The volume occupied by the air gap was corrected for by one of two methods:

1. "Micing" the sample accurately and computing the volume of the air space between the sample and cylinder,

2. Extrapolating the data taken after the sample had expanded into the wall back to room temperature.

In practice, the second method was more accurate because of the difficulty in measuring the inside diameter of the cylinder and outside diameter of the plastic samples. The extrapolation method consisted of using a third degree polynomial to extrapolate back to room temperature. Density was then calculated by subtracting out the apparatus expansion from the corrected total expansion.

The resulting difference was then the volume expansion of the sample minus the volume expansion of the cell due to diameter change.

The procedure used for d_s determination of Teflon*, 6-10 Nylon, and P.E.O. was essentially the same as described above with the exception that the piston cylinder arrangement consisted of the microwave sample cell as the cylinder and a brass rod with a 0.28 inch diameter hole drilled through the center (to clear the inner conductor) as a piston. This measurement provided d_s directly after corrections for the cell and gasket had been taken into account.

The gasket thickness, d_t, was determined at room temperature using a 0-1.0000 inch micrometer. This length was then corrected for temperature from Teflon* density data which had been obtained earlier in the aforementioned apparatus. The relative dielectric constant of Teflon* was determined using the microwave apparatus at room temperature, 100°C, 150°C, 200°C and 250°C. Both dielectric constant and density were obtained from a 2 inch sample which was taken from a 1 foot rod of Teflon*. This insured that the data obtained would not be sensitive to manufacturing techniques used in making the Teflon* rod. All the gaskets were then made from the remaining 10 inches of the aforementioned rod. Figure 19 contains graphs of both the relative dielectric constant, ε_3', and the volume expansion of Teflon* as a function of temperature. It was found that ε_3' was independent of frequency between 0.25 and 8 GHz and that ε_3'' was negligible. It can also be seen that the dielectric constant of Teflon* is related to the density by the formula

$$\left(\frac{\varepsilon' - 1}{\varepsilon' + 2}\right)_T = \left(\frac{\varepsilon' - 1}{\varepsilon' - 2}\right)_{25°} \quad X \quad \frac{\overline{V}_{25°}}{\overline{V}_T} \tag{65}$$

which is a rearrangement of the Clausius-Mosotti equation (7).

The length of the sample cell as a function of temperature was necessary as a consequence of Equation 57. This was found by measuring x_o' and λ_1 as a function of temperature, using the empty short-circuited cell. From this information, it was then possible to find L_{cell} as a function of temperature. This procedure is outlined in the Appendix. As more than one cell was used, the L_{cell} data are presented as part of the description of the respective cells.

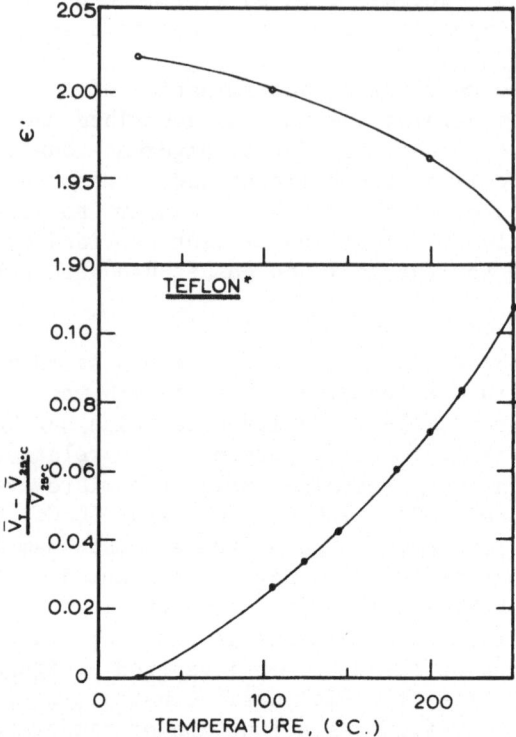

Figure 19. Relative dielectric constant and relative specific
            volume as a function of temperature of polytetrafluoro-
            ethylene.

Microwave Sample Cell: The cell constructed is illustrated
in Figure 20. This cell consisted of three major parts: (1) the
upper section containing the outer conductor, (2) the short circuit
and center conductor, and (3) the bottom cap. The upper section
was constructed by jacketing the bottom of a 9 inch length of
General Radio precision outer conductor, Type 900-9509, with a 6
inch length of 1.5 inch diameter brass rod, using metal-to-metal
interference shrink fit. The upper three inch section of outer
conductor was machined down, resulting in a 0.650 inch diameter
neck with three cooling fins attached to a G.R. 900-AP connector.
The thin neck wall and cooling fins were constructed in order that
the top of the cell could be kept cool. Excessive heating might
have damaged the Teflon* gasket in the coaxial connector of the
slotted line.

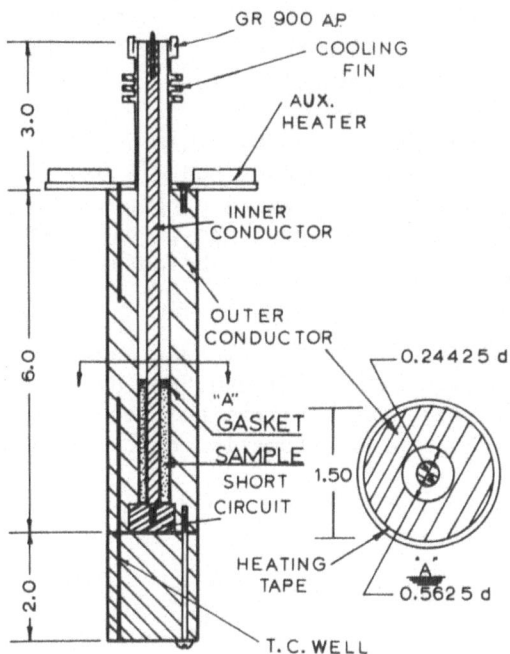

Figure 20. Cell used to study all polymers in microwave frequencies.

        The center conductor was made from G.R. 900-9507 precision
inner conductor rod. The bottom of this conductor screwed into a
hole drilled in the center of a 0.873 inch diameter, 0.505 inch
long, copper-plated rod, which acted as the short circuit. The
short circuit assembly was placed in a 0.875 inch diameter hole
bored into the end of the upper section, whose face was perpendicular to the outer conductor wall. This centered the inner conductor
to within 0.001 inch of the center line of the outer conductor on
a parallel with the walls. The short was firmly seated by forcing
it into the upper section seat by bolting on the cap.

        Thermocouple wells were drilled to a depth of 5 inches. Tem-
perature gradients could be measured in the sample area by with-
drawing the thermocouples and measuring temperature as a function
of distance. No measurable gradients were found in the sample area
when the auxiliary heater thermocouple was within 1°C of the control thermocouple at 250°C.

The resulting cell was a high precision coaxial cell with a characteristic impedance of 50.00 ± 0.50% ohms. The walls of both conductors were silverplated with a surface finish of better than 8 micro inches A.A. to minimize line loss.

The length of the inner conductor was 21.91 cm. This conductor had been fabricated in such a manner that it was 0.001 inch shorter than the outer conductor. This was done to prevent possible damage of the slotted line inner conductor due to differential thermal expansion arising from the fact that the outer conductor was being cooled by a fan.

The electrical length of the cell was constant to within ± 0.01 cm for frequencies ranging from 0.5 to 8.0 GHz, which is of the same order of error expected from the accuracy of the scale and vernier. This length as a function of temperature was found to be

$$L_{cell} = 21.91 \ (\pm \ 0.01) + 4 \times 10^{-4} \ (T - 25°C) \qquad (66)$$

The linear expansion of samples could be measured directly in this cell by using an annual piston. This was done by heating the cell with sample in it on a separate stand and monitoring the linear expansion by the method previously described. After the cell had reached equilibrium, it was then transferred to the slotted line and attached. To prevent the slotted line from bearing the full weight of the cell, a spring was attached to the cap and in turn to a ring stand below the cell. This then loaded the cell from the bottom.

When using molten polymers it was necessary to place a Teflon* gasket between the piston and the sample. This annular gasket was usually about 0.10 inch thick with an O.D. of $0.5615 \ ^{+0.0000}_{-0.0010}$ and an I.D. of 0.246 inches. The length dimension could be accurately machined to within 0.001 inch. All gaskets were machined from the same piece of Teflon* on which tests had been previously conducted to determine its coefficient of linear expansion and relative dielectric constant.

The gasket was deemed necessary for three reasons.

1. The unrestrained polymer would not form a flat interface upon melting.

2. The piston used to measure the expansion could not be withdrawn without some of the molten polymer adherring to its surface.

3. Air attacked the interface of the polymer forming cross-linked gels on degradation which have been reported to affect dielectric properties or to cause monomer formation and gassing and bubbling.

The presence of this gasket caused a discontinuity in the characteristic impedance of the line and, as a result, Equation 45 was developed to account for the residual reflections. This equation was tested by making two runs with 6-10 Nylon at 25°C; first with a gasket 0.368 cm long and second, without this gasket. The same nylon sample was used in both tests. The table below summarizes the results of these two tests.

Frequency (GHz)	With Gasket		Without Gasket	
	ε'	ε''	ε'	ε''
4.0	No Convergence		2.772	0.026
3.2	2.774	0.030	2.787	0.029
2.5	2.779	0.030	2.778	0.030
2.0	2.794	0.032	2.787	0.054
1.60	No Convergence		2.807	0.033
1.25	2.789	0.039	2.809	0.033
0.79	2.801	0.038	2.838	0.039

It can be seen that there is good agreement between ε', better than 1% on the average. The agreement between ε'' should not be compared due to the fact that loss data at this low of loss are too unreliable to be accurate. The ε'' data given should only have been reported to one significant digit for these low values. The parameter most sensitive to reflections would be ε', and the agreement between the gasket and no gasket values indicate that the reflections have been adequately accounted for.

ACKNOWLEDGMENT

The research reported here was supported by the National Science Foundation.

NOMENCLATURE

Symbol	Definition	Units
a	inside diameter of coaxial line	cm
A	area	cm^2
b	inside diameter of coaxial line	cm
c	velocity of light, 2.998 x 10^8	m/sec
C	capacitance	pf
C'	specific conductor capacitance	pf/cm
CC	complex coefficient defined by Equations 39 and 45	
C_{mt}	bridge capacitance reading with an empty sample cell connected	pf
C_o	geometric capacitance (capacitance with $\epsilon' = 1.000$)	pf
C_p	parallel-model capacitance reading of the bridge	pf
C_{px}	parallel-model capacitance of the sample	pf
C_{sh}	parallel cap of a shunt capacitor	pf
d_s	sample length	cm
d_t	gasket thickness	cm
D	dissipation factor	
e	voltage	volt
E	voltage magnitude	volt
E'	E exp($j\omega t$)	volt
E_{max}	maximum voltage of a standing wave	volt
E_{min}	minimum voltage of a standing wave	volt
f	frequency = $\omega/2\pi$	Hz
f_{max}	frequency at which maximum loss appears	Hz
G	conductance	mho
G'	specific insulator conductance	mho/cm
i	current	amp
I	current magnitude	amp
I'	periodic current	amp
j	$\sqrt{-1}$	

Symbol	Definition	Units
L	length of transmission line	cm
L	inductance	H
L'	specific inductance	H/cm
r_o	reflection coefficient	
R	resistance	ohm
R'	specific conductor losses	ohm/cm
R_p	parallel resistance at the bridge terminals	ohm
R_{px}	parallel resistance of the sample	ohm
S	standing wave ratio, E_{max}/E_{min}	
t	time	sec
T	temperature	°C or °K
tan δ	dissipation factor	
V	voltage	volt
$\overline{V}$	specific volume	cm^3/g
x	distance	cm
x_o	distance from sample or gasket to first voltage minima	cm
x_o'	distance of first measured minima from the slotted line	cm
Y	admittance	mho
Z	impedance	ohm
Z_o	characteristic impedance	ohm
α	attenuation constant, real part of γ	cm^{-1}
α'	dielectric decay function	sec^{-1}
β	phase constant, imaginary part of γ	cm^{-1}
ε	relative dielectric constant	
ε'	real part of $\varepsilon*$	
$\varepsilon*$	complex dielectric constant = $\varepsilon' - j\varepsilon''$	
ε''	loss, imaginary part of $\varepsilon*$	
γ	propagation constant, $\alpha + j\beta$	cm^{-1}
δ	phase angle	radian

Symbol	Definition	Units
ω	angular frequency = $2\pi f$	radian/sec
*	registered trademark	

REFERENCES

1. R. H. Boyd and C. H. Porter, Symposium on Dielectric Properties of Polymers, American Chemical Society, Los Angeles, 1971; Polymer Preprints, 12, 1, 135 (1971).
2. DIELECTRIC MATERIALS AND APPLICATIONS, A. von Hippel, ed., Technology Press of MIT and John Wiley and Sons, New York, 1954.
3. N. G. McCrum, B. E. Read and G. Williams, ANELASTIC AND DIELECTRIC EFFECTS IN POLYMERIC SOLIDS, John Wiley and Sons, New York, 1967.
4. HANDBOOK OF CHEMISTRY AND PHYSICS, 47th Edition, The Chemical Rubber Company, Cleveland, Ohio, 1966.
5. CHEMICAL ENGINEERS' HANDBOOK, 4th Edition, R. H. Perry, ed., McGraw-Hill Book Company, New York, 1963.
6. G. P. Weeg and G. B. Reed, INTRODUCTION TO NUMERICAL ANALYSIS, Blaisdell Publishing Company, Waltham, Massachusetts, 1966.
7. C. P. Smyth, DIELECTRIC BEHAVIOR AND STRUCTURE, McGraw-Hill Book Company, New York, 1955.

APPENDIX 1: COMPUTER PROGRAMS

Program 1 (TANPLT)

Definition: A program which will calculate all solutions to Equation 49 between specified limits of ε' and ε''. The program first calculates values of the real and imaginary parts of Equation 49 and determines whether these values, as a function of $\alpha_2 d_s$ and $\beta_2 d_s$, are positive or negative. When both functions change sign at the same point, this represents a solution or a pole. The values of $\alpha_2 d_s$ and $\beta_2 d_s$ found by this method are then fed to a Newton iteration in the complex plane, which results in improved values of these variables which, in turn, are used to calculate ε' and ε''. The output is optional, including: (1) a plot of the solutions; (2) estimated values from scanning this plot; and (3) the converged values of the Newton iteration.

Input variables are defined:

1. D = sample length at T, cm (d_s).

2. ELL = sample holder length relative to slotted line "0" at T, cm (L_{cell}).

3. EZERO = no function, place holder, leave blank.

4. T = temperature, °C (T).

5. EMAX = cuts off iteration after program exceeds this value, leave blank, then EMAX = 1000.

6. DT = thickness of Teflon* gasket at T, cm (d_3).

7. E2REL = relative dielectric constant of Teflon* at T (ε'_3).

8. ELMX = upper limit expected on loss.

9. ELMN = lower limit expected on loss.

10. EMX = upper limit expected on ε'.

11. EMN = lower limit expected on ε'.

12. ANA = number of intervals in $\alpha_2 d_s$ direction scanned (100 max).

13. ANB = number of intervals in $\beta_2 d_s$ direction scanned (100 max).

14. APRNT = print option; leave blank to print out graphical solution; put in the number, 1.0, to suppress this printout.

15. ANWT = Newton iteration option; put in number 1.0 to suppress Newton iteration; otherwise leave blank.

16. F = oscillator frequency, GHz.

17. SWR = standing wave ratio, uncorrected.

18. XMIN = slotted-line reading of first minima measured, cm (x_o').

19. XNODE = slotted-line reading of a second minimum measured, cm (x_n').

20. WVLNG = wavelength; if XNODE is not on the S.L. scale, leave XNODE blank if this is entered; if it is not, leave this space blank (λ_1).

21. EEST = estimated value of ε'; leave blank, not used in present version.

22. ELOST = same as above for loss.

23. W3DB = width of minimum, if measured, cm (Δx_{3DB}).

NOTE: When in doubt as to what to enter, leave blank.

```
C
C
C                            TANPLT
C
C     PROGRAM FOR CALCULATING DIELECTRIC DATA FROM COAXIAL MICROWAVE
C     DATA OBTAINED USING AN AIR FILLED SLOTTED LINE.PROGRAM WILL ALSO
C     CORRECT FOR LINE LOSSES IF THE LINE LOSS IS FITTED TO A SECOND -
C     ORDER POLYNOMIAL WITH TAN DELTA(DISSIPATION FACTOR OF EMPTY SLOT-
C     TED LINE AND SAMPE CELL)AS A FUNCTION OF LOG FREQUENCY.  PROGRAM
C     IN ADDITON WILL CALCULATE FOR THE CASE WHERE A LOSSLESS BEAD IS
C     PLACED IN FRONT OF THE UNKNOWN SAMPLE.  THIS BEAD IS OF LENGTH D2
C      AND RELATIVE DIELECTRIC CONSTANT OF E2REL. ENTRY OF THESE TWO
C     VARIABLES IN READ STATEMENT 20 WILL AUTOMATICALLY TAKE THIS BEAD
C     INTO EFFECT
C
C
C     AA= 1ST TERM IN LINE LOSS POLYNOMIAL
C     BB= SECOND TERM IN LINE LOSS POLYNOMIAL
C     CC= THIRD TERM IN LINE LOSS POLYNOMIAL
C     DT = LENGTH OF TEFLON PLUG (CM)
C     E2REL = DIELECTRIC CONSTANT OF TEFLON PLUG AT TEMP OF SAMPLE
C     EEST=ESTIMATED DIELECTRIC CONSTANT
C     W3DB= WIDTH OF MINIMUM, FROM MIN TO TWICE MIN ON BOTH SIDES
C     D= SAMPLE LENGTH (CM)
C     ELL=DISTANCE FROM ENT OF S.L. TO SHORT (LENGTH OF SAMPLE CELL)
C     SWR=VOLTAGE STANDING WAVE RATIO
C     EL1=WAVE LENGTH (CM)
C     XMIN= LOCATION OF MINIMA NODE (CM)
C     T=TEMPERATURE (CENTIGRADE)
C     WVLNG=WAVELENGTH(ENTERED ONLY IF XNODE IS NOT ENTERED)
C     XNODE=LOCATION OF SOME MININA(AS FAR AWAY FROM XMIN AS POSSIBLE
C     F=FREQUENCY(GHTZ)
C
      DIMENSION ASA(100),BSA(100)
      DIMENSION IAA(100,100),IBB(100,100)
      DIMENSION IB(100),IA(100),IC(10),AV(16)
      COMPLEX Z,CTANH,C,ZP,ZN,ZD,X,FX
      DATA IC/1H ,1H.,1H+,1H-,1H1,1H2,1H3,1H4,1H5,1H6
    1 FORMAT(7F10.5)
    4 FORMAT(F6.3,' GHTZ',F9.3,' GHTZ',F13.3,' CM',F15.4,F15.2,F10.2,F15
     1.3)
    6 FORMAT(8X,'FREQUENCY',13X,'WAVE LENGTH', 8X,'VSWR',12X,'X0', 9X,'X
     1NODE',8X,'VSWR(CORR)',/,3X,'INPT',10X,'CALC',/)
    8 FORMAT(1X,///)
    9 FORMAT(5F10.5,2F5.2,F10.5)
   50 FORMAT(1H1)
   40 FORMAT(1X,'TEMPERATURE =',F6.1,' C',/,1X,'SAMPLE LENGTH =',F8.4,'
     1CM',/,1X,'HOLDER LENGTH =',F8.4,' CM',40X,'FMAX =',F6.2,///)
   54 FORMAT(1X,'REPEATED ROOTS')
   71 FORMAT(/,1X,'LENGTH OF TEFLON INSERT =',F6.3,'CM.     DIELECTRIC C
     1ONSTANT (TEFLON) =',F6.3//)
  162 FORMAT(8X,'LOGF = ',F5.3,//)
   73 FORMAT(/,1X,' CONVENTIONAL RUN, NO TEFLON BEAD USED AS RESTRAINT',
     1//)
  429 FORMAT(16A5)
  430 FORMAT(/,25X,16A5/)
      READ 429,(AV(JLL) ,JLL= 1,16)
   20 READ (5,1) D,ELL,EZERO,T,EMAX,DT,E2REL
```

```
  900 READ 901,ELMX,ELMN,EMX,EMN,ANA,ANB,APRNT,ANWT
  901 FORMAT(6F10.5,2F5.2)
C    1 OPTIONS IF APRNT IS .NEQ. TO 0, THE PLOT WILL BE SUPRESSED
C    2 IF ANWT IS .NEQ. TO 0, THE NEWTON ITERATION FOR FINDING ROOTS
C      USING STARTING VALUES FROM THE SCAN PROGRAM WILL BE SUPPRESED
C      ELMX= MAXIMUM LOSS EXPECTED FOR SCAN PROGRAM
C      ELMN= MINIMUM LOSS EXPECTED FOR SCAN PROGRAM
C      EMX= MAXIMUM DI EXPECTED FOR SCAN PROGRAM
C      EMN= MINIMUM VALUE OF DI EXPECTED IN SCAN PROGRAM
C      ANA= NO OF INTERVALS SCANNED FOR AD IN SCAN PROGRAM
C      ANB= NO OF INTERVALS SCANNED IN SCAN PROGRAM
      JPRNT=APRNT+.1
      JNWT=ANWT+0.1
      IF (JPRNT .EQ. 0) CALL NOSKIP
      FTMAX=SQRT(1.+(ELMX/EMN)**2)
      FTMIN=SQRT(1.+(ELMN/EMX)**2)
      FAMAX=SQRT(0.5*EMX*(FTMAX-1.))
      FAMIN=SQRT(0.5*EMN*(FTMIN-1.))
      FBMAX=SQRT(0.5*EMX*(FTMAX+1.))
      FBMIN=SQRT(0.5*EMN*(FTMIN+1.))
      WRITE(6,50)
      PRINT 430,(AV(JL), JL=1,16)
      JCODE =E2REL +.0001
      IF(E2REL .LT. 1.) E2REL=1.0
      IF(JCODE .EQ. 0 ) GO TO 70          @ NO INSERT IN SAMPLE CELL
      PRINT 71,DT,E2REL
      GO TO 72
   70 PRINT 73
   72 DX=0
      DXX=0
      IF(EMAX .LT. 1.) EMAX = 1000.
      WRITE(6,40),T,D,ELL,EMAX
   19 READ(5,9)F,SWR,XMIN,XNODE,WVLNG,EEST,ELOST,W3DB
C-------ENTER WVLNG IF XNODE IS NOT AVAILABLE
C-------ENTER W3DB INSTEAD OF SWR IF DESIRED METHBOD OF SWR CALC.
C-----ENTER EEST IF GOOD ESTIMATE OF DIELECTRIC CONSTANT IS AVAILABLE
C-------ENTER ELOST IF GOOD ESTIMATE OF LOSS FACTOR IS AVAILABE
      AA=-3.24
      BB=-0.442
      CC=0.296
      IF (F.LT. .01) GO TO 20   @ LAST CARD MUST BE ZERO FOR REPEAT
      K=1
      IF.(WVLNG .GT. .01) GO TO 32
      ALMDA=30./F
      J=2.*ABS(XNODE-XMIN)/ALMDA +0.50 @ 0.50 ASSURES CORRECT ROUND OFF
      AJ=J
      EL1=ABS(XNODE-XMIN)*2./AJ
      GO TO 31
   32 EL1=WVLNG
   31 FCAL=30./EL1
      IF (W3DB .LT. .0001) GO TO 33
      THETA = 3.14159*W3DB/EL1
      SWR=SQRT(2.-(COS(THETA))**2)/SIN(THETA)
   33 S=1./SWR
      XZERO=XMIN+ELL-D-DT
      EPSI1=.001
      EPSI2=.0005
```

```
      B1=2.*3.14159/EL1
C-------CORRECTS SWR FOR LINE LOSSES
      ZXL=ALOG10(FCAL)
      TNWAL=10.**(AA+BB*ZXL+CC*ZXL*ZXL)
      S=S-TNWAL*XZERO*B1/2.
      SWRC=1./S
      IF(JCODE .NE. 0) GO TO 74
      Q=B1*D*(1.+(S*TAN(B1*XZERO))**2)
      C1=-(S**2-1.)*TAN(B1*XZERO)/Q
      C2=S*(1.+TAN(B1*XZERO)**2)/Q
      C=CMPLX(C1,C2)
      GO TO 97
C--------TEFLON INSERT IMPEDANCE CALCULATION
   74 SD2=SQRT(E2REL)
      COF=1./(B1*D*SD2)
      X1=TAN(B1*XZERO)
      X2=TAN(B1*SD2*DT)
      R=1./S
      G=R*R+X1*X1
      RZN=COF*((R*R-1.)*X1/G+X2/SD2)
      AIZN=COF*(1.+X1*X1)*R/G
      RZD=1./SD2-(R*R-1.)*X1*X2/G
      AIZD=-R*(1.+X1*X1)*X2/G
      ZN=CMPLX(RZN,AIZN)
      ZD=CMPLX(RZD,AIZD)
      C=ZN/ZD
   97 WRITE(6,8)
      WRITE(6,6)
      ALF=9.+ALOG10(FCAL)
      WRITE(6,4),F,FCAL,EL1,SWR,XMIN,XNODE,SWRC
      PRINT 162,ALF
  617 CP=TAN( B1*XZERO)/(B1*D)
C-------START OF SCAN PROGRAM
      B1D=B1*D
      XADM=D*B1*(FAMIN)
      XAD=XADM
      DAD=D*B1*(FAMAX-FAMIN)/ANA
      XBDM=D*B1*FBMIN
      DBD=D*B1*(FBMAX-FBMIN)/ANB
      XADMX=D*B1*FAMAX
      XBDMX=D*B1*FBMAX
      XBD=XBDMX
      JRA=ANA+0.1
      JIB=ANB+0.5
      PRINT 599,ELMX,ELMN,EMX,EMN,B1D,XADM,XADMX
  599 FORMAT(1X,'LOSS(MAX) =',F5.2,5X,'LOSS(MIN) =',F5.2,/,1X,'DIEL(MAX)
     1 =',F5.2,5X,'DIEL(MIN) =',F5.2,15X,'B1D=',F10.6,5X,/,16X,F10.5,80X
     2,F10.5,//)
      DO 600 JB=1,JIB
      DO 601 JA=1,JRA
      X=CMPLX(XAD,XBD)
      FX=CTANH(X)+C*X
      RFX=REAL(FX)
      AIFX=AIMAG(FX)
      IF(RFX.GE.0.) GO TO 602
      IA(JA)=IC(4)
      IAA(JA,JB)=0
```

```
        GO TO 603
    602 IA(JA)=IC(3)
        IAA(JA,JB)=1
    603 IF(AIFX .GE. 0) GO TO 604
        IB(JA)=IC(1)
        IBB(JA,JB)=2
        GO TO 605
    604 IB(JA)=IC(2)
        IBB(JA,JB)=3
    605 XAD=XAD+DAD
    601 CONTINUE
        IF(JPRNT .GT. 0 ) GO TO 608
        EPC=XBD*XBD/(B1D*B1D)
        PRINT 606,XBD,EPC,(IB(JV),JV=1,JRA)
    606 FORMAT(1X,F6.3,F6.2,3X,100A1)
        PRINT 607,(IA(JV),JV=1,JRA)
    607 FORMAT(1H+,15X,100A1)
    608 XAD=XADM
        XBD=XBD-DBD
    600 CONTINUE
C--------PICKS UP VALUES FROM MATRIX AND ESTIMATES ROOTS
        JLZ=1
        JRZ=JRA-1
        JIZ=JIB-1
        DO 820 JB=1,JRZ
        DO 821 JA=1,JIZ
        IS=IAA(JA,JB)
        IF(IS-IAA(JA+1,JB)) 822,823,822
    823 IF(IS-IAA(JA,JB+1)) 822,824,822
    822 IW=IBB(JA,JB)
        IF(IW-IBB(JA+1,JB)) 825,826,825
    826 IF(IW-IBB(JA,JB+1)) 825,827,825
    827 IF(JA.EQ.1) GO TO 828
        IF(IW-IBB(JA-1,JB) ) 825,828,825
    828 IF(JB.EQ. 1) GO TO 829
        IF(IW-IBB(JA,JB-1))825,829,825
    829 GO TO 824
    825 ASA(JLZ)=XADM+JA*DAD
        BSA(JLZ)=XBDMX-JB*DBD
        JLZ=JLZ+1
    824 GO TO 821
    821 CONTINUE
    820 CONTINUE
        PRINT 850
    850 FORMAT(//,35X,'ESTIMATED',20X,'CALCULATED',/,5X,'AD',8X,'BD',18X,
       1'DI',7X,'LOSS',18X,'DI',7X,'LOSS',7X,'NO WAVELNGTHS',/)
        DO 830 JLX=1,JLZ
        IF(ASA(JLX).LT. .001) GO TO 840
        AD=ASA(JLX)
        BD=BSA(JLX)
        EGRPH=(BD*BD-AD*AD)/(B1*B1*D*D)
        ELGRPH=2.*BD*AD/(B1*B1*D*D)
        PRINT 841,AD,BD,EGRPH,ELGRPH
    841 FORMAT(2F10.5,10X,2F10.5)
        IF(JNWT .GT. 0) GO TO 830
C--------START OF OPTIONAL NEWTON ITERATION
     24 Z=CMPLX(AD,BD)
```

```
      DO 2 I=1,10
      ZP=Z-(CTANH(Z)+C*Z)/(1.-CTANH(Z)**2+C)
      IF(ABS( REAL(ZP) / REAL(Z)-1.).GT.EPSI1) GO TO 2
      IF(ABS(AIMAG(ZP) /AIMAG(Z)-1.).LT.EPSI2) GO TO 3
    2 Z=ZP
   17 PRINT 5
    5 FORMAT(1H+,55X,'NO CONVERGENCE')
      GO TO 830
C--------LOSS AND DIELECTRIC CONSTANTS ARE CALCULATED FROM ALPHA AND BD
    3 D1=(AIMAG(ZP)/(B1*D))**2*(1.-(REAL(ZP)/AIMAG(ZP))**2)
      D2=2.*(AIMAG(ZP)/(B1*D))**2*(REAL(ZP)/AIMAG(ZP))
C--------WALL CORRECTION FOR LOSS AT THE WALL
      IF (D1 .LT. .01 ) GO TO 1013
      TAND=D2/D1
      EL2=EL1/SQRT(0.5*D1*(1.+SQRT(1.+D1*D1/(D2*D2))))
      ANWLG=D/EL2
      B2D=2.*3.14159*D/EL2
 1013 TB2=TAN(B2D)
      TANCOR=TAND-TNWAL-TNWAL*TB2/(B2D*(1.+TB2*TB2)-TB2)
      D2COR=D1*TANCOR
      PRINT 26,D1,D2COR,ANWLG
   26 FORMAT(1H+,60X,2F10.5,F15.3)
  830 CONTINUE
  840 GO TO 19
C     SAMPLE  DATA SET FOLLOWS AFTER LISTING
      END
```

```
CELCON MELT RUN,SHORT SAMPLE,SECOND RUN AT 28.4, OCTOBER 16,1970
3.233     21.97                  188.4                0.293      1.964
 1.5       0.        5.           2.         50.       50.
8.0        1.50     11.59        56.705
7.1        1.72     10.59        57.015
6.6        2.88     10.72        56.11
6.3        2.86      9.93        57.475
5.6        1.525    11.17        56.59
5.0        1.511    10.62        55.525
4.5        4.48     10.87        57.18
4.0        7.98     11.40        56.26
3.2        2.66     11.30        53.33
2.5                 16.675       58.65                           0.266
2.0                 11.49        56.51                           0.196
1.6                 10.44        57.09                           0.268
1.25      17.8       8.755       56.505
1.00      22.8      19.11        48.925
0.
```

Program 2 (RESON)

Description: The computer program used to calculate ε' and ε'' from Equations 15 and 16. This method uses the iterative equations defined by Equations 21 and 22. Initial values of ε' and ε'' are obtained by setting $x = 0$ in Equations 15 and 16.

Variables are defined (in order of appearance) as:

1. CO = geometric capacitance of sample cell, pf (C_o).

2. D = electrical length of transmission line connecting the 1690-A to the sample interface (d_s).

3. TEMP = temperature, °C (T).

4. F = frequency, MHz (f).

5. DCD = ΔCd (see Equation 22), pf.

6. DCV = ($\Delta C_{v-s} - \Delta C_{v-mt}$), pf.

7. VTUNE = V_{max}/V, dimensionless (m).

8. EEST = estimated ε', not necessary; leave blank if not needed.

9. ELOST = estimated ε'', not necessary; leave blank if not needed.

10. CODE - options:

    a. enter 0 for regular calculation; calculates ε' and ε'' from variables; EEST and ELOST are not needed;

    b. enter 1 for EEST to be used as first estimate of ε';

    c. enter 2, and EEST is used as the value of ε' throughout; DCD is not required;

    d. enter 3, and EEST and ELOST are used to start off ε' and ε'' iterations.

```
C
C                              RESON
C
C-----PROGRAM TO CALCULATE PARAMETERS FROM RESONANT CELL
      REAL L,LAPROX,LNEW
      DIMENSION FF(30),EE(30),EL(30),FL10(30)
   30 READ 1,CO,D,TEMP
      PRINT 100,CO,D,TEMP
      KLM=0
  100 FORMAT(1H1,1X,'CO =',F 8.3,/2X,'D =',F 8.3,' CM',/,2X,,'TEMP =',F 6
     1.1,' C.',///)
    1 FORMAT(3F10.5)
   29 READ 2,F,DCD,DCV,VTUNE,EEST,ELOST,CODE
C     F = FREQUENCY,MHTZ
C     CO = VACUMM CAPACITANCE,PF
C     D = DISTANCE FROM CAPACITOR TO CELL
C     DCD = CHANGE IN DRUM CAPACITANCE
C     DCV = CHANGE IN VERNIER,DETUNING CHANGE
C     VTUNE= VMAX/VMIN AT DETUNE
C     EEST = ESTIMATED DIELECTRIC CONSTANT
C     ELOST = ESTIMATED LOSS FACTOR
C     CODE = TYPE OF OPERATION,SEE BELOW
    2 FORMAT(7F10.5)
      PRINT 2,F,DCD,DCV,VTUNE,EEST,ELOST,CODE
      J=CODE+.01
      IF(F .LT. .001) GO TO 307
      KLM=KLM+1
      PI=3.1416
      C=2.998*1.E10
      ZO =50.
      W=2.*PI*F*1.E6
      A=W*CO*ZO*1.E-12
      B1=W/C
      ALF=ALOG10(F)+6.
      X=SIN(B1*D)/COS(B1*D)
      AX=A*X
      XDA=X/A
      AXX=A*X*X
      EAPROX=DCD/CO+1.
      XX=X*X
      LAPROX=DCV/(2.*SQRT(VTUNE*VTUNE-1.)*CO)
      IF (J.EQ. 0) GO TO 10  @ REGULAR CALCULATION
      IF(J .EQ. 1) GO TO 11  @ EEST IS USED AS FIRST ESTIMATE OF DI
      IF(J .EQ. 2) GO TO 11  @ EEST USED TO CALULATE LOSS,NO DCD GIVEN
      IF(J .EQ. 3) GO TO 13  @EEST AND ELOST USED TO START ITERATION
   10 E=EAPROX
      L=LAPROX
      GO TO 14
   11 E=EEST
      ENEW=EEST
      L=LAPROX
      GO TO 14
   13 E =EEST
      L=ELOST
      GO TO 14
   14 PRINT 15,E,L,J
```

```
 15 FORMAT(1X,'E =',F10.5,'      L =',F10.8,'      CODE =',I2,/)
    DO 50 K=1,15
    DEN=(1.-AX*E)**2 +AX*AX*L*L
    LNEW=LAPROX*DEN/(1.+XX)
    IF(ABS((LNEW-L)/L) .LT. .0001) GO TO 16
    IF (J .EQ. 2) GO TO 50
 16 IF(J .EQ. 2) GO TO 51
    CORF=1.-AX*E+XDA/E-XX-L*L*AX/E
    ENEW=(EAPROX -1. +(1.-AX+XDA-XX)/(1.-AX)**2)*DEN/CORF
    IF(ABS((ENEW-E)/E) .GT. .0001) GO TO 52
    IF(ABS((LNEW-L)/L) .LT. .0001) GO TO 51
 52 L=LNEW
    E=ENEW
 50 CONTINUE
    PRINT 60
 60 FORMAT(1X,'DID NOT CONVERGE         ')
 51 PRINT 70,F,ALF,ENEW,LNEW,J
    FL10(KLM)=ALF
    EE(KLM)=ENEW
    FF(KLM)=F
    EL(KLM)=LNEW
 70 FORMAT(1X,'FREQ =' F5.2,' MHTZ    LOG10(F) =',F6.3,'    DIELECTRI
   1C CONSTANT =',F8.3,'    LOSS =',F10.7,'    CODE =',I2////)
    GO TO 29
307 PRINT 400,TEMP
400 FORMAT(1H1,//////////,6X,'TEMPERATURE =',F5.1,' C',// 8X,'FREQUENC
   1Y',42X,'LOG10(F)',/)
    PRINT 500,(FF(I),EE(I),EL(I),FL10(I),I=1,KLM)
500 FORMAT(7X,F6.2,' MHTZ',9X,F6.3,10X,F6.3,10X,F5.3)
    GO TO 30
    END
```

APPENDIX 2:   DERIVATION OF RESONANT-CELL EQUATION

The resonant circuit used could be well approximated by the
circuit shown in Figure 21.   Shown is the load separated from the
measuring capacitors, C_d and C_v, by two lengths of transmission
line.   The first portion, having length "d" and characteristic
impedance "$Z_o 2$," is the sample cell.   The second piece, having
length "s" and characteristic impedance "$Z_o 1$," is a section of
coaxial air line extending from the sample interface to the measuring
circuit and the voltmeter.   The sample cell is terminated by an open
circuit, resulting in Z_L being infinite.   Substituting this impedance into

$$Z = Z_o \frac{Z_L + Z_o \tanh (\gamma d)}{Z_o + Z_L \tanh (\gamma d)} \tag{A-1}$$

results in

$$Z_a = \frac{Z_o}{\tanh (\gamma d)} \tag{A-2}$$

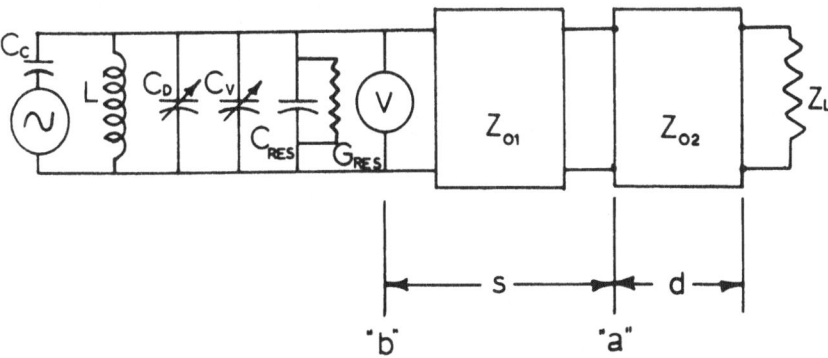

Figure 21.   Schematic representation of resonant circuit.

For small "d," this reduces to

$$Z_a = 1/(G' + j\omega C')d = 1/(G + j\omega C_p) \qquad (A-3)$$

This expression can be expressed in terms of the complex-dielectric constant and as an admittance,

$$Y_a = 1/Z_a = j\omega C_o(\epsilon' - j\epsilon'') \qquad (A-4)$$

The admittance at point "a" is then transformed to point "b" by use of

$$Y_b = Y_o1 \frac{Y_a Z_o1 + j \tan (\beta_1 s)}{1 + j Y_a Z_o1 \tan (\beta_1 s)} \qquad (A-5)$$

Substitution for Y_a in Equation A-4 and separation of the real and imaginary parts of Y_b result in the expressions,

$$Y_{br} = REAL (Y_b) = Y_o1 \frac{A\epsilon''(1 + x^2)}{DEN} \qquad (A-6)$$

and

$$Y_{bi} = IMAG (Y_b) = Y_o1 \frac{A\epsilon' - A^2\epsilon'^2 x + x - A\epsilon'x^2 - A^2\epsilon''^2 x}{DEN} \qquad (A-7)$$

where $A = \omega C_o Z_o1$, $x = \tan (\beta_1 s)$, $Y_{o1} = 1/Z_o1$, and DEN $= (1 - A\epsilon'x)^2 + (A\epsilon''x)^2$.

The total admittance at b is:

$$Y_B = - \frac{j\omega}{L} + j\omega(C_v + C_d + C_{res}) + jY_{bi} + Y_{br} + G_{res} \qquad (A-8)$$

At resonance, with material in the sample cell, the sum of the imaginary parts of Equation A-8 is equal to zero,

$$1/\omega L = \omega(C_{d1} + C_v + C_{res}) + Y_{bi} \qquad (A-9)$$

Comparing values of C_d at constant L, ω, C_{res}, and C_v, both with sample in the sample cell and with an empty cell, results in

$$\frac{C_d 2 - C_d 1}{C_o} = \frac{\varepsilon' - A\varepsilon'^2 x + \frac{x}{A} - \varepsilon' x^2 - A\varepsilon''^2 x}{DEN} - \frac{1 - Ax + \frac{x}{A} - x^2}{(1 - Ax)^2} \qquad (A-10)$$

At resonance, the admittance at b is equal to

$$Y_B = Y_{br} + G_{res} \qquad (A-11)$$

due to the fact that all the imaginary parts sum to zero. Detuning the circuit from voltage V_{max} to voltage

$$V_M = V_{max}/M \qquad (A-12)$$

by changing the vernier capacitor from C_v to $(C_v + \Delta C_v/2)$ results in an admittance of magnitude (see Figure 22),

$$Y_M = (Y_{br} + G_{res})^2 + (\Delta C_v/2)^2 \qquad (A-13)$$

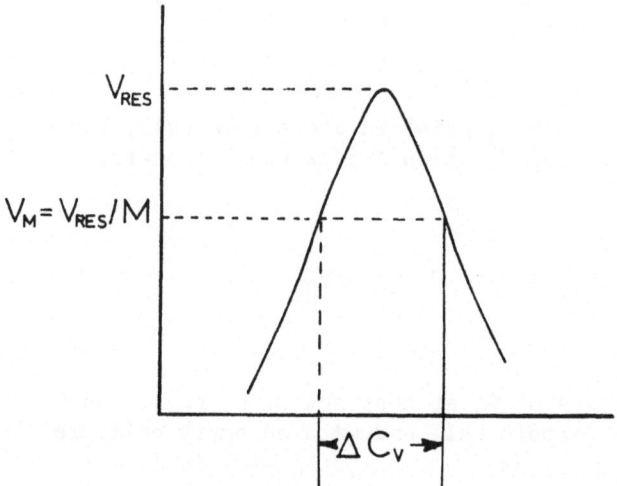

Figure 22.   Voltage around resonance as a function of the vernier
capacitor, C_v, value.

Provided that the current, I, is constant, which requires a small
value of C_c, the ratio of the voltages can also be expressed in
terms of the admittances,

$$\frac{Y_{B-M}}{Y_{B-res}} = \frac{V_{max}}{V_M} = M \tag{A-14}$$

Comparing values of ΔC_v for both the case of a sample-filled sample
cell and the case of an air-filled sample cell, at constant C_d, ω,
L, and C_{res}, gives the final expression:

$$\frac{\Delta C_{v-sample} - \Delta C_{v-empty}}{2C_o \sqrt{M^2 - 1}} = \varepsilon'' \frac{1 - x^2}{DEN} \tag{A-15}$$

APPENDIX 3:   LINE-LOSS CORRECTIONS

The derived equations used in microwave work assume that the wall resistance, R', is negligible. This, in general, is a good assumption if silverplated coaxial line is used. In some cases, however, at low frequencies during the study of material of low loss, the wall losses become a significant portion of the measured loss.

The wall loss can be measured by short circuiting the sample cell and measuring the width of minimum at some voltage node at distance "d" from the short. For the TEM mode, that mode of propagation encountered in coaxial line in a nonmagnetic material, the wall loss is then simply related to the measured parameters by the relationship

$$\tan \delta_w = \frac{\Delta x_{3DB}}{d}$$

(A-16)

The standing-wave ratio measured during an experiment arises from two sources: (1) reflections from the sample interface and (2) reflections caused by wall losses. As a result, the measured value of the standing-wave ratio must be corrected to eliminate the wall-loss contribution. This correction is made by the relationship,

$$\frac{1}{SWR_{corrected}} = \frac{1}{SWR_{meas}} - \frac{\beta_1 x_o \tan \delta_w}{2}$$

(A-17)

where x_o is defined by Equation 57. The value of $SWR_{corrected}$ is the appropriate value to use in the equations derived in the body of this work. The calculated value of the loss must also be corrected to account for: (1) line losses over the length of coaxial line containing the sample, d_s; and (2) the effect of the line loss on the sample-interface impedance.

These corrections are only important in low-loss situations; and, as a result, it is possible to use the low-loss approximations to calculate the dissipation factor. The calculated dissipation factor is corrected by the formula,

$$\tan \delta2_{corr} = \tan \delta2_{meas} - \tan \delta_w -$$

$$\frac{\tan (\delta_w) \tan (\beta_2 d_s)}{\beta_2 d_s [1 + \tan^2 (\beta_2 d_s)] - \tan (\beta_2 d_s)} \qquad (A-18)$$

The second term on the right in the above expression represents the wall loss, and the last term is the correction to the impedance at the interface. The corrected value of ε_2'' is then found by multiplying the calculated value of ε_2' by $\tan \delta2_{corr}$.

In practice, this correction was made in the computer program written for analyzing the data. Tan δ_w was measured as a function of frequency in the short-circuited cell. The values obtained were then fitted to a second-order polynomial of the form,

$$\log_{10} (\tan \delta_w) = A + B(\log_{10} f) + C(\log_{10} f)^2 \qquad (A-19)$$

This equation was then used to calculate $\tan \delta_w$, which, in turn, was used to correct $SWR_{measured}$ and the value of $\tan \delta_2$ found.

APPENDIX 4:   DETERMINATION OF L_{CELL}

The electrical length of a sample cell was measured by the following procedure:

1.   The electrical length of the short-circuited sample cell was estimated from the formula,

$$L_e = \sum_j \sqrt{\varepsilon_j}\, L_j$$

where ε_j = relative dielectric constant of substance in line, and L_j = physical length of section containing substance j, cm.

NOTE: The electrical length of a short-circuited line containing no Teflon* gaskets was just the physical length, $L_e = L$.

2.   The oscillator was set at the desired frequency and allowed to stabilize.

3.   The short-circuited sample cell was attached to the slotted line.

4.   The values of x_o' and x_n' were measured by the method outlined in the text and $\lambda_1/2$ calculated.

5.   The number of 1/2 wavelengths between x_o' and the short circuit was estimated from the formula

$$n \simeq \frac{2(x_o' + L_e)}{\lambda_1}$$

The correct value of n was the closest integer to that value calculated from the above relationship.

6.   The electrical length of the cell was calculated from the formula,

$$L_{CELL} = \frac{n\lambda_1}{2} - x_o'$$

This value is relative to the slotted line 0 and is the value used in the body of this work.

DIELECTRIC RELAXATION IN SEGMENTED POLYURETHANES

S. B. Dev, A. M. North and J. C. Reid

Department of Pure and Applied Chemistry, Strathclyde

University, Glasgow, United Kingdom

ABSTRACT

Dielectric measurements between 10^{-5} and 10^{+6} Hz have been made on one commercial (B. F. Goodrich Estane 5702 F1) polyurethane, and on polyethylene oxide and polypropylene oxide methylene bis phenyl diisocyanate polyurethanes. Four loss processes are observed and are ascribed to bulk conductance, interfacial polarization at the electrodes, Maxwell-Wagner-Sillars polarization of an occluded phase, and orientation polarization of the flexible chain segments. The processes due to migration of charge carriers give rise to large complex dielectric constants at very low frequencies. These processes are enhanced by the absorption of water. The orientation process in the polyether polyurethanes is compared with the same process in pure polyethers and is equivalent to 'normal' movement of a reduced number of the possible ether groups per chain segment. The temperature dependence of this process has been examined using a simple free volume model. For polyethylene oxide polyurethanes the ratio of critical volume to volume occupied at a reference temperature is estimated to be 0.6, compared with 0.4 in pure polyether.

A procedure is described for the evaluation of the low frequency complex dielectric constant from d.c. transient measurements. The technique fits a series of overlapping curves (parabolae or exponentials) to arbitrarily spaced data, so permitting evaluation of the Fourier transform without the inaccuracies introduced by the more conventional use of evenly spaced data.

INTRODUCTION

The mechanical characteristics of thermoplastic polyurethane elastomers are the result of a submicroscopic phase separation in which urethane-containing chain sections form 'hard' reinforcing domains in a matrix containing the flexible chain sections. Both the 'hard' and 'flexible' sub-phases consist of polar chain segments. Consequently studies of dielectric relaxation should, in principle, be able to evaluate the characteristics of the molecular motion in each of the phases. Of particular interest would be the way in which any imperfection of the molecular segregation might affect the characteristic motion of the moving groups.

A preliminary examination (1) of three polyether polyurethanes failed to detect dielectric phenomena which could be ascribed to orientation polarization of chain dipoles. Instead the dielectric properties of the system were swamped by a large low frequency polarization and loss. This was ascribed to Maxwell-Wagner-Sillars (2) interfacial polarization of a non-spherical conducting occluded phase. Careful purification and annealing of the samples reduces the low frequency polarization and losses to a point where it is possible to distinguish a number of distinct dielectric phenomena. One of these has the characteristics of orientation polarization of unit dipoles attached to a polymer backbone. In this paper are reported observations made on a commercial polyurethane elastomer (B. F. Goodrich Estane 5702 Fl), and on five polyurethanes prepared from polyether diol and methylene bis phenyl diisocyanate. The repeat units of these were

$$-CH_2-\bigcirc-NHCOO(CH_2CH_2O)_5OCNH\bigcirc- \qquad (E\ 200)$$

$$-CH_2-\bigcirc-NHCOO(CH_2)_4OOCNH\bigcirc-CH_2\bigcirc-NHCOO(CH_2CH_2O)_{15}$$

$$OCNH-\bigcirc- \qquad\qquad (E\ 600)$$

and

$$-CH_2-\bigcirc-NHCOO(CH_2)_4OOCNH\bigcirc-CH_2\bigcirc-NHCOO(CH_2CHO)_n$$

$$OCNH-\bigcirc-CH_3$$

n = 7, 15, 30   (P 450, P 1000 and P 2000)

For convenience, the samples are named with the initial letter and the molecular weight of the polyether diol residue.

EXPERIMENTAL

Materials

The polyethylene oxide 200 polyurethane (E 200) was prepared as described previously (1). The remaining polymers were chain extended with butane 1,4 diol. Samples were reprecipitated first from chloroform into methanol and then from N,N dimethyl formamide into water. The samples were dried under vacuum at 60°C, pressure moulded (1) and then annealed at 60°C. In this temperature region volatile impurities (principally water) were removed, the enormous Maxwell-Wagner-Sillars interfacial polarization was reduced due to either impurity removal or decreased asymmetry of the occlusions, and degradative side or crosslinking reactions were negligible. Water reabsorption was carried out by standing samples in an atmosphere of controlled temperature and relative humidity for a period of four weeks.

Dielectric Measurements

Below 10^{-2} Hz a d.c. transient technique has been employed to measure charge and discharge currents (3). The large d.c. conductivity component made the charge current an inexact measure of current transients and observations were restricted to the discharge current. Between 10^{-1} and 3×10^{2} Hz a.c. measurements were made on a Scheiber bridge (4) and between 10^{2} and 10^{6} Hz measurements were made on transformer ratio arm bridges as described (1) previously.

Fourier Transform of the Current Decay Function

In the present case, we are concerned with the computation of ε' and ε'' from the transient d.c. measurements. The complex dielectric constant ε^{*} is related to ε' and ε'' by its Fourier transform, namely

$$\varepsilon^{*} = F\ [\phi(t)] \tag{1}$$

where F is the half-range Fourier transform of the current decay function $\phi(t)$. Writing out in full and separating the real and imaginary parts in Equation 1, we get

$$\varepsilon' = \int_0^\infty \phi(t) \cos\omega t \, dt \qquad\qquad (2)$$

$$\varepsilon'' = \int_0^\infty \phi(t) \sin\omega t \, dt \qquad\qquad (3)$$

Although ε' and ε'' are given above as separate expressions, they are not independent of each other since they are related by the Kramers-Kronig transform. In practice, however, Equations 2 and 3 are evaluated independently.

The object, therefore, is to compute ε' and ε'' over a wide range of frequency. Usually, $\phi(t)$ is read at different times, t, from the experimentally obtained current or charge decay curve. The problem, therefore, reduces to calculating ε' and ε'' when $\phi(t)$ is given in tabular form.

The usual procedure is to represent $\phi(t)$ by an analytic function and then carry out an analytical integration. One empirical law that has been widely used is due to Hamon (5) which is a power law of the form

$$\phi(t) = at^{-n} \qquad\qquad (4)$$

The empirical law has certain limitations and therefore must be modified - especially for very short or very long times. Another restriction is on the exponent, n, which should preferably be in the range 0.3 - 1.2. Moreover, the Hamon approximation can be used to compute ε'' only. A similar treatment exists for calculating ε' and has been given by Adamec (6). However, this is valid only for a restricted range of n.

In the simplest type of relaxation, i.e. Debye relaxation, where the process is characterized by a single relaxation time, τ, $\phi(t)$ takes the form

$$\phi(t) = \phi_0 \exp(-t/\tau) \qquad\qquad (5)$$

Very few materials obey such an ideal relaxation law in practice
and it is then usual to describe the system in terms of a distribution of relaxation times. Analytically this may be represented as

$$\phi(t) = \sum_{i=1}^{n} \phi_i \exp(-t/\tau_i) \qquad\qquad (6)$$

Fitting sums of exponentials to a set of data has been the subject
of thorough investigations (7-9). The point to emerge is that no
physical significance can be attached to the coefficients and expo-
nents of such a series because of the inherent nature of mathematical ambiguities involved in such curve fitting (10).

Techniques for the numerical evaluation of Equations 2 and 3
are available (11), but almost invariably they deal with analytic
functions or closely spaced data. Although they are for a finite
Fourier integral, these can be used even when the upper limit of

integration is ∞. Basically, the integral $\int_0^\infty$ is written as the sum

of $\int_0^{t_n}$ and $\int_t^\infty$, where t_n is suitably chosen. The function $\phi(t)$ is

such that the contribution from the second integral is negligible.

Should this not be the case, it is possible (12) to represent $\int_{t_n}^\infty$

by an asymptotic series in this range and the resulting integral is

summed with $\int_0^{t_n}$.

In connection with work on dielectric measurements by the
transient d.c. technique, it becomes necessary to evaluate Equations
2 and 3 for a wide range of frequencies or where $\phi(t)$ may extend
over as many as five or six decades of log t. In such cases, it
is impossible to sample data at close intervals because of the large
number of points that would be necessary for each set of data. On
the other hand, increasing the time interval would necessarily
result in the loss of information concerning the initial stage of
the experiment. Ideally, therefore, data should be sampled at
frequent intervals at the beginning, but at greater intervals with
increasing time. Approximately, this means reading off $\phi(t)$ at
t_0, kt_0, k^2t_0, ... $k > 1$ (but preferably < 2), t_0 being the
time of the first reading.

Not much work has been done on numerical integration of arbitrarily spaced data. Details are given in References 11 and 13.
The method of cubic spline for finite integrals on equally spaced
data at close intervals has been described and the difficulty of

tackling data with non-uniform portioning has been pointed out (14).

It was, therefore, decided to develop a technique which would Fourier transform a set of arbitrarily spaced data irrespective of the nature of the decay function. To compute ε' and ε'' with an accuracy that is comparable to or better than what is achieved in the reading of $\phi(t)$ was considered sufficient.

We describe below the technique of "Overlapping Parabolas" for obtaining the Fourier transform, but compare results to those obtained from linear (successive points are assumed to be on straight lines) and multiple exponential fits. No rigid rule is laid down as to the spacing of the data except that they are distributed roughly in geometric progression. Assume that $\phi(t)$'s are given at abscissae corresponding to $a = t_0 < t_1 < t_2 < t_3.....$ $<t_{n-1} < t_n = b$. The integral to be evaluated is given by

$$\varepsilon^* = \int_{t_o}^{t_n} e^{-i\omega t} \phi(t) \, dt \tag{7}$$

$$= \sum_{i=0}^{n-1} \int_{t_i}^{t_{i+1}} e^{-i\omega t} \phi(t) \, dt \tag{8}$$

Let

$$P_2 (t_{i-1}, t_i, t_{i+1}) = a_i t^2 + b_i t + c_i \tag{9}$$

be the quadratic polynomial that interpolates $\phi(t)$ at three consecutive points t_{i-1}, t_i, t_{i+1}. Then for $i = 1, 2, \ldots n - 2$, we use the approximation

$$\int_{t_i}^{t_{i+1}} \phi(t) \, e^{-i\omega t} \, dt \approx \frac{1}{2} \int_{t_i}^{t_{i+1}} \{P_2 (t_{i-1}, t_i, t_{i+1})$$

$$+ P_2 (t_i, t_{i+1}, t_{i+2})\} \times e^{-i\omega t} \, dt \tag{10}$$

$$\frac{a_i + a_{i+1}}{2} \int_{t_i}^{t_{i+1}} t^2 e^{-i\omega t} \, dt + \frac{b_i + b_{i+1}}{2} \int_{t_i}^{t_{i+1}} t \, e^{-i\omega t} \, dt$$

$$+ \frac{c_i + c_{i+1}}{2} \int_{t_i}^{t_{i+1}} e^{-i\omega t} \, dt \tag{11}$$

This technique combines integration as well as smoothing. Over the first and last intervals, no smoothing is carried out, i.e.

$$\int_{t_0}^{t_i} \phi(t) \, e^{-i\omega t} \, dt \approx \int_{t_0}^{t_1} P_2 \, (t_0, t_1, t_2) \, e^{-i\omega t} \, dt \tag{12}$$

and

$$\int_{t_{n-1}}^{t_n} \phi(t) \, e^{-i\omega t} \, dt \approx \int_{t_{n-1}}^{t_n} P_2 \, (t_{n-2}, \, t_{n-1}, \, t_n) \, e^{-i\omega t} \, dt \tag{13}$$

The coefficients a_i, b_i and c_i can be evaluated by Lagrangian interpolation or by any standard technique of matrix inversion. The integrals

$$\int_{t_i}^{t_{i+1}} t^k \, e^{-i\omega t} \, dt \qquad\qquad (k = 0, \, 1, \, 2)$$

are, of course, exactly integrable and are evaluated in a straight-forward manner either (i) by integration by parts or, preferably, (ii) differentiation under the sign of integration starting with

$$\int_{t_i}^{t_{i+1}} e^{-i\omega t} \, dt = - \frac{1}{i\omega} \, e^{-i\omega t} \, \Big|_{t_i}^{t_{i+1}}$$

The linear case is a particular example of the quadratic case.

An example of the use of the method is given in Figure 1.   In
this example a sample of estane has been allowed to absorb 0.5%
water so that low frequency polarization and losses were large.
The sample also showed a large d.c. conductance. The discharge
current-time data were transformed to the ε' and ε'' curves shown,
and the data have the characteristic form of a relaxation process
symmetric in log frequency. By comparison ε'' data evaluated using
the Hamon approximation have a misleading distortion at low frequencies. The method can be applied to much more complex systems
where the current-time curve cannot be represented by a simple
exponential power law, by a small sum of exponentials, or by a
simple polynomial.

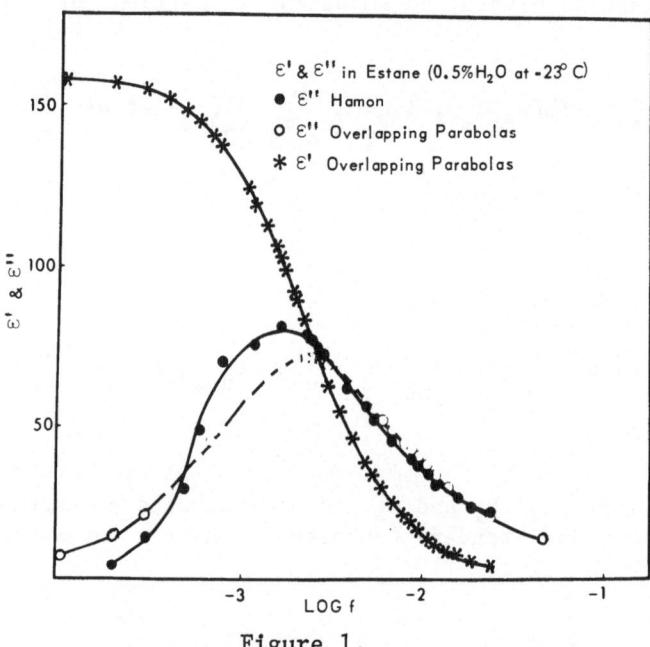

Figure 1.

RESULTS AND DISCUSSION

The principal features of the dielectric loss observations for
the polyurethanes are illustrated in Figures 2 to 8. In these
figures the data obtained by transformation of the d.c. transient
discharge current (curves A in each case) are separated from and
given a different scale from the a.c. measurements (curves B). The
data of curves B as initially obtained and illustrated in these
figures do not smoothly join the data of curves A. Four general
regions can be detected.

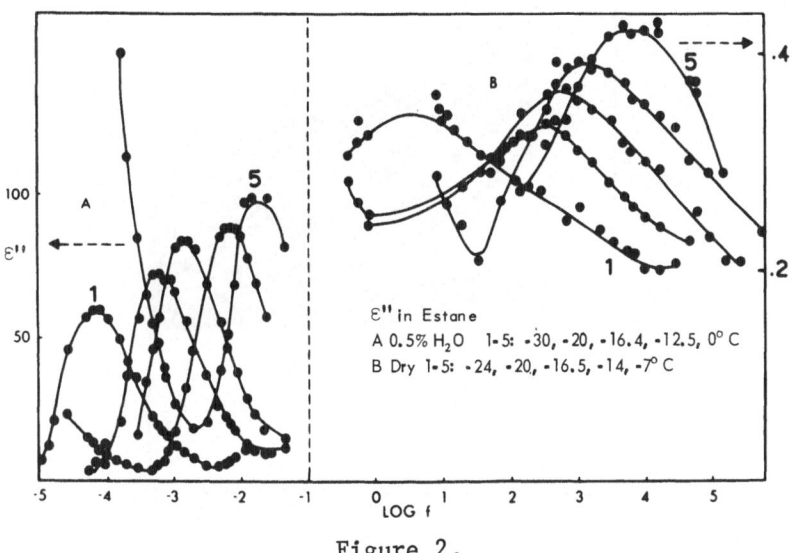

Figure 2.

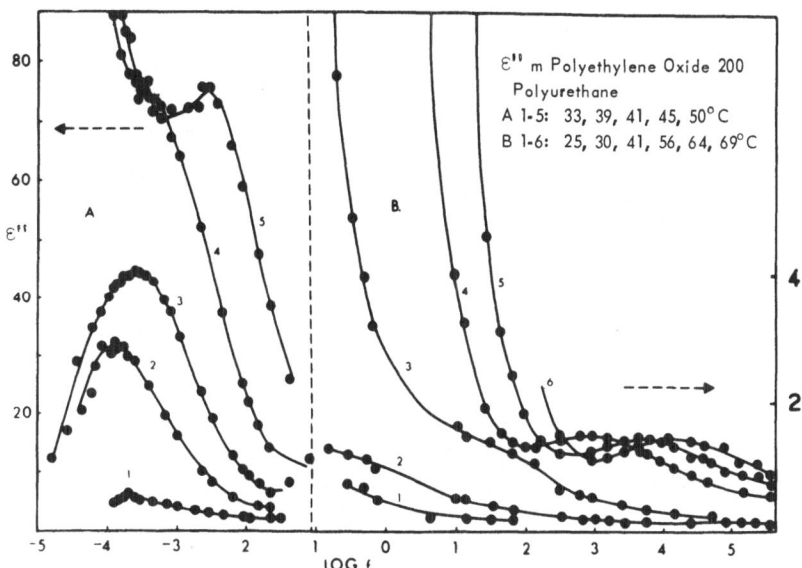

Figure 3.

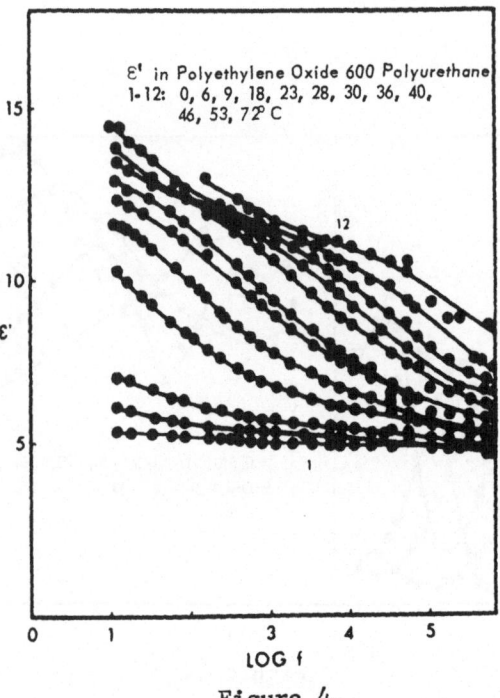

Figure 4.

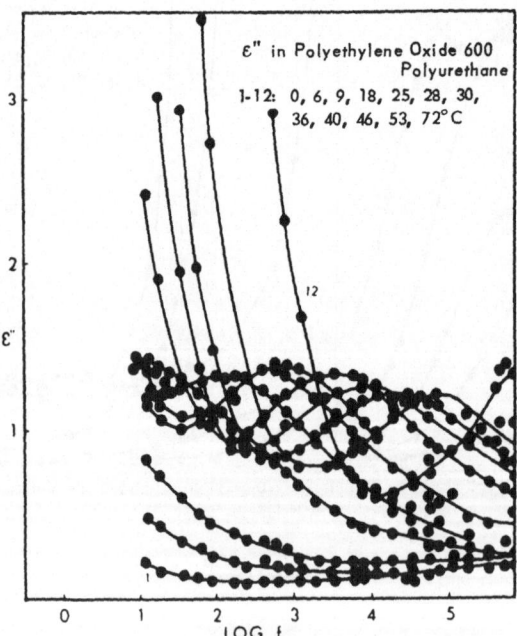

Figure 5.

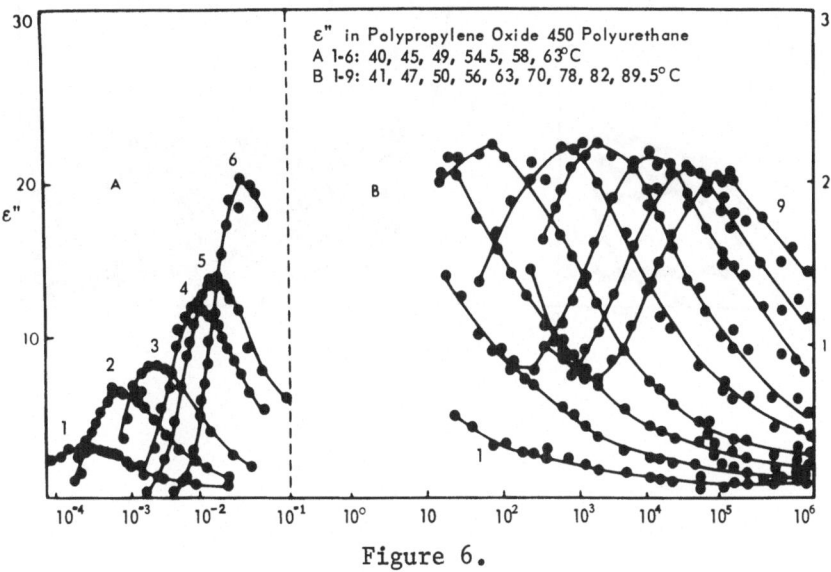

Figure 6.

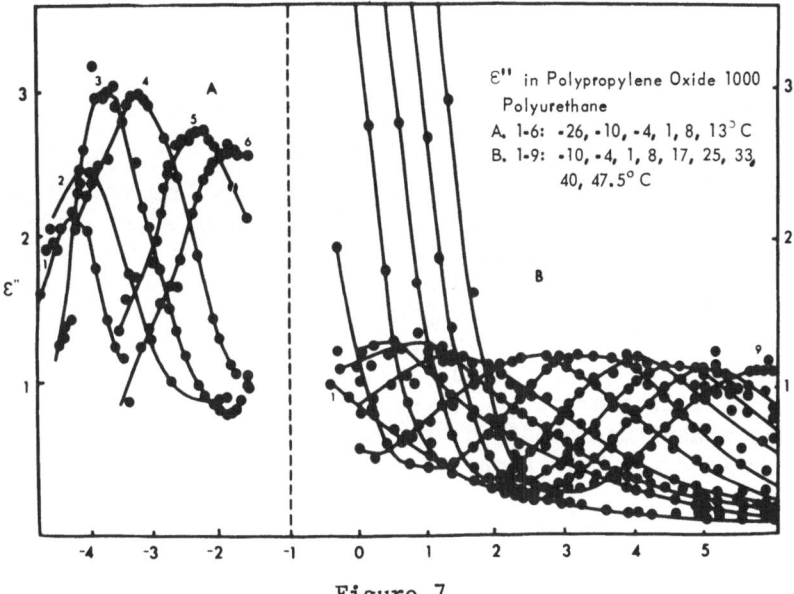

Figure 7.

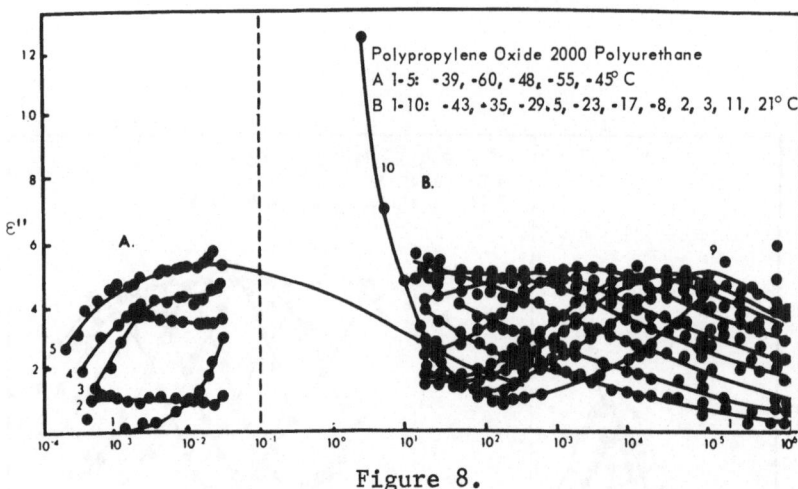

Figure 8.

(a) At the lowest frequencies and highest temperatures
(curves A of Figures 2,3), a large loss process exists which is
very dependent on temperature, electrode material, charging voltage
and water absorption by the sample. In no case has a low frequency
diminution in this loss been apparent. We have every reason to
believe that (despite experimental uncertainties estimated to be
as high as 40%) this is a real phenomenon, and is not a manifesta-
tion of base line drift in the recording equipment. We emphasize,
however, that transformation of a spurious discharge current,
recorded as apparently time invariant, would lead to a false
impression of ε'' rising continuously with decreasing frequency.

(b) At low frequencies (curves A), an important loss process
exists which is dependent on temperature, but which is independent
of electrode material and charging voltage. Reproducibility of
this process is often poor, and it is very dependent on the
annealing prehistory and dryness of the sample.

(c)   At audio frequencies (curves B), a large a.c. loss is
observed which is not detectable in discharge current measurements (curves A). The process has an amplitude inversely proportional to frequency, and, if subtracted from curves B, allows
them to extrapolate smoothly onto the corresponding curves A. The
process is only slightly dependent on electrode material at the
lowest frequencies, and is affected by temperature and moisture
absorption.

(d)   At audio and radio frequencies (curves B), an a.c. loss
is observed which is dependent on temperature, is independent of
electrode material, polarizing voltage and is only very slightly
affected (by movement in the frequency plane) by moisture absorption.

A.   The Lowest Frequency Process. This process, in common with
(b) and (c) above, is impurity dependent, relatively slow, and
involves polarization magnitudes much larger than would be possible
for any dipolar orientation process. It is reasonable, therefore,
to look for their origin in some phenomenon involving long range
migration of charge carriers (impurity conduction effects). An
important feature of this process is that it is very dependent on
the nature of the electrode-sample contact. This dependence
could be either a frequency shift or an amplitude change (measurements are capable of superposition along either axis). The phenomenon is largest (or occurs at highest frequencies) when the
electrode-sample contact is efficient as with evaporated gold
electrodes, and is less pronounced when electrodes are pressed onto
the sample during the moulding process. However, perhaps the most
important aspect of this discharge phenomenon is that the magnitude
(or frequency) of the loss is almost exactly inversely proportional
to V/T where V is the charging voltage.

At this juncture, one can do little more than speculate on the
molecular origin of the phenomenon. However, it seems that a general
explanation might lie in the discharge of charge carriers, trapped
at the sample-electrode interface during conduction, across some
energy barrier. The voltage could act by either decreasing the
number of trapped charges or increasing the barrier height by some
double layer formation (and so decreasing the discharge rate). This
latter explanation requires double layer decay to be slower than
charge transport across the interface, suggesting that different
species might be responsible for the long range transport and barrier formation. Electrode contact efficiency would then act by
increasing the discharge rate. Such a phenomenon has been reported
(15) in observations on the behavior of thin insulating films.

B.   Maxwell-Wagner-Sillars Interfacial Polarization. When a
dielectric material of low conductivity contains an occluded phase
of differing (conventionally higher) conductivity, electric charge

can migrate through the conducting body to the interface, so causing a polarization manifested as an increase in apparent dielectric permittivity (2). For spherical inclusions of permittivity equivalent to a polar organic substance in a medium of low (2-4 relative to free space) permittivity the increase in capacitance (and hence apparent permittivity) is by factors of about 2 - 4. However, if the shape of the occluded phase is non-spherical, the effect may be larger by many powers of ten.

Although this effect is often considered as affecting only the apparent permittivity of a heterophase sample, there is also an effect on the apparent loss factor at certain frequencies. This arises because there is an ohmic loss as current flows in the conducting phase to produce the polarization charges at the interfaces with the insulating phase. The time required for this charging process to occur is governed by the geometry of the system and the conductivity of the conducting phase. Thus the Maxwell-Wagner relaxation frequency for metal inclusions is typically at optical frequencies which is why metal-filled plastics can be used as microwave lenses. On the other hand, the relaxation frequency is below 10^3 Hz when the conductivity of the inclusions is below 10^{-8} Ω^{-1} m^{-1}.

It must be emphasized that this effect is not the same as a.c. conduction due to charge carrier migration through the whole sample. The second low frequency process observed here is almost certainly Maxwell-Wagner-Sillars polarization of an occluded phase. The magnitude suggests (2) that the occlusions have the form of prolate ellipsoids or cylinders rather than spheres.

The temperature dependence of the process is of particular interest. Both the amplitude and the frequency of the process increase with increasing temperature. This is in line with a phenomenon depending on both the number and mobility of charge carriers in a poorly conducting medium. Interestingly, the effect becomes markedly reduced at the glass transition temperature of the continuous rubbery phase, and the amplitude temperature coefficients roughly parallel those for bulk conductivity (Table I). The conclusion must be that the region of charge localization lies in the diffuse interphase boundary region where the electrical characteristics resemble those of the polyether phase.

C. The Bulk Conduction Process. The process observed at the lowest a.c. frequencies, but not in discharge measurements, must be simple bulk conductivity through the continuous phase. In support of this is the frequency dependence and the fact that the process is markedly reduced below the main glass transition temperatures of the rubbers as measured by differential scanning calorimetry.

TABLE I

Polymer	"Activation Energy" for Magnitude of M-W-S Loss KJ mol^{-1}	"Activation Energy" for d.c. Conduction KJ mol^{-1}
E 200	138	90
P 450	64	55
P 1000	175	175

D. The Highest Frequency Process. This loss process is the
only one of the four for which the amplitude (as well as the characteristics mentioned earlier) is in line with dipole orientation.
In each case the temperature at which the critical frequency is
about 10^0 Hz corresponds to the major glass transition temperature.
The Arrhenius plots are non-linear (Figure 9), but activation
energies between 10^2 and 10^4 Hz are reported in Table II. These
compare with 210 KJ mole^{-1} reported (16) for the α transition
(over the same frequency range) of polyethylene oxide and 130
KJ mole^{-1} for (17) polypropylene oxide respectively. Increasing
the length of the polyether segment always results in a lower
activation energy.

An alternative interpretation of the temperature dependence of
the process can be made by combining the critical free volume expression of Cohen and Turnbull (18) with the Williams-Landel-Ferry picture of the thermal expansion of free and occupied volume. The
result gives the ratio of the critical free volume required for
the movement, v_c, to the occupied volume, v_o, at a reference temperature, T_o. This latter is the temperature at which the total
volume-temperature line, extrapolated from above the glass transition, intersects the occupied volume-temperature line. The new
data are plotted according to

$$(\log a_T)^{-1} = \frac{2.3 \, \Delta\alpha \, v_o}{\gamma \, v_c} \left\{ (T_1-T_o) + \frac{(T_1-T_o)^2}{(T_2-T_1)} \right\}$$

TABLE II

Polymer	Temperature °C	$\varepsilon_o - \varepsilon_\infty$ α Process	Fuoss Kirkwood Distribution Parameter, β	Acti- vation Energy α Process KJ mole^{-1}	Number of Ether Groups Per Segment	Equiva- lent Freely Rotating Ethers
E 200	56	7.8	0.43	214	5	3.4
E 600	46	7.3	0.30	198	15	5.9
P 450	63	8.8	0.43	193	7	4.0
P 1000	17	7.2	0.35	159	15	8.3
P 2000	-8	3.9	0.27	126	30	8.2

where a_T is the frequency shift factor from constant T_1 to T_2, $\Delta\alpha$ is the change in expansion coefficient at the glass temperature and $\gamma \sim 1$ allows for overlap of free volume.

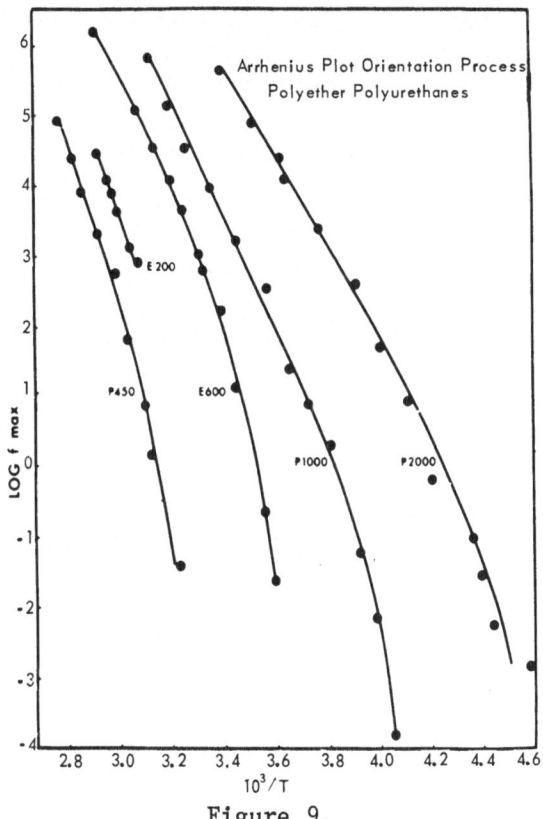

Figure 9.

As an example data for polyethylene oxide 600 polyurethane are illustrated in Figure 10. The two polyethylene oxide polyurethanes give a value of v_c/v_o of 0.6, compared with 1.3 for the α-transition of polyoctyl methacrylate (19) and 0.4 for bulk polyethylene oxide (20). Comparable plots for the propylene oxide polymers were less significant owing to a large extrapolation error in the intercept. However, there seems little doubt that the dielectric process involved is equivalent to an α transition in the polyether chain backbone, but that rather more critical free volume is required than in the pure polyether.

The magnitude of the dielectric relaxation for this process has been analyzed using the Kirkwood expression so as to obtain the number, n, of dipoles, moment μ, orienting in each repeat unit of the chain. The Kirkwood correlation parameter g was calculated from the effective dipole moment (21,22) of a CH_2OCH- group in

pure polyether. The results were equivalent to "normal" rotation
of less than the total possible number of ether groups (μ = 1.30D)
per repeat unit of polyether polyurethane as formulated in the
Introduction (Table II). The conclusion is that orientation of
the very polar urethane moieties is almost non-existent in this
process and that movement of the polyether units is restricted
by attachment to the urethane-rich domains.

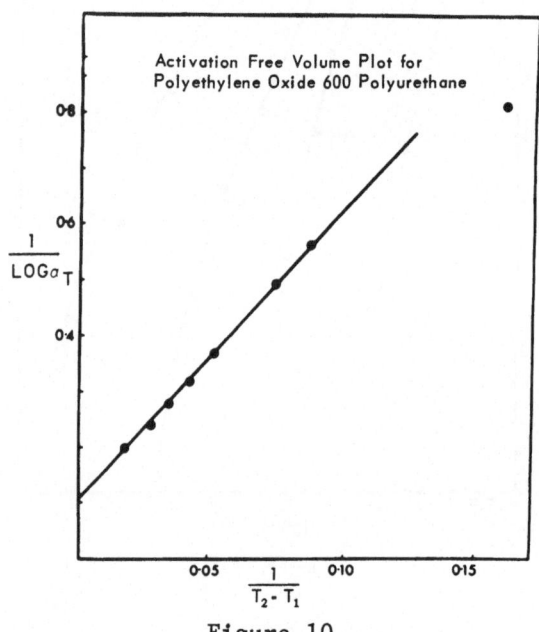

Figure 10.

CONCLUSIONS

Dielectric observations on three samples of polyurethane
elastomers show four distinct processes. Three of these are
ascribed to mobility of conduction charge carriers being (in order
of increasing frequency) polarization by trapping at the sample-
electrode interface, polarization of the two-phase interfaces,
and normal conduction. The highest frequency process has all the
characteristics of an α relaxation but corresponds to movement of
only small polyether sections in each repeat unit. The activation
parameters for this movement are similar to those in pure poly-
alkylene oxide.

REFERENCES

1. A. M. North, J. C. Reid and J. B. Shortall, Eur. Polymer J. $\underline{5}$, 565 (1969).
2. R. W. Sillars, Proc. Roy. Soc., London A169, 66 (1939).
3. W. Reddish, J. Polymer Sci. C14, 123 (1966).
4. D. J. Scheiber, J. Res. Nat. Bur. Standards C65, 23 (1961).
5. B. V. Hamon, Proc. Inst. Elec. Eng. $\underline{99}$, Monograph 27 (1952).
6. V. Adamec, Proc. Inst. Elec. Eng. $\underline{116}$, 1119 (1969).
7. K. Bonke, A.S.E.A. Research $\underline{8}$, 49 (1964).
8. B. H. Worseley and L. C. Lax, Biochim. and Biophys. Acta. $\underline{59}$, 1 (1962).
9. F. B. Hilderbrand, INTRODUCTION TO NUMERICAL ANALYSIS, McGraw Hill, Maidenhead, 1956, pp. 380-1.
10. C. Lanczos, APPLIED ANALYSIS, Sir Isaac Pitman, London, 1957.
11. P. J. Davis and P. Rabinowitz, NUMERICAL INTEGRATION, Blaisdell Publishing Company, Waltham, Massachusetts, 1969, pp. 272-280.
12. C. J. Tranter, INTEGRAL TRANSFORMS IN MATHEMATICAL PHYSICS, Methuen, London, 1966, p. 72.
13. D. Secrest, J. Siam. Numer. Anal. B2, 52 (1964).
14. B. Emarsson, B.I.T. $\underline{8}$, 279 (1968).
15. F. Stern and C. Weaver, J. Phys. C: Solid State Physics, $\underline{3}$, 1736 (1970).
16. Y. Ishida, M. Matsuo, S. Togami, K. Yamafugi and M. Tahayanagi, Koll. Z. $\underline{183}$, 74 (1962).
17. G. Williams, Trans. Faraday Soc. $\underline{61}$, 1564 (1965).
18. M. H. Cohen and D. Turnbull, J. Chem. Phys. $\underline{31}$, 1164 (1959).
19. J. D. Ferry, VISCOELASTIC PROPERTIES OF POLYMERS, John Wiley, New York, 1961, p. 213.
20. T. M. Connor, B. E. Read and G. W. Williams, J. Appl. Chem. $\underline{14}$, 74 (1964).
21. J. Marchal and H. Benoit, J. Chim. Phys. $\underline{52}$, 818 (1955).
22. G. D. Loveluck, J. Chem. Soc. 4729 (1961).

DIELECTRIC AND MECHANICAL RELAXATIONS IN ETHYLENE-ACRYLIC ACID AND

ETHYLENE-METHACRYLIC ACID COPOLYMERS

W. J. MacKnight and F. A. Emerson[*]

Chemistry Department and Polymer Science and Engineering

Program, University of Massachusetts, Amherst

ABSTRACT

The dielectric relaxation behavior of ethylene-carboxylic acid copolymers has been studied over the frequency range of 50 Hz to 10^4 Hz and over the temperature range of -160°C to + 100°C. The copolymers studied consisted of four acrylic acid and two methacrylic acid copolymers of varying acid content. In the temperature and frequency region investigated, two dielectric dispersions occur and these are labelled β' and γ in order of decreasing temperature. The β' dispersion may be quantitatively accounted for on the basis of the orientation of free carboxylic acid groups. The concentration of these groups is assessed from temperature dependent infrared studies. It is suggested that the γ relaxation also involves free carboxyl group motion but quantitative correlations are not possible in this case.

INTRODUCTION

The bulk properties of ethylene polymers containing carboxylic acid side groups have excited considerable interest in recent years. Dynamic mechanical studies have been carried out on acrylic acid copolymers (1) and mechanical and dielectric relaxation studies on methacrylic acid copolymers (2,3,4). The general features of the relaxation behavior of all the copolymers are similar. Two main mechanical and dielectric relaxation regions exist, labelled β' and γ in order of decreasing temperature. For example, a 4.1 mole percent methacrylic acid copolymer was observed to exhibit the

[*]Present Address: Naval Weapons Center, China Lake, California

237

dielectric γ relaxation at -120°C at the same frequency (2). The
effect of acid concentration on the mechanical relaxation behavior
of the acrylic acid copolymers has been studied (1) and it has been
found that the temperature of the β' relaxation increases with
increasing acid content. The effect of acid concentration on the
γ relaxation was not reported, and the corresponding effects on the
dielectric relaxation behavior have also not been studied.

In view of the dependence of the mechanical β' relaxation on
acid content, it is of interest to establish the extent of participation of the carboxylic acid groups in the dielectric β' process.
It is known that the acrylic and methacrylic acid groups in these
copolymers exist very largely in the form of hydrogen bonded
dimers (1,5) at the temperature of the β' relaxation and thus the
question is whether the small concentration of free carboxylic
acid groups present can account for the observed magnitude of the
β' relaxation.

In the present paper we report both the mechanical and dielectric
relaxation behavior of acrylic and methacrylic acid copolymers of
varying acid contents. A comparison of the relaxation behavior
determined by the two methods and an analysis of the dielectric
relaxation magnitudes using the free carboxyl group concentration
determined from infrared studies leads to the conclusion that the
β' dielectric relaxation can be quantitatively accounted for on
the basis of reorientation of free carboxylic acid side groups.

Table I summarizes some of the physical properties of the
copolymers studied. It can be seen that most of the copolymers are
partly crystalline although their degrees of crystallinity are low
relative to that of low density polyethylene and decrease with
increasing acid content. The problem of a quantitative measurement
of the crystallinity of these copolymers is complicated by several
factors, including lack of agreement between various experimental
techniques and the large role played by thermal history. The crystallinity values in Table I were determined on films annealed at
90°C for 12 hours by the infrared method described by Read and
Stein (6), based on the 1894 cm^{-1} absorption. All the copolymers
contain hydrocarbon branches on the ethylene sequences and their
concentrations are reported in Table I as the number of methyl
groups per 100 methylene groups.

The problem of the distribution of the carboxylic acid side
groups along the chain is of considerable importance to the discussion of the properties of the copolymers. All the copolymers studied
were prepared by a free radical, high pressure copolymerization of
ethylene and either acrylic acid or methacrylic acid. Because of
the considerable disparity in chemical nature existing between the
reacting species, it might be argued that the acid units should have

a tendency to exist as blocks. However, available evidence suggests that the acid groups are, in fact, randomly distributed (7),
and we accordingly adopt this view here.

EXPERIMENTAL

The ethylene-acrylic acid copolymers (samples 1A-5A) were
obtained from the Tennessee Eastman Company while the ethylene-
methacrylic acid copolymers (samples 6M and 7M) were obtained from
the duPont Company. These latter were obtained in the form of
partially neutralized sodium salts which were converted to the
acid form by a technique previously described (5). Sample 6 was
the base copolymer for other studies reported from these laboratories (2,3,4,5). The copolymers were purified by dissolving in
a mixed solvent of p-xylene and tetrahydrofuran (90/10 v/v) and
precipitating in methanol. The precipitated copolymers were dried
in a vacuum oven to constant weight. The dried materials were
compression molded into films at 160°C and 10,000 psi in a Carver
press. Samples were removed from the press and cooled to room
temperature in air. Further thermal conditioning was imposed on
all samples consisting of annealing the films at 90°C for 12 hours.
The acid contents of the copolymers, reported in Table I, were
determined by oxygen analysis.

INFRARED MEASUREMENTS

Infrared analysis was used to determine (1) the number of
methyl groups per 100 CH_2's reported in Table I, (2) the volume
fraction crystallinity, also reported in Table I and discussed in
the introduction, and (3) the concentration of free carboxylic acid
groups as a function of temperature. The instrument employed was
a Perkin Elmer IR 257 spectrometer.

The 1378 cm^{-1} band, attributed to symmetric deformation of
methyl groups, was the basis for the determination of the degree of
branching according to a well-known analytical procedure. This was
previously carried out on sample 6 (5) and sample 6 was used as a
standard for the other copolymers. The number of methyl groups per
100 methylenes reported in Table I is corrected for the methyl
groups present in the methacrylic acid units in the cases of
samples 6 and 7.

Temperature dependent infrared studies were conducted in a
specially constructed environmental chamber (8). Films were placed
between salt plates and placed in the chamber which was continuously
purged with nitrogen gas. Spectra were recorded approximately every
10°C and temperature control was maintained within $\pm$ 0.5°C.

TABLE I

Properties of Ethylene-Acrylic Acid and
Ethylene-Methacrylic Acid Copolymers

Sample[a]	Melt Viscosity Cp. at 190°C	$\dfrac{CH_3}{100CH_2}$	$\dfrac{COOH}{100CH_2}$	ρ/cm³ [b] g/cm³	Volume % Crystallinity[c]
1A	10.5×10^4	4.1	1.16	0.923	26.8
2A	12.0×10^4	4.4	1.55	0.926	24.8
3A	30.0×10^4	4.7	2.00	0.930	24.5
4A	18.0×10^3	3.1	2.57	0.036	14.7
5A	2.8×10^2	4.5	4.82	0.952	0.0
6M	–	2.6	2.1	0.934	9.8
7M	–	2.8	3.37	0.953	8.8
8	–	1.4	–	0.926	44.4

[a]Samples 1A – 5A are acrylic acid copolymers. Samples 6M and 7M are methacrylic acid copolymers. Sample 8 is low density polyethylene.

[b]Determined by displacement in methanol.

[c]Determined using the 1894 cm^{-1} infrared absorption.

MECHANICAL AND DIELECTRIC MEASUREMENTS

Dynamic mechanical response was studied using a Vibron DDV II dynamic viscoelastometer (Toyo Instrument Company). This is a forced vibration device operating in tension and measurements were made over the temperature range of -160°C to +100°C. The frequencies employed were 3.5, 11 and 110 Hz.

Dielectric behavior was studied with a General Radio Capacitance Measuring Assembly (Type 1620A). Capacitance, tan δ, or conductance were measured at 50, 100, 200, 500, 1000, 2000, 5000 and 10,000 Hz. Measurements from room temperature to -160°C were made using a three terminal cell (Balsbaugh Type LD-3) with 53 mm diameter electrodes. Temperature variation in this range was achieved by regulating the flow of dry nitrogen gas through a copper coil immersed in liquid nitrogen and control to $\pm$ 0.5°C was achieved. Measurements from room temperature to 100°C were made using a specially constructed two terminal stainless steel cell with 53 mm diameter electrodes and teflon insulation. Temperature variation and control in this range was achieved by an oil bath regulated to $\pm$ 0.1°C. 50 mm diameter aluminum foil was placed on the surface of the films by means of a thin layer of silicone grease to improve contact with the electrodes.

A General Radio Megohmmeter (Type 1862-A) was used to measure d.c. resistance in the range of $10^{12} - 10^6$ ohms. These data were in turn used to correct the dielectric loss data for d.c. conductivity.

Values of the dielectric loss tangent, tan δ, the dielectric constant, ε', and the dielectric loss factor, ε", were calculated in the usual manner (9) and necessary corrections for the use of the aluminum foil, edge effects, tan δ readings greater than 0.1 and d.c. conductivity were made when necessary (9).

RESULTS AND DISCUSSION

Infrared evidence is the basis in large measure for the statement made in the Introduction that the carboxylic acid side groups exist as hydrogen bonded dimers and that, above the glass transition temperature, a temperature dependent monomer dimer equilibrium exists. The equilibrium constant and associated thermodynamic parameters characterizing this equilibrium have been determined by studying the temperature dependence of the absorption coefficients of two different infrared bands (1,5). These two methods determine the association and dissociation constants respectively. In the method of Otocka and Kwei (1), the concentration of hydrogen bonded dimer species is measured using the absorption of the 935 cm^{-1} band, assigned to out of plane bending of the hydrogen bonded hydroxyl group. The success of this method rests on the determination of the extinction coefficient of the 935 cm^{-1} band. This is accomplished by using samples of varying thickness and making the additional assumption that all the carboxylic acid groups are in the dimer form at room temperature or just above the glass transition temperature of the material. The temperature dependence of the peak absorbance of the 935 cm^{-1} band then provides the dimer concentration as a function of temperature. The association content follows as

$$Ka = \frac{[(COOH)_2]}{[COOH]^2} \tag{1}$$

where $[(COOH)_2]$ is the concentration of dimers and $[COOH]^2$ is the concentration of monomers, found by difference.

In the method employed in these laboratories previously (5), the dissociation constant, which is, of course, the reciprocal of the association constant, is determined using the absorption of three bands. These are the 1700 cm^{-1} band assigned to hydrogen bonded carbonyl stretching, the 1750 cm^{-1} band assigned to free carbonyl group stretching, and the 3540 cm^{-1} band assigned to free hydroxyl group stretching. The concentration of free acid groups is determined by taking the ratio of the absorbance between the 3540 cm^{-1} band and the 1750 cm^{-1} band at a temperature where a base line can be established for the latter and assuming this ratio to remain constant at other temperatures. The extinction coefficient for the 1750 cm^{-1} band is taken as 1/2 the value of the 1700 cm^{-1} band which was determined on the basis of the known concentration of carboxylic acid groups in the copolymer.

Table II lists the Ka's and Kd's at 25°C for the samples of this investigation determined by the two methods described. It is immediately apparent that the Kd's of Table II are not the reciprocals of the Ka's but are, in fact, approximately two orders of magnitude smaller. This means that the free carboxyl group concentration calculated at any given temperature will differ by one order of magnitude depending on whether Ka or Kd is used. It is thus of first importance to determine the reasons for the discrepancies noted in order to use the true value for the dimerization equilibrium. The method employed for the determination of Kd has been criticized (1) on the basis that analysis reveals the presence of at least four separate absorptions at 1750, 1735, 1705 and 1690 cm^{-1}. The band at 1735 cm^{-1} apparently represents the presence of adventitious oxidation products introduced by the polymerization process. No such interference exists in the 935 cm^{-1} region. Thus, although conclusive evidence is not available for a decisive choice, the Ka determination is adopted here and values of the free carboxyl group concentration utilized in what follows are obtained from this method.

TABLE II

Characteristics of the Carboxylic Acid
Monomer Dimer Equilibrium

Sample	Ka (25°C) cm^3/mole	Kd (25°C) mole/cm^3
1A	61.0 x 10^4	1.86 x 10^{-8}
2A	25.3 x 10^4	5.37 x 10^{-8}
3A	20.4 x 10^4	1.48 x 10^{-8}
4A	18.1 x 10^4	0.44 x 10^{-8}
5A	6.7 x 10^4	3.55 x 10^{-8}
6M	17.2 x 10^4	28.8 x 10^{-8}

Figures 1 and 2 represent the temperature dependencies of the storage, E', and loss, E", moduli at 110 Hz of all the copolymers studied. In addition, a similar plot of the mechanical behavior of low density polyethylene is included in Figure 1 for the sake of comparison. The general features of these curves are similar to those reported previously (1,4). The low density polyethylene exhibits the familiar α, β, and γ relaxations in order of decreasing temperature while the acid copolymers exhibit the β' and γ relaxations described in the Introduction. Two overlapping peaks exist in the β' region of the samples of low acrylic acid content, namely samples 1A, 2A and 3A, and these consist of the β' relaxation itself, raised in temperature due to the crosslinking effect of the hydrogen bonded acid dimers, and the polyethylene α relaxation, lowered in temperature due to the modification in crystal morphology introduced by the presence of the acid groups (1,10). In contrast to the strong dependence of the β' relaxation peak temperature on acid content, the γ peak occurs at the same temperature for all copolymers and for low density polyethylene. Table III summarizes the temperature locations of the β' and γ relaxations in the acid copolymers and includes the activation energies obtained from plots of log frequency versus 1/T.

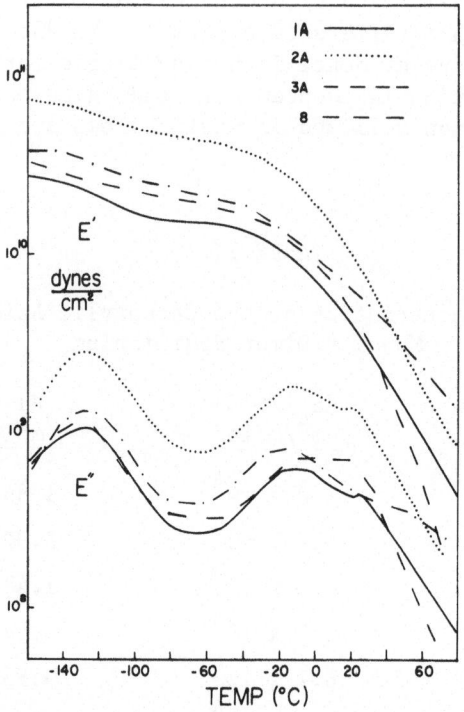

Figure 1.   Temperature dependence of E' and E'' at 110 Hz for
samples 1A – 3A and 8.

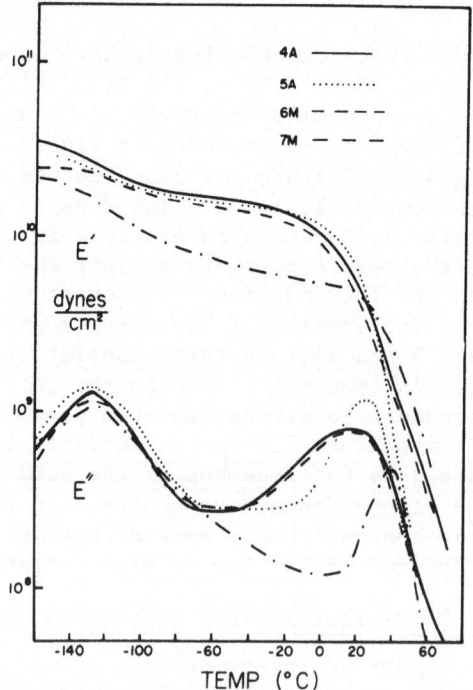

Figure 2.   Temperature dependence of E' and E'' for samples 4A,
5A, 6M and 7M.

TABLE III

Characteristics of the β' and γ Mechanical Reactions
for the Copolymers Studied at 110 Hz

Sample	β' T_{max} (°C)	β' ΔE (Kcal/mole)	γ T_{max} (°C)	γ ΔE (Kcal/mole)
1A	-10	-	-128	18
2A	- 1	-	-127	13
3A	5	-	-128	15
4A	19	54	-127	16
5A	25	60	-128	19
6M	17	60	-124	18
7M	39	93	-130	13
8	-20	120	-126	13

Figure 3 is a plot of the temperature dependence of the
dielectric loss factor, ε", at three frequencies for sample 7M. The
behavior of sample 7M is typical of all the copolymers in that only
two dielectric relaxation regions are observed. These correspond
to the mechanical β' and γ relaxations (2). The α relaxation is
not observed in the dielectric relaxation data. This behavior is
in contrast to polar derivatives of polyethylene in which the polar
groups may be incorporated into the crystal phase. Among the most
widely studied of these latter are ethylene-carbon monoxide
copolymers (11) and oxidized polyethylene (12), both of which contain carbonyl groups attached to the chain backbone. These
materials exhibit well defined dielectric relaxations corresponding
to the mechanical α relaxation of polyethylene. Inasmuch as the β'
is an amorphous phase relaxation and the γ relaxation is also known
to occur largely in the amorphous phase (13), it may be concluded
that the active dipolar species in the acid copolymers reside in
the amorphous phase. This immediately leads to the tentative conclusion that the free carboxylic acid groups are responsible for
the observed dielectric behavior since they are excluded from the
polyethylene crystal lattice (1,2,4,5,10,13). Table IV summarizes
the dielectric results for all the copolymers. The activation
energies quoted in Table IV were obtained in the same manner as
the mechanical activation energies of Table III, i.e. from the
slopes of plots of log frequency versus 1/T. The increases in the
magnitudes of both the β' and γ relaxations with increasing acid
content reported in Table IV also tend to reinforce the conclusion
that free carboxyl groups are the active polar species responsible
for the dielectric relaxation results. The temperature of the

dielectric β' relaxation is approximately the same for all the
copolymers with the exception of 7M, contrasting sharply with the
behavior of the mechanical β' relaxation (Table III). This difference is rationalized on the basis that the free carboxyl groups
cannot contribute to the dielectric β' relaxation until the glass
transition temperature has been reached and until a critical concentration of free carboxyl groups is established on the basis of
the temperature dependent monomer-dimer equilibrium existing above
the glass transition temperature. The relatively close temperature
correspondence between the mechanical and dielectric γ relaxation
temperatures, on the other hand, reflects the short range localized
motions underlying this relaxation. Presumably the small concentration of free carboxylic acid groups "frozen in" at the glass
transition temperature (i.e. the temperature of the mechanical β'
peak) undergo reorientation in the γ relaxation region and are
responsible for the dielectric relaxation behavior observed.

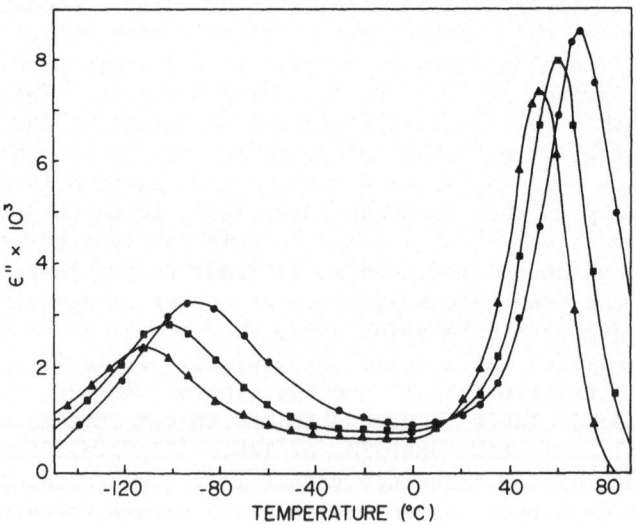

Figure 3.   Temperature dependence of ε" for sample 7M (▲, 100 Hz;
          ■, 1 KHz; ● 10 KHz).

TABLE IV

Characteristics of the β' and γ Dielectric Relaxations
for the Copolymers Studied at 100 Hz

Sample	β'			γ		
	Relative ε''	T_{max} (°C)	ΔE (Kcal/mole)	Relative ε''	T_{max} (°C)	ΔE (Kcal/mole)
1A	1.00	24	39	1.00	-126	15
2A	1.06	30	42	1.11	-119	16
3A	1.66	22	39	1.77	-117	19
4A	2.19	27	38	2.39	-114	16
5A	6.37	29	38	6.25	-114	16
6M	1.00	29	54	1.00	-120	10
7M	2.07	54	70	2.58	-112	13

The dielectric and mechanical activation energies correspond quite well for the γ relaxation but are widely disparate for the β' relaxation. Inasmuch as the mechanical activation energies were obtained on the basis of only three frequencies (3.5, 11 and 110 Hz), caution must be exercised in any conclusions drawn from the data. However, the much higher mechanical activation energies for the β' relaxation compared to the dielectric values are in accord with the assignment of the mechanical β' relaxation to microbrownian segmental motion accompanying the glass transition and the dielectric β' relaxation to reorientation of free carboxyl groups.

The concentration of free acid groups above the glass transition may be calculated using the approximate relation (2)

$$N_f = \left[\left(\frac{N_t}{2Ka}\right)\right]^{1/2} \tag{2}$$

where N_f represents the number of free acid groups/cm^3 and N_t represents the total number of acid groups/cm^3. This latter quantity is given by

$$N_t = N_A \, \rho/M \tag{3}$$

where N_A is Avogadro's number, ρ is the density and M is the molecular weight of the copolymer repeat unit. From the dielectric relaxation data it is possible to calculate $N\mu^2$, where N is the number of dipolar groups/cm^3 and μ is the dipole moment of each group, on the basis of the Onsager equation (14)

$$N\mu^2 = \frac{3kT}{4\pi} \left(\frac{2\epsilon_R + \epsilon_u}{3\epsilon_R}\right) \left(\frac{3}{\epsilon_u + 2}\right)^2 (\epsilon_R - \epsilon_u) \tag{4}$$

where ϵ_R is the relaxed (low frequency) and ϵ_u is the unrelaxed (high frequency) values of ϵ', the real part of the complex dielectric constant. The difference $(\epsilon_R - \epsilon_u)$ is obtainable according to

$$(\varepsilon_R - \varepsilon_u)_{T_{max}} = \frac{2\Delta E}{\pi R} \int_0^\infty \varepsilon'' \, d(1/T) \tag{5}$$

where ΔE is the activation energy and R is the gas constant. The integral in Equation 5 is evaluated graphically from plots of ε'' vs $1/T$. The procedure is indicated for the case of sample 6M in Figure 4. An additional assumption is then necessary in order to obtain ε_R and ε_u separately for use in Equation 4. Inasmuch as the dispersions investigated are roughly symmetrical, it seems logical to assume:

$$\varepsilon' = \frac{\varepsilon_R + \varepsilon_u}{2} \tag{6}$$

at the temperature of maximum ε''.

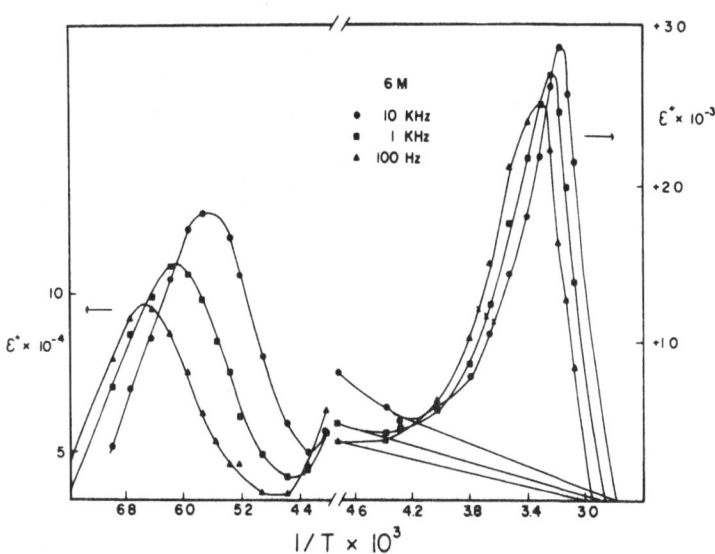

Figure 4. ε'' vs $1/T$ for sample 6M at various frequencies. The base line used in estimating the area of the β' relaxation is shown.

The results of the calculations embodied in Equations 2 - 6 are summarized in Table V for the β' relaxation. The value of μ for the free carboxyl group is taken as 1.7 D (16). The values of ε_R-ε_u are frequency dependent, increasing with increasing frequency (Table V). This is probably due to the increase in free carboxyl group concentration with temperature. The ratios of $N\mu^2$ to $N_f\mu^2$ reported in Table V, calculated at 100 Hz, indicate approximately 30 to 60 percent of the observed magnitude of the β' relaxation may be accounted for by reorientation of free carboxyl groups. In view of the approximations and assumptions inherent in the calculations, it may be concluded that the free carboxyl groups are solely responsible for the dielectric β' relaxation. The Onsager equation does not take orientation correlations of dipoles into account and thus the successful application of the Onsager equation to the acid copolymers as noted above implies that the carboxylic acid side groups are sufficiently well spaced along the chains so that the motion of a particular group is not affected by the presence of other groups. This provides an additional argument in favor of the supposition that the acid groups are randomly distributed and are not present as blocks.

In the case of the γ relaxation, it is not possible to compare directly values of $N\mu^2$ and $N_f\mu^2$ since the monomer-dimer equilibrium is "frozen" at the glass transition temperature and the infrared technique is not sufficiently sensitive to detect directly the small concentration of free carboxyl groups present at the temperature of the γ relaxation. Nevertheless, Table VI summarizes the calculations for the γ relaxation and it may be seen that $N\mu^2$ increases with increasing acid content in a similar manner to the β' relaxation although the absolute values of $N\mu^2$ for the γ relaxation are at least an order of magnitude less than the corresponding β' values. Thus, these results are consistent with the assignment of the γ relaxation to the reorientation of free carboxyl groups.

CONCLUSIONS

The β' and γ dielectric relaxations observed in the ethylene-acrylic acid and ethylene-methacrylic acid copolymers studied both originate in the motion of free carboxylic acid groups in the amorphous phase.

In the case of the β' relaxation, this motion occurs as a consequence of the long range microbrownian segmental motion accompanying the glass transition and is quantitatively relatable to the concentration of free carboxylic acid groups present estimated from infrared studies.

TABLE V

Calculated Values of $\varepsilon_R - \varepsilon_U$, ε_R, ε_U, $N\mu^2$, and $N_f\mu^2$ for the β'

Dielectric Relaxation of Ethylene-Methacrylic Acid and Ethylene-Acrylic Acid Copolymers

Sample	(Hz)	$\varepsilon_R - \varepsilon_U$	ε_R	ε_U	$N\mu^2 \times 10^{16}$ (ergs)	$N_f\mu^2 \times 10^{17}$ (ergs)	$N_f\mu^2/N\mu^2$ x100
1A	10^4	.0204	1.6875	1.6671	1.41		
	10^3	.0202	1.6939	1.6737	1.36	3.50	29
	10^2	.0192	1.7044	1.6852	1.25		
2A	10^4	.0292	1.8196	1.7907	2.01		
	10^3	.0282	1.8271	1.7989	1.80	7.63	46
	10^2	.0269	1.8274	1.8006	1.67		
3A	10^4	.0364	1.8007	1.7641	2.38		
	10^3	.0339	1.2102	1.7734	2.09	10.3	57
	10^2	.0296	1.8218	1.7922	1.80		
4A	10^4	.0397	1.2427	1.2031	2.59		
	10^3	.0356	1.8613	1.8517	2.24	10.7	50
	10^2	.056	1.8556	1.8200	2.16		
5A	10^4	.1185	1.7782	1.6598	8.37		
	10^3	.0966	1.7733	1.6767	6.60	25.0	46
	10^2	.0844	1.7622	1.6778	5.49		
6M	10^4	.0609	2.0794	2.0186	3.50		
	10^3	.0637	2.0949	2.0313	3.56	12.0	35
	10^2	.0655	2.1072	2.0418	3.56		
7M	10^4	.1509	2.2204	2.096	9.10		
	10^3	.1360	2.2370	2.1010	8.22	46.2	64
	10^2	.1278	2.2439	2.1161	7.22		

TABLE VI

Calculated Values of $\varepsilon_R-\varepsilon_U$, ε_R, ε_U, $N\mu^2$, and $N_f\mu^2$ for the γ Dielectric Relaxation of Ethylene-Methacrylic Acid and Ethylene-Acrylic Copolymers

Sample	Frequency (Hz)	$\varepsilon_R-\varepsilon_U$	ε_R	ε_U	$N\mu^2 \times 10^{18}$ (ergs)
1A	10^4	.00112	1.6204	1.6194	4.28
	10^3	.00122	1.5985	1.5973	4.46
	10^2	.00107	1.5737	1.5727	3.66
2A	10^4	.00101	1.6751	1.6741	3.88
	10^3	.00103	1.6465	1.6455	3.77
	10^2	.00095	1.6267	1.6258	3.31
3A	10^4	.00206	1.7192	1.7172	7.78
	10^3	.00212	1.7020	1.6999	7.55
	10^2	.00230	1.6872	1.6848	7.88
4A	10^4	.00269	1.7756	1.7730	10.3
	10^3	.00282	1.7457	1.7429	10.1
	10^2	.00288	1.7252	1.7214	9.81
5A	10^4	.00659	1.6497	1.6431	26.6
	10^3	.00807	1.6256	1.6176	30.4
	10^2	.00775	1.6246	1.5968	28.2
6M	10^4	.00235	1.8846	1.8822	8.33
	10^3	.00216	1.8498	1.8476	6.98
	10^2	.00187	1.8247	1.8229	5.82
7M	10^4	.0208	2.150	1.9942	7.08
	10^3	.0200	1.9700	1.9500	6.51
	10^2	.0172	1.9439	1.9267	5.35

In the case of the γ relaxation, this motion occurs as a consequence of the local reorientations of a few chain backbone units including free carboxyl groups present as a result of being trapped in the glassy matrix. It is not possible to independently estimate the concentration of free acid groups responsible for the γ relaxation.

ACKNOWLEDGMENTS

The authors are grateful to the National Science Foundation for partial support of this research. We thank Dr. Carl Wooten of the Tennessee Eastman Company and Dr. Ruskin Longworth of duPont for providing samples. Thanks are also due to Dr. Longworth for helpful discussions concerning the distribution of acid groups along the chains.

REFERENCES

1. E. P. Otocka and T. K. Kwei, Macromol. 1, 244 (1968).
2. B. E. Read, E. A. Carter, T. M. Connor and W. J. MacKnight, Br. Polymer J. 1, 123 (1969).
3. P. J. Phillips and W. J. MacKnight, J. Polymer Sci. A2, 8, 727 (1970).
4. W. J. MacKnight, T. Kajiyama and L. McKenna, Polymer Eng. and Sci. 8, 267 (1968).
5. W. J. MacKnight, L. W. McKenna, B. E. Read and R. S. Stein, J. Phys. Chem. 72, 1122 (1968).
6. B. E. Read and R. S. Stein, Macromol. 1, 116 (1968).
7. R. Longworth, private communication.
8. M. Yang, Ph.D. Thesis, University of Massachusetts, 1971.
9. N. G. McCrum, B. E. Read and G. Williams, ANELASTIC AND DIELECTRIC EFFECTS IN POLYMERIC SOLIDS, Wiley, New York, 1967, Chapter 6.
10. W. J. MacKnight, L. W. McKenna and B. E. Read, J. Appl. Phys. 38, 4208 (1967).
11. P. J. Phillips and R. S. Stein, ONR Tech. Report No. 119, Project No. NONR 056-378, June, 1969.
12. Reference 6, Chapter 10.
13. L. W. McKenna, T. Kajiyama and W. J. MacKnight, Macromol. 2, 58 (1969).
14. L. Onsager, J. Am. Chem. Soc. 58, 1486 (1936).
15. B. E. Read and G. Williams, Trans. Faraday Soc. 57, 1979 (1961).
16. C. P. Smyth, DIELECTRIC BEHAVIOR AND STRUCTURE, McGraw Hill, New York, 1955.

A COMPARATIVE STUDY OF DIELECTRIC BEHAVIOR OF POLYETHYLENE AND
CHLORINATED POLYETHYLENE

S. Matsuoka, R. J. Roe and H. F. Cole*

Bell Telephone Laboratories, Murray Hill, New Jersey

ABSTRACT

Dielectric studies were made at audio frequency ranges on
linear and branched polyethylenes and linear polyethylene chlorinated to concentrations ranging from 0.1 to 4 mole percent. For
the polymers with low levels of chlorine concentration, chlorine
atoms are found predominantly at chain ends, whereas at higher
levels of concentration, they become more randomly distributed
along the chain. These features make the chlorinated polymers a
convenient set of model compounds to study molecular dynamics of
linear and branched polyethylene.

The intensity of the γ relaxation increases and that of the α
relaxation decreases with temperature approximately following the
exponential 1/T type relationship, rather than following a 1/T
dependency. The sum of the α, β and γ peaks, however, follows the
Onsager 1/T relationship and is proportional to the chlorine concentration among all of the bulk crystallized samples. However,
solution crystallization always reduced the loss several times.

Some molecular aspects of the α and γ processes are discussed.
For the α transition, our results seem consistent with the model of
Hoffman, Williams and Passaglia (15) in which the motion of a dipole
near the crystalline surface is coupled with the motion of the
crystalline interior.

*Present Address: Bell Telephone Laboratories, Denver, Colorado

INTRODUCTION

We report here the results of dielectric studies on linear
polyethylene samples that were chlorinated to various mole fractions.
Samples with relatively low chlorine concentrations were studied,
i.e., up to ca. 4 atoms of chlorine per 100 methylene units, so
that they were morphologically comparable to commercially available grades of branched polyethylene. The technique of attaching
a polar atom on practically nonpolar polyethylene to make accurate
dielectric measurements feasible has been previously tried by
Reddish (1) and Ishida (2). In both cases, the polymer was
slightly oxidized. The chlorine atoms in our study were introduced
not only as a probe for dielectric measurements, but also as a
means of altering the morphology of the semicrystalline sample in
a defined manner. A chlorine atom can be considered as a model
for a short branch in branched polyethylene, with an added advantage of having a large dipole moment from whose concentration a
variation in the loss intensity for each polymer can be studied.

This dielectric study was actually preceded by detailed thermodynamic and morphological studies of the same chlorinated samples
that were crystallized in solution, as well as from the melt, and
these findings will be published elsewhere (3).

CHLORINATED POLYETHYLENE SAMPLES

Chlorinated samples were prepared and kindly supplied to us by
Dr. E. P. Otocka of this laboratory. They were prepared by passing
chlorine gas through a solution of Marlex 6050 in tetrachloroethy-
lene (at 105°C) for various lengths of time and dried in vacuo.
The samples thus obtained were in a fluffy powder form. They were
subsequently either compression molded or solution crystallized
from xylene.

The overall chlorine content was analyzed by the x-ray fluorescence technique as well as by a conventional wet analysis technique.
Infrared studies showed that chlorination is accompanied by a
decrease in vinyl unsaturation. Since the vinyl unsaturation is at
the chain ends, the chlorinated segments are presumably concentrated
at chain ends for the samples with very low levels of chlorination,
e.g., 0.2 mole percent. At higher levels of chlorination, e.g.,
3 mole percent, on the other hand, the chlorine content is well
beyond the number of such vinyl unsaturations available, so that
most of the chlorine atoms are expected to be distributed randomly
throughout the polymer chains.

Since chlorinated segments would be difficult to fit into the
crystalline cell of straight polyethylene, these segments would

either tend to be excluded from the crystalline regions and/or influence the size of the crystalline unit cell. The related properties are listed in Table I.

The crystalline lattice dimensions were determined from x-ray diffraction photographs obtained with a Guinier-deWolff camera. The unit cell parameters, a, b, and c, were determined by a least square fit to the positions of 22 observed diffraction lines. A noticeable lattice expansion was observed with an increase in the chlorine concentration. The dependence of a, b, and c dimensions on the chlorine concentration is shown in Figure 1. It is primarily the a dimension that expands with increasing chlorine concentration. Swan (4) has shown previously that short branches, such as propylene as co-mers, affect the a dimension in a similar way. The a dimension is also most sensitive to a temperature change.

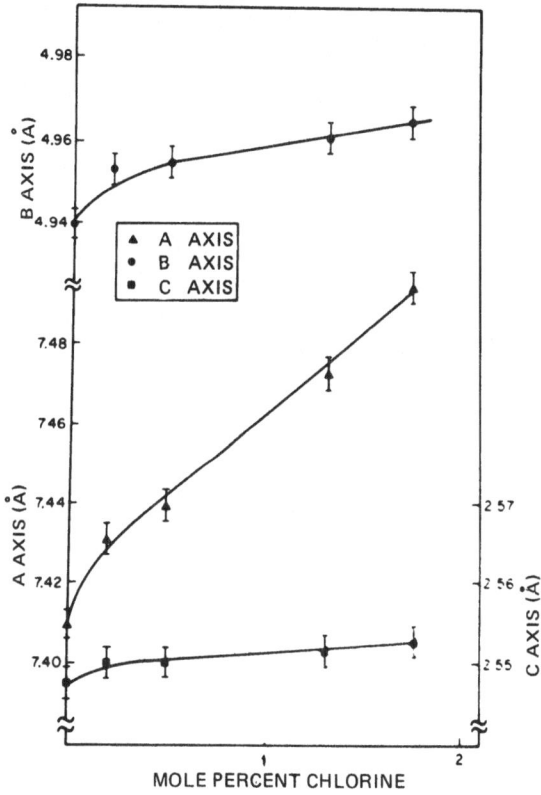

Figure 1. Relationship between the crystalline unit cell dimensions, a, b, and c, and chlorine concentration.

S. MATSUOKA, R. J. ROE AND H. F. COLE

TABLE I

Sample No.	Mole % Cl	Lattice Vol., Å^3	Mole % Cl in Crystalline Regions	Mole % Cl in Amorphous Regions	Lattice Heat of Fusion
Marlex 6050	0.00	93.26	0.00	0.00	67 cal/g
CL-1	0.19	93.88	0.05	0.75	63
CL-2	0.49	94.02	0.11	2.1	58
CL-6	1.35	94.61	0.34	5.3	52
CL-4	1.76	94.90	0.54	6.9	50
CL-10	3.19	96.40	0.4	13	41

The column termed the lattice heat of fusion in Table I is the thermodynamic heat of fusion obtained by the melting point depression with an added solvent. It decreases with the increasing concentration of chlorinated segments because of the increasing lattice dimension. The primary effect of chlorine is to change the properties of the crystalline unit cell, rather than to decrease the degree of crystallinity by leaving the crystalline regions unchanged. Here the degree of crystallinity on the conventional two-phase basis is not a realistic nor relevant physical parameter. In fact, if we define the degree of crystallinity as the ratio of the apparent heat of fusion (by DSC) to the thermodynamic heat of fusion mentioned above, it is about 0.85 for all samples.

The chlorine content inside and outside the crystalline regions is very important in interpreting the dielectric data. The values shown in Table I were obtained from the lattice specific volume, the specific volume of the amorphous state extrapolated from the specific volume of the melt, and the overall specific volume of the sample with the degree of crystallinity, as determined by the ratio of the apparent heat of fusion and the thermodynamic heat of fusion. A marked tendency for the crystalline regions to reject the chlorine is evident from the relative chlorine concentration of only about one twentieth in the crystalline regions, as compared to that in the amorphous regions. This would not mean that all of the loss maxima observed in this study are transitions only in the amorphous regions. For example, it may be due to the motion of the crystalline interior coupled to the motion of the dipole at the chain fold.

The chlorine atoms and the short branches have many common features with respect to both the morphological and dielectric properties of polyethylene. For linear polyethylene the vinyl unsaturation at the chain end is slightly polar so that our slightly chlorinated polymer, with most of the chlorine atoms at chain ends, should be a good model compound to compare.

DIELECTRIC RESULTS AND DISCUSSION

Dielectric Measurement

For dielectric measurements the signal, generated by a Hewlett-Packard H20-200CD Wide Range Oscillator with a nominal frequency range from 5 Hz to 600 KHz, was passed through a general radio 1615-A capacitance bridge. For signal detection, a Princeton Applied Research Model HR-8 Lock-In Amplifier was used. A Balsbaugh Model LD-3 Three-Terminal Sample Holder was used to hold a sample

about 50 mils thick and 2.5 inches in diameter. The sample holder
was covered with a custom-fitted thermal insulation cover made of
styrene foam. This assembly was placed inside two plastic bags.
Dry nitrogen was introduced into this assembly after having passed
through a coiled copper tube immersed in liquid nitrogen, warm
water, or a heated oven, depending on the temperature range of
the test. Typically, measurements were made repeatedly at 100 Hz
1000 Hz, 10,000 Hz, and 100,000 Hz as the temperature of the sample,
measured by two thermocouples directly in contact with the sample,
was allowed to increase from the starting temperature of -160°C.

Loss Data for Chlorinated Polymers

In Figures 2, 3, and 4 are shown the dielectric loss tangent
vs. temperature curves for the samples of the low, intermediate,
and high chlorine contents, respectively. The tan δ is shown here
instead of the ε'', merely because our data were obtained in this
form. The analyses of the intensity of relaxation, the frequency,
and the temperature of each transition, etc., were carried out on
ε''.

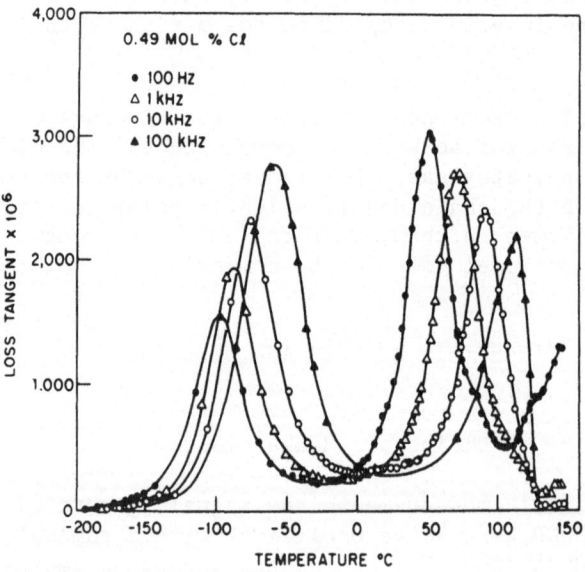

Figure 2.   Dielectric loss tangent vs. temperature of bulk-
            crystallized specimen with chlorine concentration of
            0.49 mole percent - only the α and γ transitions are
            observed.

The loss curves for the sample with a low chlorine content shown in Figure 2 exhibit only the α and γ peaks, while those for the high chlorine content shown in Figure 4 exhibit the β peak as well. We have already pointed out that the chlorine atoms for the sample shown in Figure 2 were mostly at the molecular chain ends, while 90% of those for the sample shown in Figure 4 were distributed along the chain. Among all of the samples tested, we noted the absence of the β peak for the samples with less than 0.5 mole % of chlorine, whereas for those samples with higher chlorine concentration, the intensity of the β peak was found to increase with the amount of chlorine.

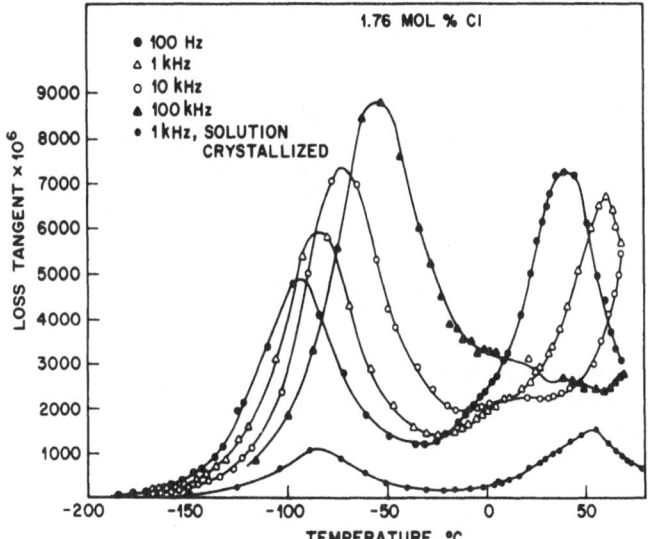

Figure 3.   Dielectric loss tangent vs. temperature of bulk-crystallized specimen with chlorine concentration of 1.76 mole percent - the β transition is marginally but distinctly observed. Data at 1 KHz for the solution-crystallized sample of the same chlorine concentration is also shown.

In Figure 3 a loss curve for the solution crystallized sample of the same chlorine concentration (checked by the x-ray fluorescence technique) is also shown. The solution crystallized sample exhibits the loss intensity of about one-fifth that of the bulk crystallized sample with the same amount of chlorine. Among the bulk crystallized samples the total loss intensity, including all of the transitions, was observed proportional to the chlorine concentration.

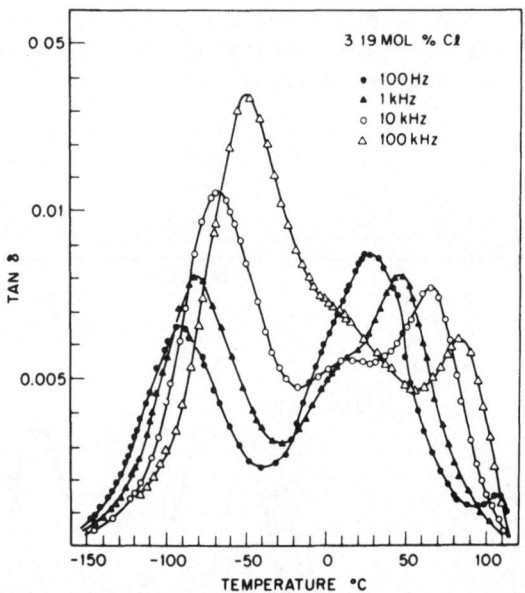

Figure 4.   Dielectric loss tangent vs. temperature of bulk-crystallized specimen with chlorine concentration of 3.19 mole percent - the three loss peaks are observed.

Both linear and branched polyethylenes (without chlorine) exhibited much lower levels of dielectric loss as compared to any of the chlorinated samples. The best grade of commercially available polyethylene with the least amount of external polar contamination showed typically the tan δ of the order of 10^{-5} at all temperatures within the frequency limits of our test. The dielectric loss for linear polyethylene is probably due to the vinyl unsaturation at chain ends, and for branched polyethylene it is due to the short branches. As evident in Figures 5 and 6, both show the multiple transitions similar, in all aspects, to those of the chlorinated samples of Figures 2, 3, and 4, with one exception in branched polyethylene in Figure 6 where the α relaxation is very weak as compared to the β relaxation.

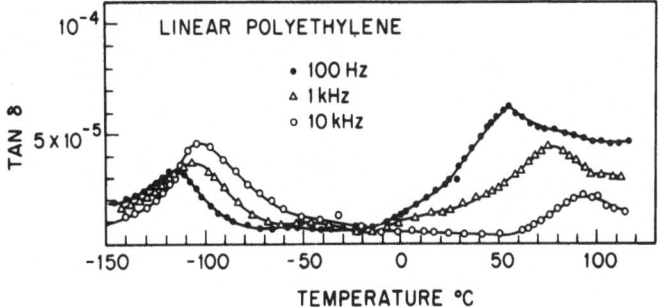

Figure 5.   Dielectric loss tangent vs. temperature of bulk-
crystallized linear polyethylene at three frequencies
and solution-crystallized at 1 KHz. Note the solution
crystallized specimen exhibits a very low loss, and
the γ transition is practically absent.

As shown in Figure 5, solution crystallization of linear
polyethylene resulted in a marked decrease of loss intensity as it
did for the chlorinated polymer shown before. The same phenomenon
was also observed in mechanical relaxation data obtained with a
Rheovibron instrument (17). The low loss in the single crystal
mats may be possibly related to the uniform orientation of the
crystallographic axes with respect to the force field, but we
have yet to perform an experiment to prove such a hypothesis.

Phenomenological Relationship Between
the α and γ Loss Maxima

Figures 2, 3, and 4 that we have discussed are in the isochronal form with temperature as the variable. Analysis of the
data in this form can lead to certain difficulties since a discussion of relaxation processes is always based on an isothermal
condition with frequency as the variable. Nevertheless, the isochronal curves provide a fair, if not quantitatively accurate,
overall view of the multiple transitions, without requiring data
over an extremely wide frequency range.

From the isochronal data we have noted (1) that both the α and
γ loss peaks occur at higher frequencies for higher temperatures,

(2) that for each sample the intensities of the α and γ loss peaks are comparable, and (3) that the intensity of the γ peak increased, and that of the γ peak decreased as temperature increased, as if to complement each other in order to maintain the sum of the two loss peaks constant.

We have found that the total intensity of relaxation according to the formula of Read, et al. (6)

$$(\varepsilon_s - \varepsilon_\infty) = \frac{2\Delta H^*}{\pi R} \int_0^\infty \varepsilon'' \, d(1/T) \tag{1}$$

is approximately independent of frequency for each polymer, even though the intensity of each relaxation varies with frequency. Here ε_s is the static dielectric permittivity, ε_∞ is the high frequency permittivity, R is the gas constant, ε'' is the dielectric loss factor, T is the absolute temperature, and ΔH^* is the enthalpy of activation. Since the α and γ processes have different values of ΔH^*, the actual evaluation of Equation 1 is carried out as a summation of the two areas under the curve using the appropriate values of ΔH^*.

The quantity $(\varepsilon_s - \varepsilon_\infty)$ is actually a very small fraction of ε_s or ε_∞ because the materials we studied have very low levels of loss. Variation of $(\varepsilon' - \varepsilon_\infty)$ with temperature, in fact, is much smaller than the variation of either ε' or ε_∞ alone, which approximately follow the Clausius-Mossotti relation.

The values of the total loss intensity obtained from Equation 1 for the various samples were found to be approximately proportional to the chlorine concentration. Following the Onsager relationship,

$$\varepsilon_s - \varepsilon_\infty = \frac{3\varepsilon_s}{2\varepsilon_s + \varepsilon_\infty} \left(\frac{\varepsilon_\infty + 2}{3} \right)^2 \frac{4\pi N}{3kT} \mu_v^2 \tag{2}$$

in which μ_v is the dipole moment and N is the concentration of the dipoles, and k is Boltzmann constant, we estimated $N \mu_v^2$ to be approximately proportional to the overall chlorine concentration, as indicated in Table II. μ_v in the last column is the effective dipole calculated by assuming all of the dipoles participated. If we instead take the value of 3 Debye for μ_v, as it is for 1-chloropropane (2.05 Debye), 2-chloropropane (2.17 Debye), and 1,3-dichloropropane (2.08 Debye), we find that about 65% of the chlorinated segments participate in the overall relaxation process. This exact number should not be taken literally, but it does suggest that most of the chlorinated segments participate in the multiple transitions.

TABLE II

Sample	Mole % Cl	$N\mu_v^2$	μ_v (Debye)
CL-1	2×10^{-3}	4.4×10^{22}	1.6
CL-2	5×10^{-3}	9.3×10^{22}	1.5
CL-4	18×10^{-3}	3.4×10^{23}	1.5
CL-10	32×10^{-3}	6.5×10^{23}	1.6

This apparently complementary relationship between the α and γ relaxation processes is summarized in the schematic diagram for $(\varepsilon'-\varepsilon_\infty)$ and ε'' shown in Figure 7. According to the simple scheme shown here, at the static frequency limit of the experiment the γ transition will not be observed and the intensity of the α transition will be equal to $\varepsilon_s-\varepsilon_\infty$, while at very high frequencies the γ relaxation will be the principal part of the overall relaxation process.

These phenomenological observations seem to apply not only to the relationship between the α and the γ transitions of crystalline polymers, but also to multi-transitions that have been observed in a wide range of crystalline and amorphous polymeric materials (7-9) and even nonpolymeric glass-forming materials (10). Williams (11) recently stressed this general nature of multiple transition processes, advocating an approach to seek a general scheme which will apply to all of these different materials. In the scheme he adopted, Williams attempts to evaluate the partial relaxation of local motions dictated by the environment which surrounds each dipole. The high frequency, low temperature transition, such as the γ transition here, is characterized by the solid-like environment.

The γ Transition

There are a number of ways to describe a solid-like environment. One of the convenient ways is to specify it by the energy level of a site model, i.e., the environment in which there is a preferred direction for a dipole with respect to its immediate neighbors such that any other orientation would have a higher energy. The energy may be due to the electrical field created by the surrounding dipoles, or the intra- and inter-molecular energies. Calculation of the temperature dependence of the loss intensity essentially follows the classical problem of Frohlich (12,13) of order-disorder transition in polar crystals, if we disregard the different nature of the environment which determines the energy

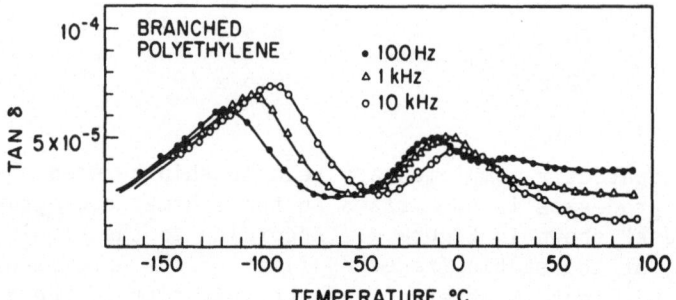

Figure 6. Dielectric loss tangent vs. temperature of bulk-
           crystallized branched polyethylene at three frequencies.
           Note the α transition is very weak.

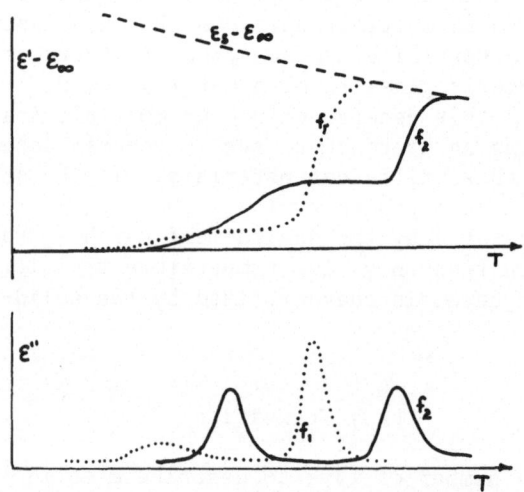

Figure 7. A schematic plot depicting relationship between the
           loss factor and the dielectric constant vs. temperature
           and frequency for the α and γ dispersion ranges.

level of each site. More specific treatment related to polymer
structure has been extensively developed by Hoffman and his coworkers (14-18). Assuming that a two-site model contains many
of the qualitative but essential features of the more complicated
structures which would lead to a multiple site model, we specify
w as the probability of finding a dipole in the higher energy
well of the two sites. Letting the evergy level difference
between the two sites be v, it follows

$$\frac{w}{1-w} = e^{-\frac{v}{kT}} \tag{3}$$

At low temperatures and v/KT >> 1, it can be shown that

$$\varepsilon_s - 1 = \frac{3\varepsilon_s}{2\varepsilon_s + 1} \, 4\pi N \left(\frac{4\mu^2}{kT}\right) 2e^{-\frac{v}{kT}} \tag{4}$$

Equation 4 describes an exponential type temperature dependence of
the loss intensity. The values of $\int \varepsilon'' \, d(1/T)$ for the γ loss peak
calculated from our data follow the same exponential 1/T relation-
ship with v $\sim$ 1 Kcal/mole, which is in the order of the enthalpy
of fusion of a polyethylene crystal.

    Variation of the intensity of the relaxation process with
temperature, in this manner, means, in approximate terms, that the
number of dipoles which participate in each transition varies with
the temperature. Hence, the intensity calculated from an isochronal
curve by $\int \varepsilon'' \, d(1/T)$ is only an approximation when v≠0. Our data
in the isothermal form cannot cover the entire loss peak, but the
area under the Rach curve can be reasonably estimated within limited
temperature ranges. The temperature dependence of the loss intensity estimated in this manner agrees with the above results from
the isochronal curves.

    The temperature dependence of the average correlation fre-
quency is another important aspect of the molecular relaxation.
Here we found that the analysis must be based on data in the isothermal form, because the isochronal form would lead to a significantly different relationship. Figure 8 shows the plot of
logarithmic frequency for loss maxima vs. 1/T. The dotted lines
are the average values of polyethylene data included in a review
article on molecular dynamics by McCall (19). Only a few of our

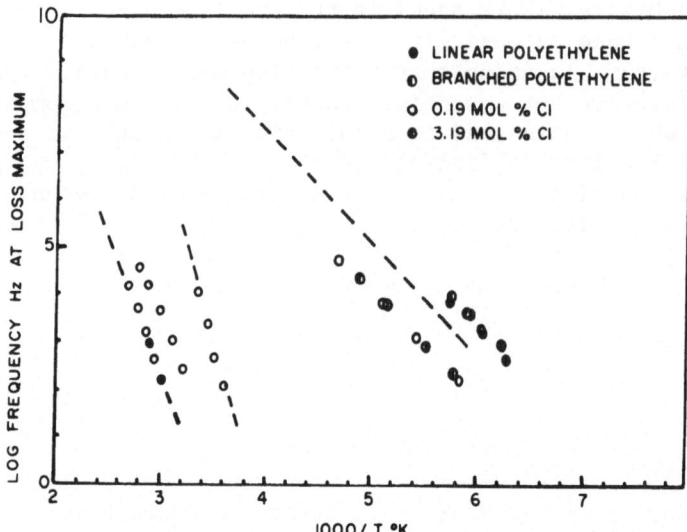

Figure 8.   Relationships between the frequency at loss maximum vs.
reciprocal temperature, often called transition map.
The dotted lines are from Reference 19.

data are shown in order to avoid confusion.   The values from all of
the chlorinated samples were found to fall on the same curve.
(This would not apply if the isochronal peaks were taken.)   The
mode of motion for the γ relaxation is probably the same for all
chlorinated polyethylenes, whether the chlorinated segments are
at the chain end or along the chain.   The dipoles are in the solid-
like environment and are mostly outside the crystalline regions,
possibly near the surface of crystals.   The activation enthalpy
was found to be 10 Kcal/mole.

The γ loss peaks for both linear and branched polyethylenes
were found much broader than those for the chlorinated samples.
Their correlation frequency vs. 1/T curves coincide with each
other, but not with the curve for the chlorinated samples.   The
activation enthalpy was found to be also 10 Kcal/mole.

A definite molecular model for the γ transition cannot be con-
clusively drawn at this stage.   However, the fairly high number of
dipoles participating in the γ process, the very close connection
with the α relaxation, and the location of most of the dipoles being

outside crystals, all seem to rule out its origin in the crystal
defects (15). The same value for the activation energy suggests
a simple mechanism that is common to all these polymers, chlorinated or otherwise, but such a relaxation process probably cannot
be independent of the neighbor molecules. Perhaps a molecular
relaxation of a crankshaft type may be suggested in which the
neighbor interaction [intra-molecular interaction when the motion
is not perfectly colinear (20) and inter-molecular interaction to
provide sufficient room for rotation] is taken into consideration,
but there may be other molecular models which will satisfy equally
well all these conditions.

α Relaxation Process

In contrast to the γ transition that can thus be classified
as the relaxation in the solid environment, the α transition in
crystalline polymers is, in some aspects, similar to the glass
transition of amorphous polymers. It is the low frequency, high
temperature transition of partially relaxed (through the γ process)
dipoles. Unlike the glass transition, however, this α process is
associated with the crystalline structure. It can be noted from
Figures 2 through 5 that the relative intensity of the α process,
as compared to the γ process, tends to be higher for more "highly
crystalline" samples. The T_{max} (obtained from the isothermal
curves) are also higher for these samples. We have maintained,
on the other hand, that the dipoles are mostly outside the crystalline region. Hence, it suggests that the α process is a coupled
motion of the chains inside the crystalline region and the dipoles
which are outside the crystal but are attached to these chains.

Hoffman, Williams and Passaglia (15) have proposed a model for
the α process supported by extensive dielectric, mechanical, and
morphological data on n-paraffins, their polar derivatives, and
polymers with defined morphological parameters. The model consists
of two overlapping mechanisms involving motions of chain folds and
reorientation (with translation) of chains in the crystalline
interior. Our results are consistent with such a model. The relationship between the dipole and the crystalline interior of the
chlorinated polymer is similar to that of paraffin-like molecules
with polar ends, such as methyl stearate studied by Broadhurst (21)
and Meakins (22). In all cases, the dipoles are at or near the
crystalline surfaces and their motion is closely coupled to the
motion of the crystalline interior. As we have shown, the crystalline lattice becomes enlarged as more chlorines are introduced, and
with it the melting point and the enthalpy of fusion will decrease.
The α transition temperature should decrease with the increasing
chlorine content, as was noted in Figure 8.

The activation energy for the correlation frequency of the α relaxation, as estimated from Figure 8, is 25 Kcal/mole for all polymers included in this work. Hoffman's model also leads to 25 Kcal/mole for polyethylene. It is the same value as obtained for the slightly oxidized linear polyethylene (2). Our data on bulk linear polyethylene show also the same value, rather than the much higher values of 40 to 160 Kcal reported elsewhere (23-25).

ACKNOWLEDGMENT

The authors wish to acknowledge Dr. E. P. Otocka for kindly supplying the chlorinated samples, Mr. G. E. Johnson for the valuable help in dielectric experiments, and Mr. H. E. Bair for advice in differential scanning calorimetry. Many valuable comments from Dr. D. W. McCall and Dr. D. C. Douglass are also gratefully acknowledged.

REFERENCES

1. W. Reddish and I. T. Barrie, IUPAC, Symposium on Macromolecules, Wiesbaden (1950), Preprint IA3.
2. Y. Ishida and K. Yamafuji, Kolloid Z., 202, 26 (1965).
3. R. J. Roe, H. F. Cole and E. P. Otocka, Polymer Preprints, 12-1, 317 (1971).
4. P. R. Swan, J. Polymer Sci., 56, 409 (1962).
5. W. Reddish, I. T. Barrie and K. A. Buckingham, Proc. of Inst. of Elec. Engr., 113-11, 1849 (Nov. 1966).
6. B. E. Read, E. A. Carter, T. M. Conner and W. J. MacKnight, Br. Polymer J., 1, 123 (1969).
7. S. Matsuoka and Y. Ishida, J. Polymer Sci., C-14, 247 (1966).
8. N. G. McCrum, B. E. Read and G. Williams, ANELASTIC AND DIELECTEIC EFFECTS IN POLYMERIC SOLIDS, John Wiley, 1967.
9. Y. Ishida, J. Polymer Sci., A2-7, 1835 (1969).
10. G. P. Johari and M. Goldstein, J. Chem. Phys., 53, 2372 (1970).
11. G. Williams and D. C. Watts, Polymer Preprints, 12-1, 79 (1971).
12. See, for example, H. Eyring, D. Henderson, B. J. Stover and E. M. Eyring, STATISTICAL MECHANICS AND DYNAMICS, John Wiley, Chapter 8, 221, 1964.
13. H. Frohlich, THEORY OF DIELECTRICS, Oxford, 1958.
14. J. D. Hoffman and H. G. Pfeiffer, J. Chem. Phys., 22, 132 (1954).
15. J. D. Hoffman, G. Williams and E. Passaglia, J. Polymer Sci., C-14, 173 (1966).
16. J. D. Hoffman, J. Chem. Phys., 23, 1331 (1955).
17. J. D. Hoffman and B. M. Axilrod, J. Res. National Bur. Std., 54, 355 (1955).
18. J. D. Hoffman, MOLECULAR RELAXATION PROCESSES, Chem. Soc. (London), Special Publication No. 20, Academic Press, London, 1966.

19. D. W. McCall, MOLECULAR DYNAMICS AND STRUCTURE OF SOLIDS, Eds. R. S. Carter and J. J. Rush, Natl. Bur. Stds. Special Publication, 301, June 1969.
20. E. Helfand, to be published; also, Bull. Am. Phys. Soc., II-16-3, 388 (1971).
21. M. G. Broadhurst, Polymer Preprints, 12-1, 128 (1971).
22. R. J. Meakins, PROGRESS IN DIELECTRICS, 3, 180, John Wiley, 1961.
23. T. Aramaki, S. Minami, F. Nagatoshi and M. Takayanagi, Repts. Progr. Polymer Phys. (Japan), 7, 237 (1964).
24. Y. Wada, H. Enjoji and H. Terada, Repts. Progr. Polymer Phys. (Japan), 5, 131 (1962).
25. S. Iwanagi and H. Nakane, Repts. Progr. Polymer Phys. (Japan), 7, 237 (1964).

MOLECULAR RELAXATIONS IN SUBSTITUTED POLYSTYRENES

R. E. Wetton

Department of Chemistry, Loughborough University

Loughborough, United Kingdom

INTRODUCTION

Interpretations of side group motions in polymer systems normally invoke cooperative processes with the main chain at the point of bonding, or hinderances of intermolecular origin (1,2). Perhaps the only exceptions to this rule are the cyclohexyl group chair-chair transition (3) and some low temperature methyl group rotations (4). In a previous study (5) of polystyrene systems, evidence was found to suggest that phenyl group rotation could only occur with cooperative motion from the main chain. From recent work by Wada (6), this cooperative motion could be the "local mode" of the main chain, but the present author differs in the main interpretation of losses in the region. Amorphous polystyrene exhibits at least two motional transitions below the main glass/rubber transition (α) at about 100°C. Previously reported (7,8) dynamic mechanical data for polystyrene over the accessible temperature range are shown in Figure 1. Labelling in order of decreasing temperature the transitions in the glassy state have been attributed (5,7,9) to the following molecular motions:

β - wide angle torsional oscillation of the phenyl group followed by complete rotation with main chain cooperation. This is a protracted process occurring between (T_g - 100°C) and T_g.

γ - small mechanical losses in the temperature range -100°C to -150°C, probably due to structural irregularities or impurities in some systems.

δ - wagging or combined wag-twist motion of the phenyl groups occurring in the liquid helium temperature range.

Even by using a combination of relaxational techniques, it is difficult to reach any unequivocal conclusion as to the precise mechanism of these complex relaxation processes. In this paper we consequently also draw conclusions concerning relaxations in unsubstituted polystyrene. It was largely in the hope of finding molecularly simple rearrangement processes that the present series of substituted polystyrenes were investigated. In particular it was hoped that "rotatable" groups, para to the main chain in the phenylene group, would be free of intra-chain effects.

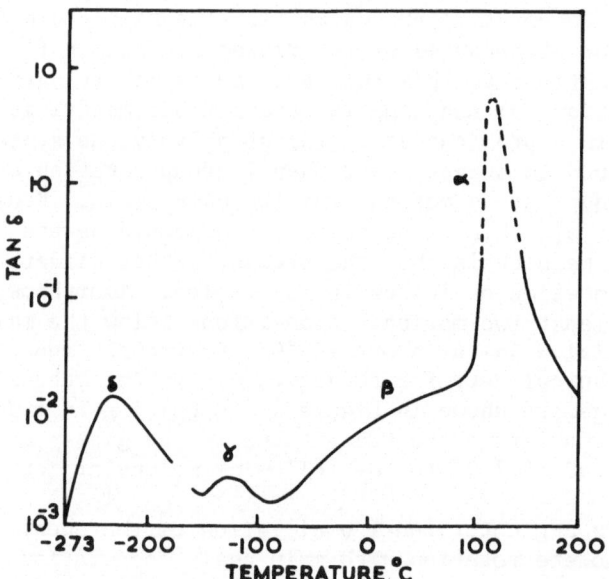

Figure 1.   Dynamic mechanical dispersion regions in commercial polystyrene according to Illers and Sinnott (8).

EXPERIMENTAL

The following polymers were synthesized from the corresponding monomers: poly 4-methoxystyrene (P4MOS), poly 4-ethoxystyrene (P4EOS) and poly 4-phenoxystyrene (P4POS). The repeat unit is thus $-CH_2-CH(Ph-O-R)-$ in each case with alteration only in the R group. Purified monomers were polymerized at 70°C in bulk with 0.01% azobisiosbutyrnitrile to approximately 50% conversion. Molecular weights were in the order of 100,000 except for some gel fraction in P4POS. Polymers were freed from impurities by normal techniques and finally freeze dried from benzene before moulding between 120°C - 140°C and cooling slowly.

Table IA summarizes the basic features of these polymers together with a schematic interpretation of their dipole disposition. Limited data on other polystyrene systems are also included and Table IB summarizes the main characteristics of these. In each case, residual solvent and monomer was thoroughly removed.

TABLE IA

Polystyrenes with Flexible Groups in the Para Positions

P4 MOS 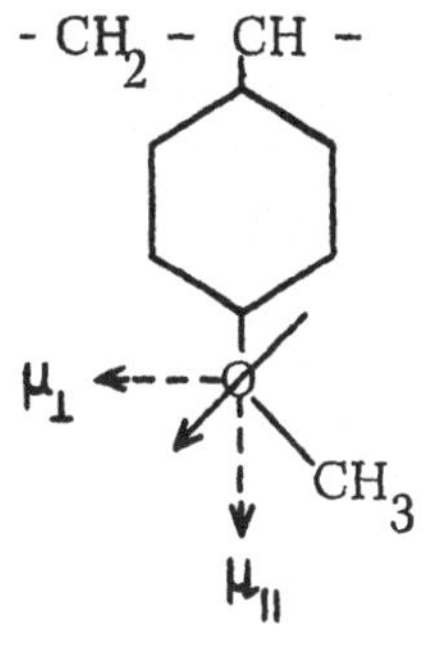	P4 EOS 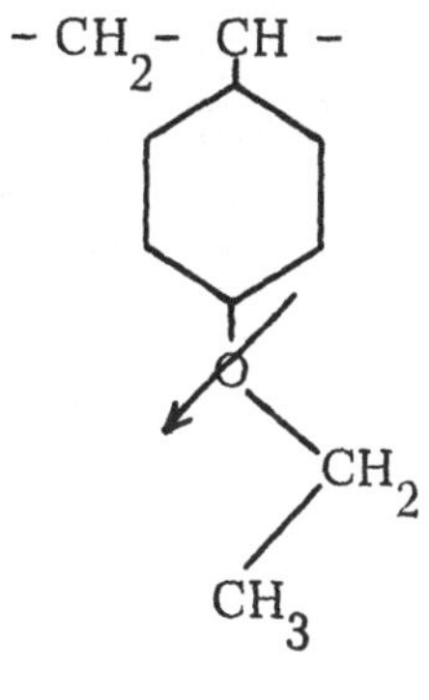	P4 POS 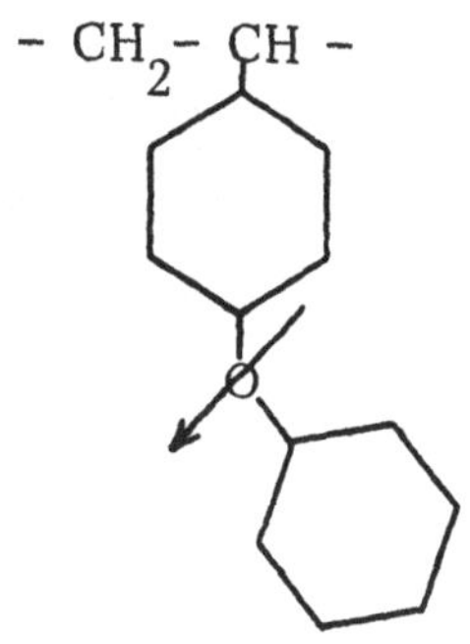

Polymerization at 70°C in bulk to 50% conversion.

Samples freeze-dried from benzene and compression moulded at 130°C.

$\overline{M}_n \simeq 100,000$ for all samples.

TABLE IB

Polymer	Initiator	Polymerization	M.W.
Anionic Polystyrene	Dilithiostylbene	Bulk, 0°C	$\overline{M}_w = \overline{M}_n = 30,000$
Anionic Polystyrene	Dilithiostylbene	Bulk, 0°C	$\overline{M}_w = \overline{M}_n = 120,000$
Commercial Polystyrene	Radical	Shell Company	$M_v = 250,000$
Poly 4-Br Styrene	Thermal, 90°C	Bulk, Low Conversion	> 100,000
Poly 3-Br Styrene	Thermal, 90°C	Bulk, Low Conversion	> 100,000
Poly 4-Cl Styrene	Thermal, 105°C	Bulk, Low Conversion	> 100,000
Poly 3-Cl Styrene	Thermal, 95°C	Bulk, Low Conversion	> 100,000
Poly 3-Me Styrene	1% AZBN, 70°C	Bulk, Three Days	—

Dielectric and vibrating reed dynamic mechanical measurements were performed using previously described instrumentation (5,10), with only minor modifications. Briefly the dynamic mechanical data were obtained by a vibrating reed resonance technique operating at a frequency in the range 200 - 400 Hz, but which varied as $(E')^{1/2}$. The dielectric measurements were performed using a three terminal cell and a transformer ratio arm bridge over the range 100 Hz - 20 KHz and using the Hartshorne and Ward resonant circuit method in the range 50 KHz - 10 MHz. In every case measurements were performed under vacuo over the temperature range -196°C to the glass transition region. The NMR data were obtained on a Varian Associates D.P.60 spectrometer operating in the wide line mode at 60 MHz with a dry N_2 gas flow controlling the sample temperature.

RESULTS AND DISCUSSION

The results of vibrating reed measurements on a number of different polystyrenes are shown as the storage component of Young's modulus and tan δ versus temperature in Figure 2. The only strong feature is the relaxation region at higher temperatures with tan δ increasing towards the α peak (main glass transition) which is expected at about 120°C (100 Hz). The protracted broad loss region extending from the α process down to about -150°C correlates with NMR narrowing (11,12) and is ascribed to torsional oscillations of the phenyl groups about their symmetry axis. By working at low frequencies, Illers, et al. (7), resolved the β loss process in the low temperature part of the α relaxation. This β peak probably indicates that complete rotation of the phenyl groups at room temperature is occurring with a relaxation time of more than 1 second. As the measuring frequency is raised the β process rapidly merges with the α process. This is the situation prevailing in the present data.

None of the present results exhibit a γ process of detectable magnitude (> 2×10^{-3} in tan δ) in the 0°C to -150°C region. Illers came to the conclusion that the γ process he observed in BASF polystyrene at -120°C, 0.5 Hz was due to structural irregularities such as head - head followed by tail - tail bonding. The flexible backbone sequence ($-CPh-CH_2-CH_2-CPh-$) would then be the origin of γ process. In this event the strength of the γ peak would be expected to vary with the method of polystyrene production. The commercial polystyrene measured in the present work is manufactured by the Shell Company and apparently is largely free of this type of structural defect. The anionic polystyrenes also then give the expected result of no γ process. Wada, et al. (6), have measured a γ peak of magnitude 10^{-2} at -70°C, 34 KHz in Asahi-Dow polystyrene, which they correlate with the γ peak of Illers and the loss reported by Baccaredda, et al. (13) at -90°C, 10 KHz on a laboratory

sample of free radical polystyrene. There are discrepancies by
factors of up to ten in the reported peak areas and by factors of
up to three in the total loss in this region. It seems to the
present author that the evidence favours either a defect or impurity
center as the cause of the very small loss sometimes observed in
this region.

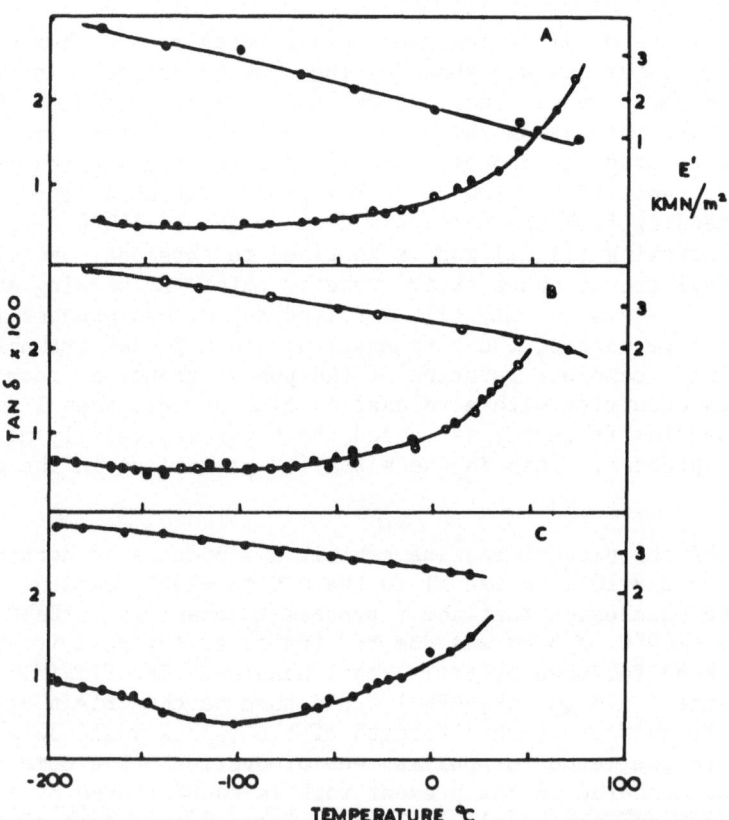

Figure 2. Dynamic mechanical data; storage component of Young's
           modulus. E' - open circles and Tan δ - filled circles
           for a series of polystyrenes. All frequencies approxi-
           mately 300 Hz. A, commercial polystyrene, Shell Company;
           B, anionic polystyrene, M_n = 120,000; C, anionic poly-
           styrene, M_n = 30,000.

The difference in behaviour between polystyrenes carrying simple substituent groups (e.g. methyl and halo) and those carrying
motionally more complex groups (e.g. methoxy and ethoxy) is immediately obvious from the data in Figure 3. A new complex range of
processes ($\gamma_1 + \gamma_2$) appears in the -200°C to 0°C region with tan δ
magnitudes large enough to distinguish these processes from those
in simply substituted polystyrenes. The data of Figure 3 agree
fairly well with that of Baccaredda, et al. (14), except that at the
higher frequency (7 KHz) used by these workers, the γ_1 process was
not resolved. We do not, however, ascribe to the Baccaredda, et
al., interpretation of this relaxation as the 'δ' process. A
molecular interpretation of these low temperature relaxations is
facilitated by comparing the mechanical with the corresponding
dielectric loss data of Figure 4. Both the low temperature relaxations involve large scale (angular) dipole relaxation. It is tempting to attribute the low temperature dielectric loss in P4MOS to
onset of complete rotational freedom of the methoxy groups. However, the complexity of the process, the dielectric strength and
dielectric activity of the β process argue against this as will be
seen later.

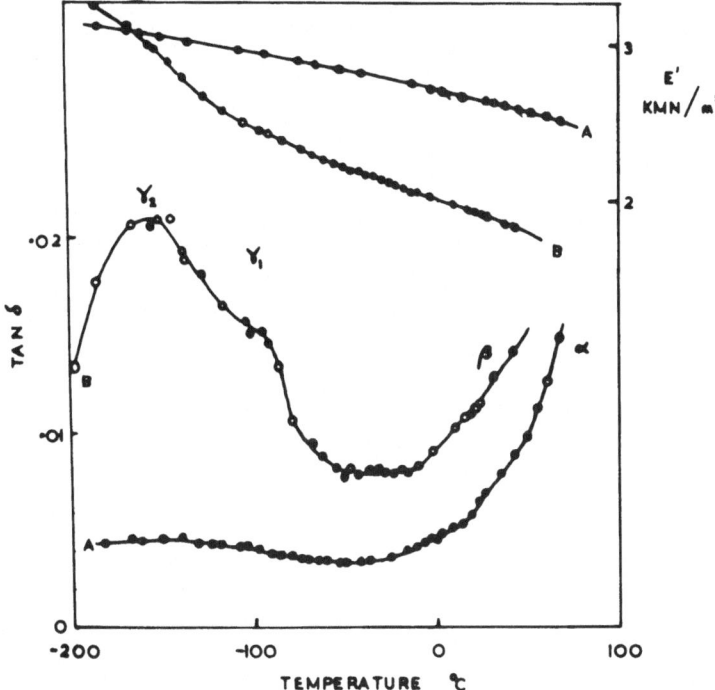

Figure 3.   Dynamic mechanical data for poly 4-methoxystyrene
            (curves B) compared with the rigidly substituted poly 4-
            methylstyrene (curves A). The open and filled points in
            the tan δ, curve A, represent data taken with temperature
            increasing and decreasing respectively.

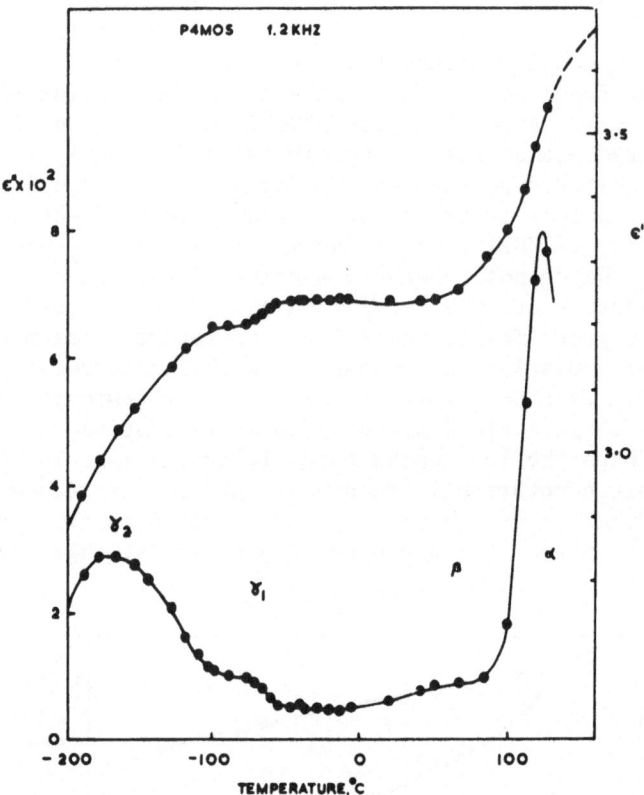

Figure 4.   Dielectric constant ε' and loss ε'' at 2 KHz for poly 4-
            methoxystyrene in the temperature plane, showing the four
            main dispersion regions.

The corresponding temperature plane data for P4 EOS are shown
in Figure 5. The lowest temperature loss is somewhat larger in
magnitude both in the mechanical and dielectric case. It could thus
be postulated that $(\gamma_1 + \gamma_2)$ were appearing as a combined transition.
However, the dielectric loss data, shown in Figure 6, in the frequency
plane although broad do not show very dramatic shape changes. Indeed only at -110°C is there clearly a second process coming in at
lower frequencies. This second process then correlates with the high
region of dielectric loss in the -80°C region of the temperature
plane plot. The activation enthalpy for both γ_2 processes agree
within experimental error, as P4MOS = 21 KJ/mole and P4EOS = 17 KJ/
mole. Accordingly we ascribe the -80°C loss to the γ_1 process
paralleling the γ_1 process in P4MOS. Strongly similar relaxation
behaviour is expected in these two polymers because of similarity of
the pendent motional groups, but the P4EOS possesses an extra degree
of internal rotational freedom and must exhibit a relaxational spectrum at least as complex as that of P4MOS.

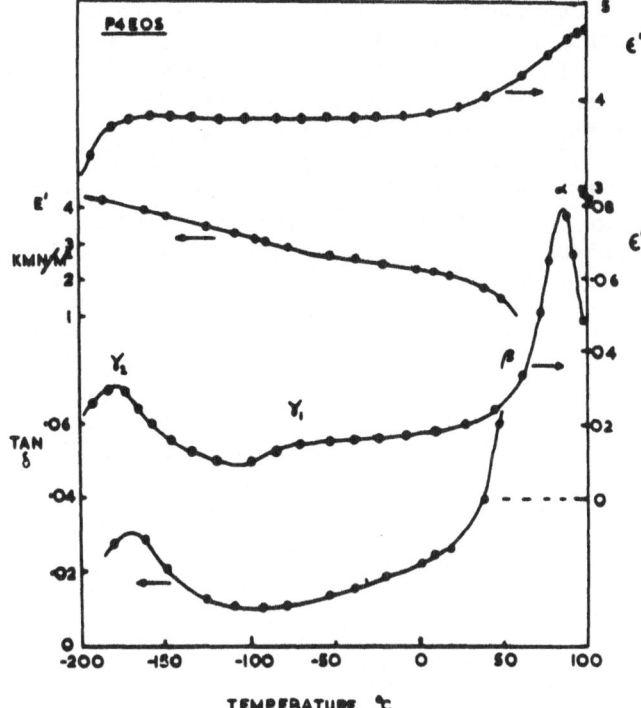

Figure 5. Temperature plane relaxation data for poly 4-ethoxysty-
rene. Filled circles - dielectric at 1 KHz, open circles - dynamic mechanical at approximately 300 Hz.

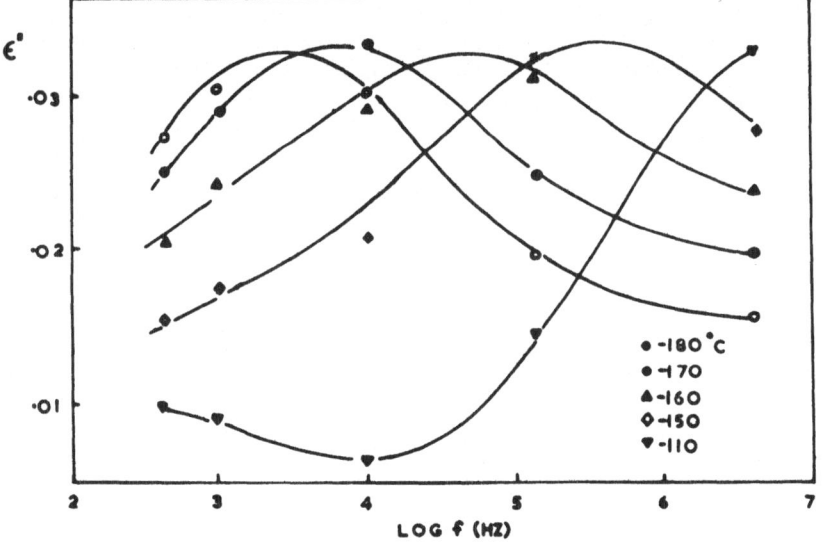

Figure 6. Frequency plane dielectric loss data for poly 4-ethoxy-
styrene in the γ dispersion at a series of constant
temperatures.

Both for P4MOS and P4EOS the dielectric data are paralleled by the dynamic mechanical data which means that all the motions are dielectrically active. Insight into the mechanism for the γ_1 and γ_2 relaxations may thus be obtained from the dielectric strength through the transition regions. Table II lists the combined strength of the γ dispersions and of the $(\alpha + \beta)$ dispersion region as assessed by the increment in ε' between the temperature listed. For completeness the data for P4POS are also listed, but the relaxation in this polymer is considerably different and will be discussed later.

TABLE II

Dielectric Strengths of Transitions

Polymer	$\Delta\varepsilon'$ (γ)	$\Delta\varepsilon'$ $(\alpha + \beta)$	Total $\Delta\varepsilon'$
P4MOS	0.4 (−200 to −60°C)	0.5 (−60 to +130°C)	0.9
P4EOS	1.1 (−200 to 0°C)	1.4 (0° to +120°C)	2.5
P4POS	0.07 (−20 to +40°C)	0.5 (+40 to +130°C)	0.57

The γ_2 process in P4EOS correlates well with the γ_2 peak in P4MOS in temperature and frequency location. Moreover, the dielectric strength of $(\gamma_1 + \gamma_2)$ as a fraction of the total is also in line with the P4MOS results. The major difference in model requirements for P4EOS is that rotation of the ethyl group about the O-Et axis cannot acquire complete freedom as in the methyl case in P4MOS. It is restricted to angular librations through about 240°. This is, however, sufficient to allow a new setting angle for the whole O-Et group about the O-Ph axis, to an angle of 55°, as with the O-Me in the γ_1 transition. Because of the requirement to librate a bulkier group the γ_1 transition in P4EOS occurs at higher temperatures than in P4MOS. The weakness of the γ_1 mechanical dispersion (0 to −80°C) suggests a conservation in the volume swept out by the librating ethoxy group.

Let us consider firstly a two site model in which the methoxy (or ethoxy) group gains rotational freedom (about O-Ph) by rotating between two equal and symmetrical minima at 180° to each other (on opposite sides of the phenylene plane). Estimating the Ph-O-Me bond angle (15) as 120° and assuming a high degree of intermolecular randomness of the oxygen dipole directions, then $\Delta\varepsilon'(\gamma) \propto g_0$ $(\mu_0 \sin 60)^2/T$ approximately for relaxation of the perpendicular dipole component. If the main (α) process relaxes the parallel component $\Delta\varepsilon'(\alpha) \propto g_0 (\mu_0 \cos 60)^2/T\alpha$. Here μ_0 represents the free

ether dipole and g_o is a factor allowing for intramolecular corre-
lations. This predicts $\Delta\varepsilon'(\gamma): \Delta\varepsilon'(\alpha)$ = 11:1, clearly in complete
contradiction to the data in Table II for P4MOS.

We thus propose the more satisfactory model in which the
methoxy (or ethoxy) groups gain oscillatory freedom between symmetrical energy minima at 30° to the normal to the phenylene plane during
the lower temperature (γ_2) transition. Model studies indicate a
clash of the methyl hydrogens with the phenylene hydrogens at angles
greater than 35°. The satisfactory Arrhenius plot for the γ_2
relaxation indicates a barrier height of 21 KJ/mole. The methyl
group rotation is hindered by close approach of its hydrogens with
the π orbitals of the phenylene, but if now the methyl group gains
rotational freedom, the whole methoxy group can oscillate to a
wider angle energy minimum at approximately 55° to the normal.
Figure 7 schematically shows the configuration of the methoxy group
at these angular minima. This activated acquisition of higher
rotational angle by the methoxy group is thus the γ_1 mechanism.
The loss peaks are not sufficiently resolved in the present work
to evaluate its activation energy, but it must clearly be greater
than the 21 KJ/mole for γ_2.

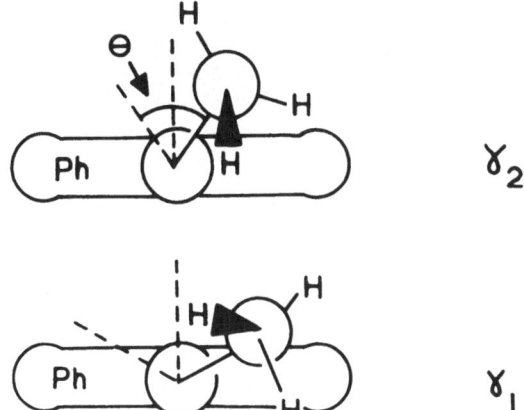

Figure 7.   Model for the γ_1 and γ_2 sites for methoxy group rotation
            about the oxygen-phenylene axis. The setting angles θ
            are assessed as 30° for the γ_2 process and 55° for the
            γ_1 process. The γ_1 position can only be achieved by
            cooperative rotation of the methyl group as indicated.

The test of these proposals, based on molecular models, must be sought in their predictions of the observed relaxation strengths, which are fairly well defined for the P4MOS system. We use the data to evaluate an effective 'g' in the Fröhlich form of the Onsager relation (16).

$$g\mu_o^2 = \frac{9kT}{4\pi N} \frac{(2\varepsilon_R + \varepsilon_u)}{\varepsilon_R (\varepsilon_u + 2)^2} \Delta\varepsilon'$$

where ε_R and ε_u are the relaxed and unrelaxed dielectric constant either side of the transition and μ_o is the dipole moment of the ether linkage. Values of $g\mu_o^2$ so obtained are shown in Table III together with the ratio g/g_o, where g_o is the value of the correlation parameter above α when ε' should be fully relaxed. We then compare this with the ratio of g/g_o predicted from the model. Thus for γ_2 we assume that the $\sqrt{g_o} \mu_o \sin(\theta/2) \sin 30°$ component can be relaxed, θ being the ether bond angle and $30°$ the site angle (Figure 7). Above the γ_1 transition the relaxed dipole component will be

$$\sqrt{g_o} \mu_o \sin(\theta/2) [f \sin 55 + (1 - f) \sin 30]$$

where f is the fractional population of the $55°$ site. If the $30°$ and $55°$ sites have equal energy then the term in the square brackets represents the arithmetic mean of the sines. In general, however, the site being occupied in the γ_1 process will be higher than that of the γ_2 process by ΔE and if there are equal numbers in each site then

$$f = e^{-\Delta E/RT}$$

It is apparent that $g_o \times$ (angular function)^2 is the effective 'g' obtained from the above models and can be compared with the experimental value. Above the $(\alpha + \beta)$ relaxation when the dipole relaxation is complete the relaxed dipole moment squared is by definition $g_o\mu_o^2$.

TABLE III

Comparison of Experimental and Predicted Relaxation Strengths

(via g/g_o) for P4MOS

Transition	$T(°K)$	$\Delta\varepsilon'$ (Trans.)	$\Delta\varepsilon'$ (Total)	$\dfrac{g\mu^2_o}{}$	g/g_o (exptl.)	g/g_o (theory)
γ_2	98	.4	.4	.26	.12	.19
γ_1	193	.04	.44	.55	.25	.33 (= .19 for $\Delta E = 2.2$ KJ)
$\alpha + \beta$	394	.5	.94	2.24	(1 by definition)	(1 by definition)

We see that the predictions from the model give quite good
agreement with experiment. The 0.33 theoretical ratio for γ_1 is
calculated on the assumption that all angular minima are of equal
energy. In fact, the sites involved in the γ_1 process will have
a higher energy than the γ_2 sites and we find that a difference in
energy of 2.2 KJ between sites will bring the γ_1 prediction into
line with experiment. Clearly also the site angles could be
regarded as parameters and adjustment of these within fairly small
limits could make the experimental and theoretical agreement exact.

Considering now the (α + β) high temperature relaxation region,
Figures 4 and 5 show this dielectric dispersion for P4MOS and P4EOS
respectively. Corresponding temperature plane data for P4POS are
shown in Figure 8. Because of limitations in the vibrating reed
method in high loss regions the dynamic mechanical data cease on
the low temperature side of the relaxation region. A basic fact of
behaviour in this region which we have pointed out previously (5)
is that in simply substituted polystyrenes, i.e. 4-Cl, 4-Me, 3-Cl,
3-Me, a dielectrically active β process is shown only in those systems which have a dipole component perpendicular to the phenylene
symmetry (rotation) axis. Curtis (17) has reported that poly 4-
chlorostyrene having a dipole lying along the rotation axis shows a
symmetrical and, for a polymer, a sharp dielectric α loss peak and
no β process, despite a wide frequency and temperature coverage. In
dynamic mechanical studies, however, in which the activity is not
concerned with dipole orientation, but only with the coupling of
molecular motion with the mechanical stress field, both the symmetri-
cally and unsymmetrically substituted polystyrenes show an α and a
β relaxation. This is very strong evidence that some form of rotational motion about the phenylene axis is responsible for the β loss
process in these systems.

Figure 9 shows normalized dielectric loss curves obtained with
the present polymer systems compared with the 3-Cl and 4-Cl substituted polymers. This allows a good comparison of line shapes in
the frequency plane with the polymers in relatively similar (compared to T_g) temperature ranges. The P4EOS data are clearly complex and the data we have presented previously (5) show that the
two processes amalgamate with increasing temperature. The loss
peaks for the other systems are all very much broader than that for
poly 4-chlorostyrene. The implication is that all the para phenyl
ether groups retain a weak dipole component perpendicular to the
phenylene rotational axis and that this component is relaxed in the
β process just prior to the main α relaxation of residual dipole
vectors (i.e. those parallel to the phenylene axis). In the case
where broad peaks are observed, the presence of the β process on the
high frequency side would be the cause of the broadening. Sensitive
low frequency measurements at lower temperatures would be expected
to resolve the β process in these cases also. The present results,

however, were confined to frequencies above 100 Hz. Even so, pronounced dielectric loss due to the β process is present in the 0° –
70°C region in the temperature plane.

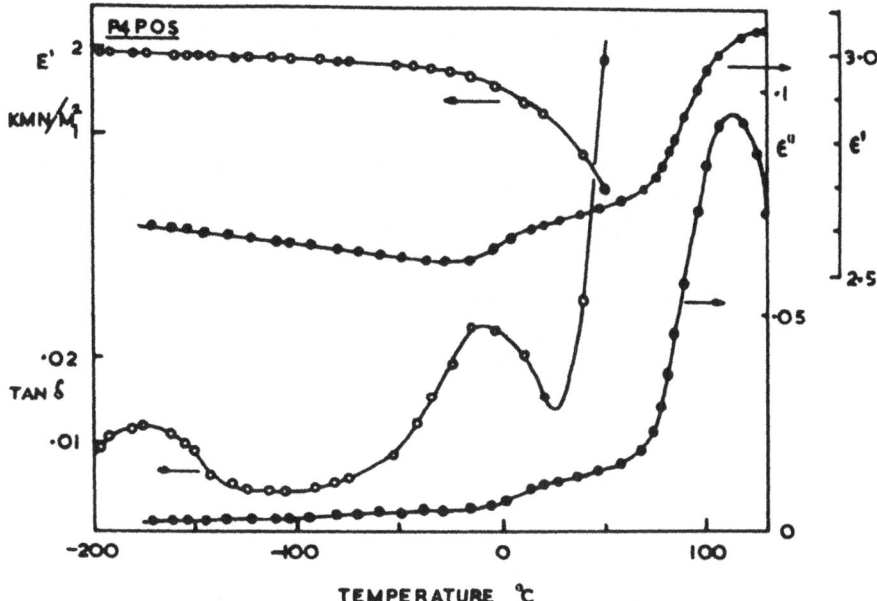

Figure 8. Poly 4-phenoxystyrene temperature plane data. Open circles – dynamic mechanical results at approximately 300 Hz.
Filled circles – dielectric results at 1 KHz.

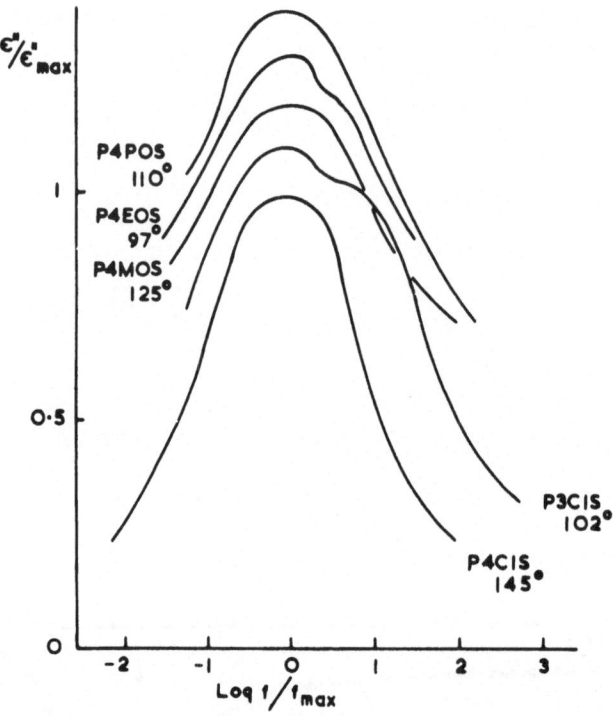

Figure 9.   Normalized dielectric loss curves (each displace 0.1 in
             turn on the ordinate scale for clarity) in the fre-
             quency plane in the region of the (α + β) main relaxa-
             tion region. The actual measuring frequency ranges are
             all similar at the temperatures indicated.

        Wada, et al. (6), have recently found very weak dielectric
activity in the β dispersion region of polystyrene itself. Polystyrene has a weak permanent dipole along the phenyl symmetry axis
and this has caused these authors to ascribe the β process to the
main chain local mode, which occurs in a number of linear polymers
prior to the main glass relaxation process (i.e. on the low
temperature/high frequency side). However, this interpretation is
against a wealth of proton (12) and fluorine (11) NMR data as well
as the dielectric data on substituted polymers. The magnitude of
the dielectric loss (ε'') in the β region is less than 10^{-4} while

the α process is about 23 x 10^{-4}. This puts the $\beta:\alpha$ dielectric
loss ratio at about 1:25. For the well investigated case of the
polyethers (18) this ratio is typically 1:5. This suggests that
the polystyrene data should be interpreted in a different way.
At the very low loss levels of the relaxation reported by Wada,
et al., in the β region, it is conceivable that oscillating dipoles
set up by the variation of induced polarization of the π bond system with rotation angle of the phenyl group are the cause of this
loss process.

The present observations of strong dielectric activity of the
β process in P4EOS and P4MOS support the librational model which
successfully explains the magnitude of the γ processes. The phenyl
group rotation will relax the remaining dipole contribution perpendicular to the phenyl rotation axis, which in terms of our model
is $\sqrt{g_0} \sin(\theta/2) \{1 - [f \sin 55 + (1 - f) \sin 30]\}$. The behavior
in the α dispersion region of the P4POS is less clear. The curves
are not as broad as the other systems in Figure 9 and a more
detailed study (19) of this region has shown that little shape
change occurs over a thirty degree temperature range. That this
is the main T_g relaxation for the polymer there can be no doubt as
it was mouldable at 140°C.

The low temperature relaxation processes in P4POS with the
very bulky phenoxy substituent are shown in Figure 8. These have
not been discussed in comparison to the processes in P4MOS and
P4EOS as they are very different. In particular, we notice that
in contrast to the other polymer systems, the glassy state relaxations in P4POS show no significant dielectric activity despite
their large mechanical strength.

The terminal phenyl group should be capable of wide angle
oscillation about its own axis without much intra-molecular constraint. Its large size, however, must impose constraints from
neighboring chains. At a higher temperature the Ph-O group can
undergo a large angle cooperative libration (through about 90°),
again without major constraints. If these types of motion can
occur in the glassy state, they should reflect a dramatic line narrowing in the proton NMR broad line spectrum. The NMR spectrum
for water free samples exhibited a single broad spectrum of line
half width shown in Figure 10 as a function of temperature. In
the same figure published data on polystyrene (12) are shown for
comparison. Motion largely of the terminal phenyl has narrowed
the line width to half the rigid lattice value by 32°C, whereas in
polystyrene the main chain bond phenyls do not achieve this until
110°C. The dielectric constant of this polymer is very low below
the γ transition at -7°C, 260 Hz (mechanically) and has increased
by 0.07 by 20°C, 500 Hz (dielectrically) without resolution of a
clear ε'' peak. As this motion is the first major dielectric

relaxation process exhibited with increasing temperature, we ascribe
it to Ph-0 angular oscillation about the phenylene-0 axis, with
cooperation from angular libration of the phenyl about its own axis.
The angle of oscillation involved for the Ph-0 motion is from model
studies, about 45°, but to give the observed relaxation strength the
angle would have to be considerably less than this. Whether or not
the terminal phenyl actually acquires free rotation during the -7°C
transition cannot be stated categorically. If the Ph-0-Ph bond
angle (20) is in reality 128°, this could be feasible.

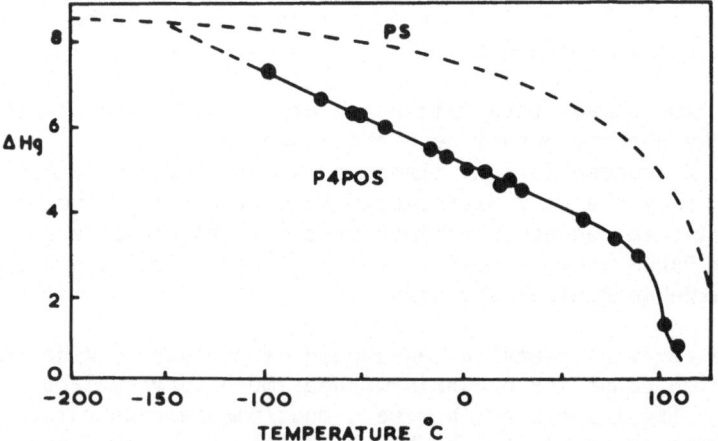

Figure 10.   Variation of NMR line widths (in Gauss) with temperature
             for poly 4--phenoxystyrene compared with the polystyrene
             results of Reference 12.

        The strong low temperature loss peak (δ) at -175°C, 300 Hz
has little or no dielectric activity and does not correlate with
any of the transition discussed for the other systems. The loss
does, however, fit into a pattern of behaviour if it is correlated
with the δ process observed in all polystyrene systems in the ultra
low temperature range. The loss peak location lies between that of
polystyrene and that of poly α-methylstyrene, but its magnitude is
greatly enhanced. It seems as though each phenyl is able to make
its own contribution of the relaxation strength.

Despite some recent excellent dielectric studies of this process by Work, et al. (9), its origin is still not resolved. The present author has measured the dielectric behaviour of some halo substituted polystyrene of very high purity. The dielectric data was less accurate than that of Work, et al., but gave good correspondence with the loss peak positions. Recent work by Morgan (12) on the same high purity samples established that the position of the δ loss process does not depend on the substituent in the series CH_3, Cl, and Br but depends only on its position of substitution in the ring (3 or 4 in this work). This clear pattern of behaviour agrees with the dielectric studies of Work, et al., but confounds the latter's interpretation, which invoked a combined wag-twist mechanism with the dipole position (2, 3 or 4) weighting the spectrum in favour of either the wag or twist component. The agreement of mechanical loss peak positions with the dielectric loss position clearly cannot be interpreted in terms of dipole vector orientation. A possible explanation of the curious, but well established, temperature order of the loss peaks, namely 3-2-4, might lie in the configurational properties of three molecules. We thus used light scattering techniques to measure the unperturbed end-to-end length of the same 3 and 4 Cl substituted polystyrene as were measured mechanically compared to polystyrene itself. These results together with T_g values we have obtained are shown in Table IV.

The molecular weight of the samples is high and yet the δ process is a relatively strong mechanical loss peak. Defects due to chain ends therefore seem to be ruled out, although curiously the low temperature dynamic mechanical loss in Figure 2 shows an increasing trend with decreasing molecular weight. If one regards a fair comparison to be between substituted polystyrenes and that polystyrene itself should not necessarily fit into the pattern, the δ peak location shows the same trends as the main T_g. Indeed the fact that the nature of the substituent in the series CH_3, Cl, Br does not change T_g or T_δ but changing the position of substitution affects T_g and T_δ similarly makes this correlation quite strong. These changes could, of course, be due to sensitivity of the chain conformation on the position of substitution. The direct evidence we have from light scattering in θ solvents does not rule this out, but correlation between the characteristic ratio (r_0^2/nl^2) and T_δ is less obvious than that between T_δ and T_g.

An in-plane phenylene ring oscillation (wag) with torsion of the skeletal carbon bonds still seems the most feasible explanation of the process in the polystyrene family of polymers. The temperature of the relaxation is sensitive to the same parameters (free volume or chain configuration) as the main T_g. In this sense the δ process seems a normal relaxational process. It is interesting in this regard that the increased mass of the Br group over CH_3 does not change the frequency/temperature location of T_δ and therefore rules

TABLE IV

Low Temperature 'δ' Process in Polystyrenes and Molecular Parameters

Polymer	$\bar{M}_n \times 10^{-5}$	$\bar{M}_w \times 10^{-5}$	(r_o^2/nl^2)	T_g (°C)	T_δ (°A) Mechanical (~1 Hz)	Dielec. (1 KHz)
Polystyrene	~2	-	10 (visc)[a]	100	40[b]	45[c]
Poly (3-Cl styrene)	6.0	12.5	15 (l.s.)	97	15[d]	19[c]
Poly (3-Br styrene)	8.7	-	expect ditto	96	15[d]	-
Poly (3-Me styrene)	-	-	expect ditto	-	15[f]	-
Poly (4-Cl styrene)	5.4	6.9	13 (l.s.)	131	40[d]	46[c]
Poly (4-Br styrene)	3.5	-	expect ditto	133	40[d]	-
Poly (4-Me styrene)	-	-	expect ditto	130	40[f]	-
Poly (2-Cl styrene)	-	-	?	~140	-	35[c]
Poly (2-Me styrene)	-	-	-	(~140)	40[f]	-
Poly (α-Me styrene)	-	-	-	192[e]	138	

Present work except as indicated: (a) P. J. Flory, (22); (b) K. M. Sinnott, (8); (c) R. D. McGammon, et al. (9); (d) R. Morgan, (21); (e) W. G. Barb, (23); (f) J. M. Grissman, et al. (24).

out any possibility of treating this as a damped resonance (far
i.r.) vibration

CONCLUSIONS

Dielectric and mechanical relaxations in polystyrene and a
number of substituted polystyrenes are only explicable in terms of
librational motions in the glassy state with complete rotational
freedom (i.e. of phenyl group) not being acquired until the β
process just prior to the main chain rotational relaxation (α).
In particular, methoxy and ethoxy groups are constrained to lie on
one side of a phenylene symmetry plane (which we have taken to be
the ring itself) even up to T_g. They do acquire librational freedom
in two distinct stages and model considerations suggest sites at 30°
and 55° to the normal to the phenylene plane. The relaxation strength
ratios predicted by this model agree well with experimental observations. It must be mentioned here that anisole is regarded as a
molecule with free internal rotation in the liquid state at room
temperature, yet in the glassy state polymer rotation is not free
until approximately T_g. The direct intramolecular environment of
the methoxy group must be the same in the polymer as in anisole and
one therefore concludes that a major part of the constraint on this
group in the polymer case must be long range interactions which are
probably inter- rather than intra-molecular.

It is concluded that the δ transition in polystyrene and simply
substituted polystyrenes is a typical relaxational process showing
the same trends as the main glass transitions of the polymers. An
in-plane phenyl wag seems the most plausible explanation and the
large mechanical loss in P4POS is explained on this basis.

ACKNOWLEDGMENTS

The author acknowledges with gratitude contributions made to
this work by Messrs. V. Bertelsen (guest workers from the University
of Copenhagen), H. Aaserud and Dr. G. S. Fielding-Russell who performed most of the relaxation studies, Dr. C. Richardson who performed the light scattering measurements and Dr. T. Connor (N.P.L.)
for his assistance with the broad-line NMR measurements. The author
is indebted to the National Physical Laboratory for a vacation consultancy during which the NMR measurements were made and to the
Monsanto Company, St. Louis, for a visiting consultancy during which
some of the low temperature 'δ' process studies were undertaken.

REFERENCES

1. J. Heijboer, Proc. Int. Conf. Non-Crystalline Solids, Delft, North Holland (Amsterdam), 231 (1965).
2. G. Williams, Trans. Faraday Soc. 62, 2091 (1966).
3. J. Heijboer, Kolloid-Z. 171, 7 (1960).
4. J. A. Sauer and R. G. Saba, J. Macromol. Sci.-Chem. A3, 1217 (1969).
5. G. S. Fielding-Russell and R. E. Wetton, MOLECULAR RELAXATION PROCESSES, Chem. Soc./Academic Press, 95, 1966.
6. O. Yano and Y. Wada, J. Polymer Sci. A2, 9, 669 (1971).
7. K. H. Illers and E. Jenckel, Rheologica Acta 1, 322 (1958).
8. K. M. Sinnott, Soc. Plastics Eng. Trans. 65 (1962).
9. R. D. McCammon, R. G. Saba and R. N. Work, J. Polymer Sci. A2, 7, 1721 (1969).
10. G. S. Fielding-Russell and R. E. Wetton, Plastics and Polymers 38, 179 (1970).
11. M. V. Vol'kenshtein, A. I. Kol'tsov and A. S. Khachaturov, Polymer Sci. U.S.S.R. 7, 324 (1965).
12. A. Odajima, J. A. Sauer and A. E. Woodward, J. Polymer Sci. 57, 107 (1962).
13. M. Baccaredda, E. Butta, V. Frosini and S. De Petris, Materials Sci. and Eng. 3, 157 (1968).
14. M. Baccaredda, E. Butta, V. Frosini and P. Magagini, J. Polymer Sci. A2, 789 (1966).
15. H. J. Bowen and L. E. Sutton, Tables of Interatomic Distances, etc., Chem. Soc. London, 1958, supplement, 1965.
16. H. Frohlich, THEORY OF DIELECTRICS, Second Edition, Oxford University Press, 1958.
17. A. J. Curtis, Soc. Plastics Eng. Trans. 82 (1962).
18. R. E. Wetton, G. S. Fielding-Russell and K. U. Fulcher, J. Polymer Sci. C, 30, 219 (1970).
19. V. Bertelsen, private communication.
20. L. Sutton and Hampson, Trans. Faraday Soc. 31, 945 (1935).
21. R. Morgan (Monsanto/Washington University Materials Program), to be published.
22. P. J. Flory, STATISTICAL MECHANICS OF CHAIN MOLECULES, Interscience, 1969.
23. W. G. Barb, J. Polymer Sci. 37, 515 (1959).
24. J. M. Crissman, A. E. Woodward and J. A. Sauer, J. Polymer Sci. A3, 2693 (1965).

THE ELECTRET PROPERTIES OF A SERIES OF CORONA-CHARGED SUBSTITUTED

POLYOLEFINS

R. A. Creswell, M. M. Perlman[*] and M. A. Kabayama

Bell Canada-Northern Electric Research Limited

Ottawa, Canada

ABSTRACT

The following polymer films were negatively corona-charged as electrets: a series of ten commercially available polyolefin films, with varying degrees of halogen substitution; an anionically polymerized polystyrene series including different end groups and carbon-filled; a polyethylene series with varying crystallinities. This method injects negative charge into a thin layer near the surface of the film. These films were studied for their electret properties using thermally stimulated charge decay (TSCD) and thermally stimulated currents (TSC), from room temperatures to just below the softening point. These are experimental techniques whereby the stored charge in a film is released as the material is heated at a uniform rate. Analyses of the TSC by previously derived theory has permitted the assignation of energies of activation of the trap sites, the charge mobility-free life-time product, trap densities and relaxation times associated with each peak. The most stable electrets were the symmetric polymers such as polytetrafluoroethylene and linear polyethylene. Total halogen substitution improved electret stability whereas partial substitution, forming less symmetric polymers, causes large decreases in stability. The phenyl group provides a deep trap site and enhances electret stability. Some of the TSC peak temperatures are at the T_g of the polymer, indicating that morphological changes give rise to charge release. In other cases, release appears due solely to thermal excitation and, in some instances, a dual process may be operative.

[*]Present Address: Department of Physics, College militaire royal de Saint-Jean, Saint-Jean, Quebec, Canada

INTRODUCTION

A theory has previously been developed to describe the thermally stimulated currents (TSC) obtained, without an external electric field, from near-surface charged insulators (1,2,3). It has been possible to explain the peaks observed and obtain their kinetics. Both the effect of heat and humidity on charge storage in some polymers have been determined. The major shortcoming in the work to date has been the inability to determine the mechanism, at the molecular level, by which the corona charge is deposited, stored and released from the insulator. Some success has occurred both in our work and elsewhere but no thorough and systematic study of a series of closely related polymers has taken place.

A series of polyolefins, Table I, consisting of polyethylene and including alkyl, aromatic and halide substitution, has been negatively corona-charged. This series of polymers is of particular interest due to their charge storage stability. One of the uses for these charged polymers (called electrets) is in the electret microphone. A stable charge, even in the presence of high humidity, is essential in this application. Some of the substituted polyolefins appear to meet this requirement (4,5,6).

Thermally stimulated currents have been obtained over the range room temperature to just below the softening point, analyzed for their kinetics, and compared. Net surface charge decay, using the vibrating electrode-equivalent bias technique (7), has been measured both at constant temperature and at a constant heating rate over the temperature range used to produce the TSC. These experiments complement the TSC data and show more directly how surface charge decays in a polymer (loss of its electret property) during an application such as the electret microphone. Both sets of data, TSC and thermally stimulated charge decay (TSCD), can then be compared.

In addition to the study of charge storage in this series of polymers having similar chemical composition, the effect of varying the degree of crystallinity in polyethylene was investigated. Finally, a polystyrene series including end group substitutions and carbon filling to alter the internal structure has been studied. All thermally stimulated current data are compared and discussed with particular emphasis being placed on actual charge storage sites in the polymers. These results also give additional insight into the conduction mechanism in these polymers.

THEORY

The expressions given here and used in the calculations have previously been derived (1,2,3). The basic expression, describing

TABLE I

Polymer Series Studied

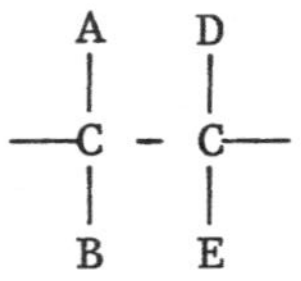

$$\begin{array}{ccc} A & & D \\ & | & | \\ -& C - C &- \\ & | & | \\ B & & E \end{array}$$

	A	B	D	E
1. Substituted Polyolefins				
Polyethylene (PE) High Density	H	H	H	H
Polyvinyl Chloride (PVCl)	H	H	Cl	H
Polyvinyl Fluoride (PVF)	H	H	F	H
Chlorinated Polyvinylchloride (Cl/PVCl)	H	Cl	Cl	H
Polyvinylidene Fluoride (PVF$_2$)	H	H	F	F
Polypropylene (PP)	H	H	CH$_3$	H
Polystyrene (PS)	H	H	C$_6$H$_5$	H
Polytetrafluoroethylene (PTFE) "Teflon"	F	F	F	F
Polyfluoroethylenepropylene (PFEP) "Teflon"	F	F	CF$_3$,F	F
Polytrifluorochloroethylene (PTFCE) "Aclar"	F	F	Cl	F

2. Polystyrenes with SO$_3$H End Groups, Also C Filled (16%)

3. Polyethylenes with Different Crystallinities, Branching

the peaks in the thermally stimulated currents obtained from near surface charged insulators, heated at a constant rate, is:

$$j = A \exp [-E_a/kT - B \int_{T_o}^{T} \exp (-E_a/kT) \, dT] \qquad (1)$$

Here, j is the externally measured current, E_a the charge trap depth, T the absolute temperature, k is a constant and

$$A = (n_{t_o} \, e\delta)^2 (\mu\tau)/(2\varepsilon d\tau_o) \qquad (2)$$

$$B = 2/(\beta\tau_o) \qquad (3)$$

n_{t_o} is the initial charge density in traps, $(\mu\tau)$ the charge mobility-free lifetime product, e the electronic charge, δ the depth of penetration of charge, ε the dielectric permittivity, d the insulator thickness, τ_o the inverse of the trap escape frequency and β the heating rate.

The current released initially, just above T_o in Equation 1, can be shown to have the form:

$$\ln j \approx \text{constant} - E_a/kT \qquad (4)$$

The activation energy E_a for the discharge process, responsible for a peak, can be obtained from a plot of $\ln j$ vs. $1/T$.

Using an approximation to the integral in Equation 1, it is possible to write:

$$j \approx A \exp [(E_a/k)(-1/T - B \{\exp (-E_a/kT)\} (E_a/kT)^{-2}] \qquad (5)$$

Using the Cowell and Woods curve fitting technique (8), with an initial low guess of E_a, then increasing it in small steps, it is possible to determine E_a and τ_0.

Equation 1 can also be written in the form (9):

$$N \left[\left(\int_t^\infty j \, dt \right) / j \right] = \ell n \ (1/\beta B) + E_a/kT \tag{6}$$

The parameters E_a and τ_0 may be determined from a straight line semi-log plot of the remaining charge divided by the current at a particular temperature vs. the inverse of that temperature. We refer to this method as the modified Bucci plot.

Values for n_{t_0} and $(\mu\tau)$ can be obtained from the expressions:

$$n_{t_0} = (Q/e\delta S)(2d/\delta) \tag{7}$$

$$(\mu\tau) = \varepsilon(\delta/d)(\delta S/Q) \tag{8}$$

where Q is the charge released from a trap and S is the surface area of the sample. The depth of corona charge penetration δ can be obtained from a separate experiment (10) involving a measure of capacitance vs frequency for charged and uncharged samples, or it can be calculated (11) from a measurement of the surface charge density σ prior to charge Q being released as a peak in the TSC:

$$\sigma = \left[\int_0^\delta \rho(x) \ (d-x) \ dx \right]/d \tag{9}$$

$$Q/S = \left[\int_0^\delta \rho(x) x \, dx \right]/d$$

$$\simeq \left[\int_0^\delta \rho(x) \, dx \right] \bar{x}/d$$

$$\simeq \left[\int_0^\delta \rho(x) \, dz \right] \delta/2d \tag{10}$$

Since

$$\sigma + Q/S = \int_0^\delta \rho(x)dx \qquad (11)$$

it follows that

$$\delta = 2d(Q/S)/(\sigma + Q/S) \qquad (12)$$

Here $\rho(x)dx$ is the charge per unit area in a cross-sectional element (1) dx. It should be noted that the charge released in the thermally stimulated current, Q, is only that charge in excess of what is initially induced on the electrode at or near the charged surface of the insulator. It is due to the motion, toward the surface, of charge in the thin surface layer.

EXPERIMENTAL

The polymer film (~25μ thick) was corona charged in air (1) by passing it on rollers, metallized side down and grounded, 0.5mm beneath a knife edge held at -4 to -6kV. The net surface charge was measured immediately after charging, and while being discharged at elevated temperature, by the vibrating electrode-equivalent bias technique (8). Values up to 140nc.cm^{-2} were obtained.

Samples were discharged at a linear heating rate of, typically, 1°C min^{-1} over the range room temperature to just below the softening point. For the thermally stimulated currents, conducting paint was used on the charged surface and the current measured with an electrometer. Peak cleaning and partial heating, as described in Reference 1, were used to isolate individual peaks and determine the extent of their overlap.

The apparatus is automated using a data acquisition system consisting of a multiplexer, analog-digital converter, and interface to a teletype that records current, temperature, time and the electrometer scale factor. These values are punched out on paper tape. The paper tape is then converted to cards, and all calculations are performed with the aid of a Control Data 3300 Computer.

RESULTS

Typical TSC and TSCD curves are shown in Figure 1. The complexity of the TSC curve is evident whereas the TSCD curve in general shows less detail. It does not always show a stepwise decrease for each TSC peak. Because of this, all discharge kinetics calculations have been carried out on the TSC peaks. However, both the total current released and total surface charge decay occur at the same temperature. Since the depth of charge penetration was found to be <5% and at one surface only, an external bias was not necessary to obtain an external current (in contrast to uniformly charged electrets). Thus, the TSC could be interpreted without consideration of current reversals; these reversals result from charge movement toward the opposite electrode. Figure 2 shows how cleaned TSC peaks are obtained by subtracting the current in a second run, after scaling, from that obtained in the first run (1). An example of the calculations carried out on the cleaned peaks in shown in Figure 3.

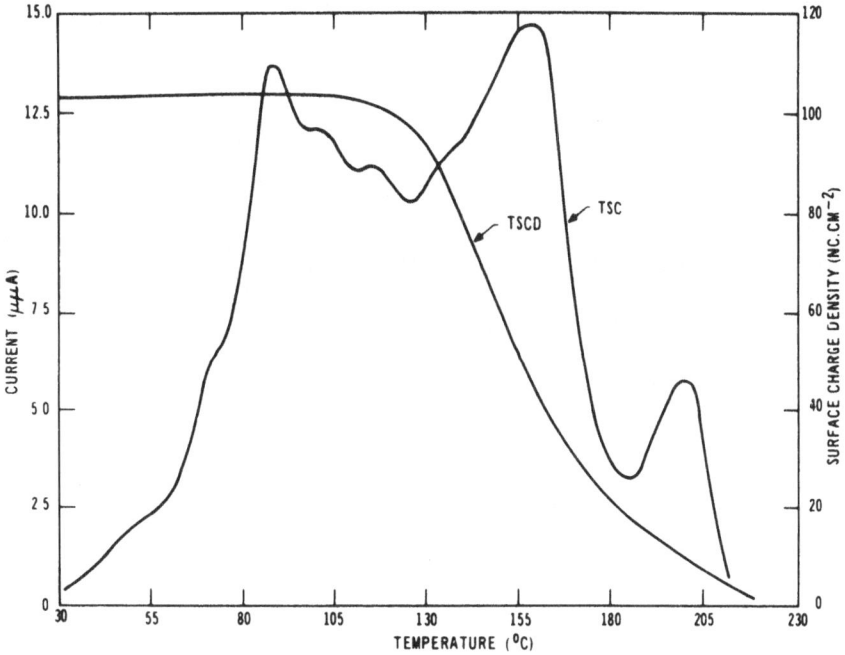

Figure 1.   Thermally stimulated current (TSC) and surface charge decay (TSCD) in negatively corona charged 25μ Teflon FEP when heated 1°C min^{-1}.

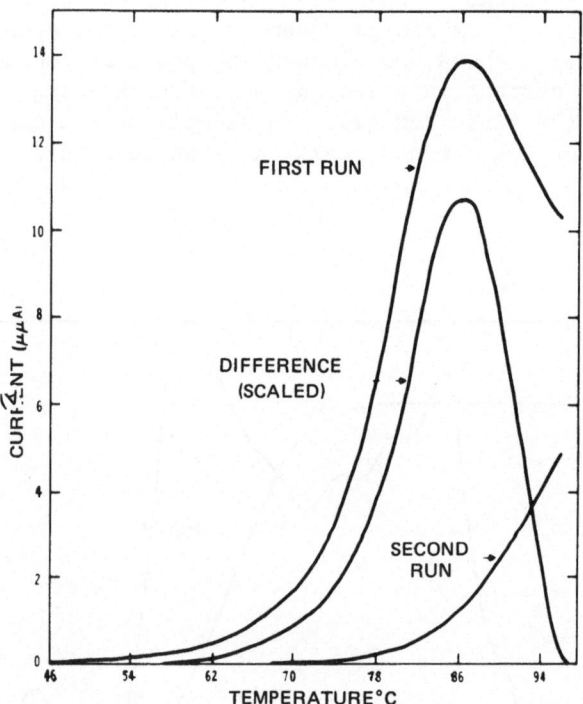

Figure 2.   Thermal current peak cleaning before kinetics calculations are carried out.

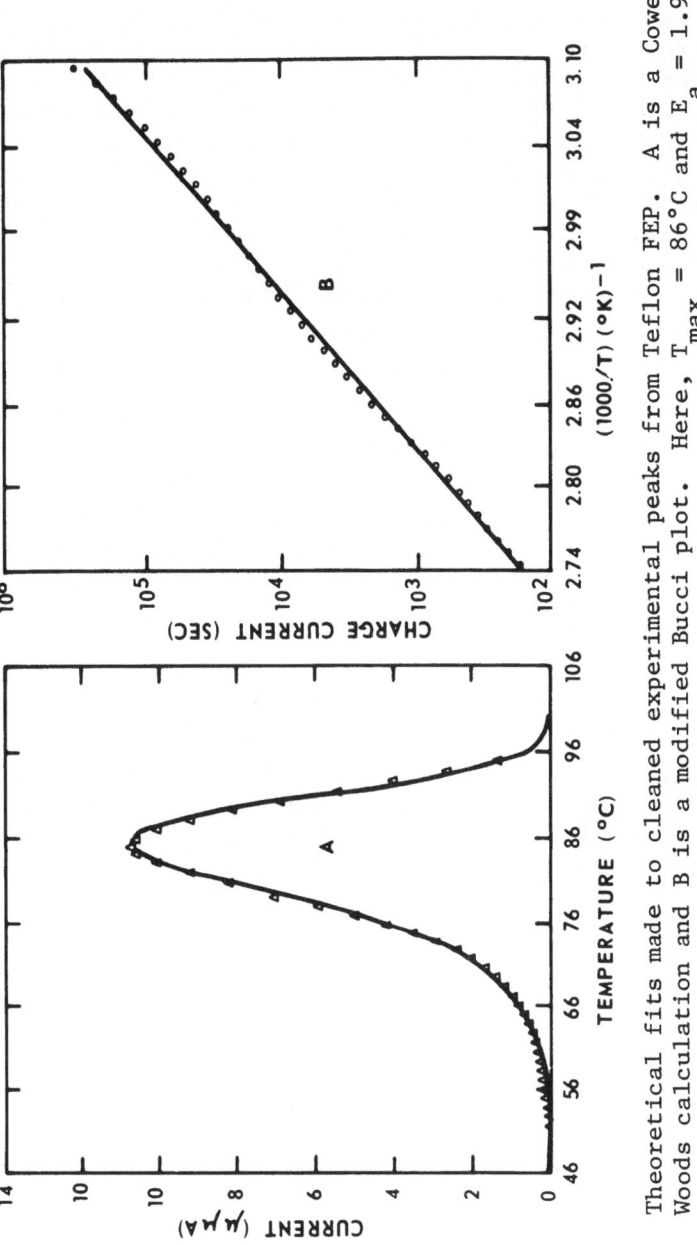

Figure 3. Theoretical fits made to cleaned experimental peaks from Teflon FEP. A is a Cowell and Woods calculation and B is a modified Bucci plot. Here, T_{max} = 86°C and E_a = 1.9 eV.

Overall TSC and TSCD curves from a series of fluorine substituted polyolefins are shown in Figures 4 and 5. A single
fluorine substitution, to produce PVF, forms a corona charged
electret which is much less stable than high density PE itself.
Total discharge, when heated at 1°C min^{-1}, occurs at about 40°C
as compared to PE(HD) at 140°C. A second fluorine substitution,
to produce PVF$_2$, forms a more stable electret that is not totally
discharged until 90°C. With total fluorine substitution (PTFE),
an extremently stable electret is formed which can be totally
discharged only at 230°C. Substitution of a single chlorine atom
in PTFE, to produce Aclar, causes a large reduction in electret
stability and total discharge occurs at 120°C, a 110°C reduction.
A similar large reduction (140 to 40°C) was seen above on substitution of a single fluorine atom in PE(HD) to form PVF.

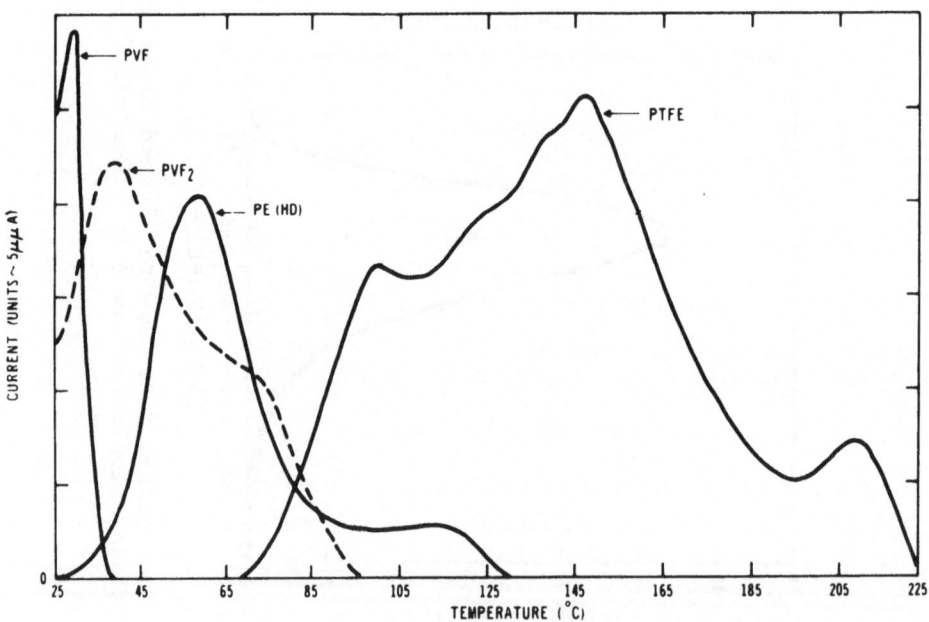

Figure 4.   Thermal current spectra from a series of corona-
            charged fluorine substituted polyolefins.

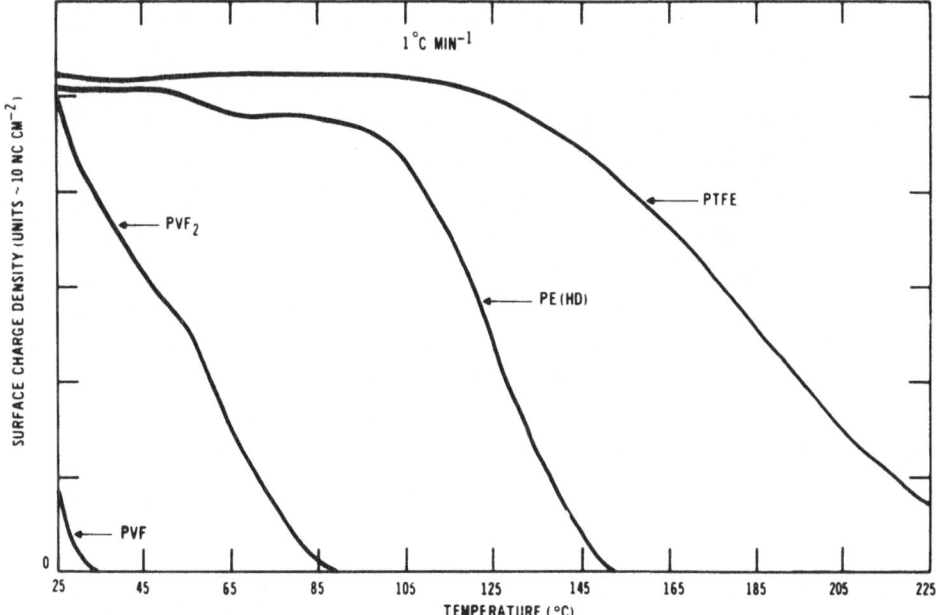

Figure 5.   Surface charge decay in a series of corona-charged
            fluorine substituted polyolefins.

        Polyolefins with equal numbers of fluorine and chlorine sub-
stitution have similar electret properties (Figure 6). In both
cases, a single substitution (PVF and PVCl) produced a very unstable electret. They are totally discharged at 40 and 55°C
respectively. A second substitution in each case (PVF$_2$ and
Cl/PVCl) produces a much more stable electret which is not totally
discharged until 100°C. The TSC characteristics are also very
similar in both cases.

        The phenyl group is advantageous in the formation of stable
electrets. Polystyrene does not become totally discharged before
reaching its softening point at 125°C and has TSC characteristics
(Figure 7) which are very similar to Mylar (1). Polystyrene prepared with a sulfonic end group has similar electret properties
to the pure anionically prepared polymer. Polystyrene which was
carbon filled, with some grafting, was too conductive to form a
stable electret.

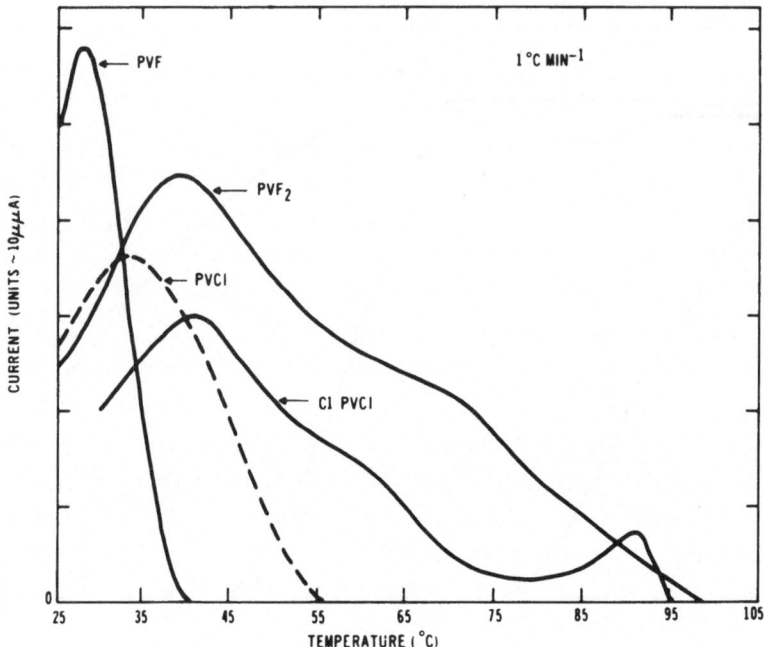

Figure 6.   Thermal current spectra from similar fluorine and
chlorine substituted polyolefins.

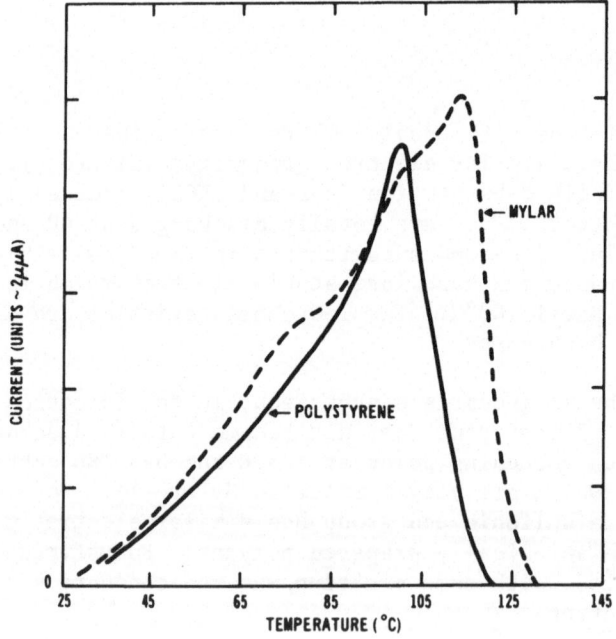

Figure 7.   Thermal current spectra from polymers containing the
phenyl group.

The degree of branching in PE has a large effect on the electret charge stability (Figure 8). High density polyethylene, PE(HD), which has the least branching, is totally discharged at 140°C. Medium and low density PE, which are much more highly branched, are totally discharged at 120 and 85°C respectively. Polypropylene, which has the CH_3 side group, lies between PE(HD) and PE(MD) and has TSCD characteristics which are common to both.

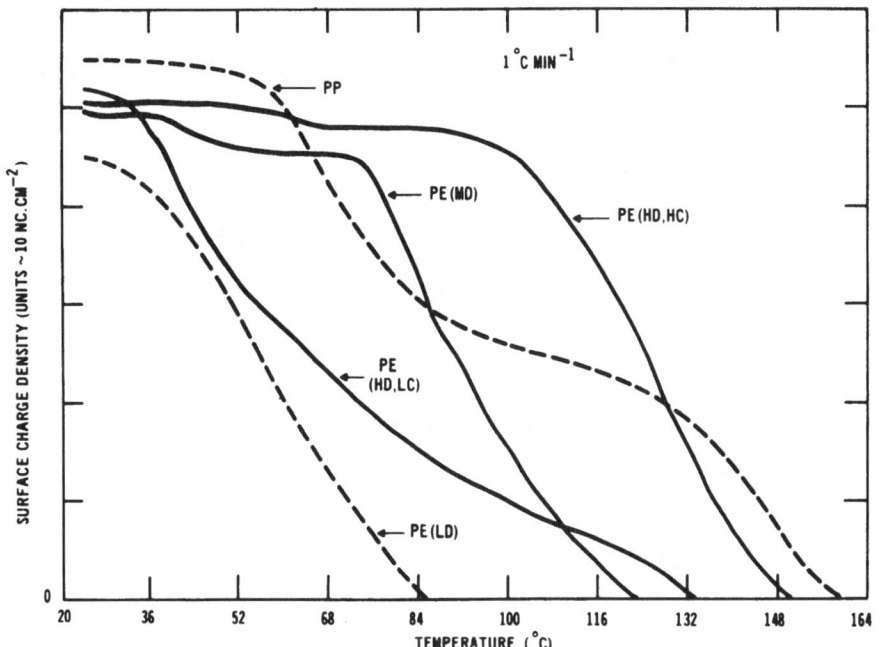

Figure 8. Surface charge decay at a constant heating rate showing the effect of branching and crystallinity in polyethylene. PE(HD), PE(MD) and PE(LD) are polyethylene of 0.965, 0.935 and 0.915g/cc density respectively. PE(HD,LC) has lower crystallinity than PE(HD,HC) due to rapid cooling.

High density polyethylene, cooled slowly from just below the softening point to allow maximum crystalline formation (largest crystallites), is indicated as PE(HD,HC) in Figure 8. It forms a much more stable electret than high density PE which is rapidly cooled to minimize crystallite formation, PE(HD,LC). The initial rate of discharge on heating is greatly increased in PE(HD,LC)

PE(HD,LC) but is less rapid beyond 60°C and total discharge does
not occur until about 130°C, just 10°C below that of PE(HD,HC).
Density measurements indicated that the total crystallinity was
reduced by only 7%; however, the effect on the average crystallite
size may have been much greater.

The kinetics calculations carried out on the clean TSC peaks
obtained from the polymer series are summarized in Table II. In
general, the activation energy, E_a, associated with the discharge
processes, lies between 1 and 2eV and is of the same order as is
associated with molecular motion or phase transitions in polymers.
The E_a did not necessarily increase with increasing temperature
maxima of the TSC peaks. Initial charge densities, n_{t_0}, are about
$10^{15} cm^{-3}$ and charge mobility lifetime products, $\mu\tau$, are about
$10^{-9} cm^2 V^{-1}$. Relaxation times, τ_0, vary a great deal because TSC
peaks were obtained from these polymer electrets over a wide
range, from near room temperature to beyond 200°C.

DISCUSSION

The most stable electrets are the symmetric polymers such as
PTFE and linear polyethylene (high density and little branching).
Disturbing the symmetry by substitution of a chlorine atom (Aclar)
in PTFE or a fluorine (PVF) or chlorine (PVCl) atom in PE greatly
reduces the electret stability. Similar decreases in stability,
but to a lesser extent, are seen with the substitution of the tri-
fluoromethyl group (PFEP) in PTFE and the methyl group (PP) in PE.
Addition of a second substituent in PE, thereby returning to a
more symmetric structure, enhances the electret stability (e.g.
PVF_2 and Cl/PVCl).

The greater stability of the equally symmetric PTFE over
PE(HD) is due to the presence of the halogen (F) atom in place of
hydrogen. It appears that the more electronegative halogen substituent provides a more stable charge storage site for negatively
charged electrets. Preliminary work has indicated that positively
charged PTFE, however, is much less stable.

In polyethylene, the symmetry of the chemical environment is
best in high density, linear polyethylene, not as good in medium
density, and worse in low density, highly branched polyethylene.
The electret stability decreases in the same order (Figure 8). The
physical packing of the chains also decreases in the same order,
and this is again reflected in a separate mechanism of charge
storage. The low crystallinity linear polyethylene, PE(HD,LC),
which was produced by rapid cooling yields a TSCD curve which
interestingly combined the lower part of the branched PE(LD) curve
with the upper part of the linear PE(HD,HC) curve.

TABLE II

Summary of Electret Properties from Thermally Stimulated Current Analysis

Polymer	Link	T_g °C	T_{max} °C	i_{max} μμA	nt_0 10^{15} CM^{-3}	E_a eV	t_0 Days	t_{25} Days	$\mu\tau$ 10^{-10} CM^2V^{-1}
TEFLON PFEP	$-CF_2-CF_2-$		86	12	2.1	1.9	1.3×10^{-29}	2.5×10^3	11
			103	7	1.2	1.9	3.8×10^{-29}	6.6×10^4	19
	$-CF_2-C-F$ $\quad\;\; CF_3$		115	a					
			135	5	1.1	1.6	6.3×10^{-23}	4.0×10^5	20
			156	8	1.6	1.4	3.5×10^{-20}	7.8×10^5	13
			200	a					
TEFLON PTFE	$-CF_2-CF_2-$	127	100	6	1.2	1.4	8.5×10^{-22}	8.4×10^2	19
			115	a					
			135	a					
			160	10	2.8	1.4	1.1×10^{-18}	1.2×10^6	7.8
			207	3	1.2	1.5	1.1×10^{-17}	4.5×10^7	1.8
ACLAR	$-CF_2-CF-$ $\quad\;\;\; Cl$	45	42	8	1.2	1.9	2.8×10^{-31}	0.3	22
			53	2	.3	1.7	7.4×10^{-32}	4.5	97
			79	2	.4	1.6	4.2×10^{-26}	1.5×10^2	67
			105	a		1.6			
POLYETHYLENE (HD)	$-CH_2-CH_2-$	-125	56	80	1.6	1.7	1.5×10^{-28}	3.0	1.8
			70	20	4.0	1.3	2.5×10^{-22}	10	5.0
			98	10	3.0	1.2	3.5×10^{-19}	1.6×10^2	7.5

TABLE II
(Continued)

Summary of Electret Properties from Thermally Stimulated Current Analysis

Polymer	Link	T_g °C	T_{max} °C	I_{max} μμA	n_{t_0} 10^{15} CM^{-3}	E_a eV	τ_0 Days	τ_{25} Days	$\mu\tau$ 10^{-10} CM^2V^{-1}
POLYPROPYLENE	$-CH_2-CH-$ $\mid$ CH_3	-20	48	50	1.1	1.2	1.3×10^{-21}	0.3	1.9
			68	35	4.8	2.3	7.0×10^{-37}	4.1×10^2	4.6
			93	3	0.9	1.4	8.1×10^{-22}	2.5×10^2	24
			111	2	0.4	1.9	9.5×10^{-27}	8.2×10^4	64
POLYSTYRENE	$-CH_2-CH-$ $\mid$ $\emptyset$	100	65	a					
			80	a					
			100	9	2.5	1.3			
PVF	$-CH_2-CHF-$	-20	29	50	4.7	1.5	3.0×10^{-27}	1.5×10^{-2}	5
POLYVINYLIDENE FLUORIDE	$-CH_2-CF_2-$	-45	40	120					
			63	65					
			100	20					

a - Not determined because of small magnitude and overlap.

The phenyl group, present in polystyrene (and in Mylar), provides a deep trap site and this polymer is not significantly discharged before reaching the softening point.

A number of the TSC peaks occur at known phase transitions in the polymers. This is the case with PTFE, PE, PS and Aclar. This fact, combined with the magnitude of the E_a associated with the peaks (1 to 2 eV), leads one to the conclusion that, in many cases, the trapping site itself is being destroyed due to increased molecular motion (12,13). In other cases, the TSC peaks do not occur at known transition temperatures and may result only from direct thermal excitation of the trapped charge. Both processes may be operative (a dual process) in still other cases.

The charge traps in polymers do not appear to arise from impurities. The same electret properties (TSC peaks, TSCD) have been observed in samples of the same polymer supplied by different manufacturers as well as film prepared from resins in our laboratory. The corona itself, however, can affect the chemical composition of the polymers. On exposure to a corona in air, two major chemical changes have been observed to take place in the polymer (14); the formation of double bonds ($\supset$C=C$\subset$) and the carbonyl group ($\supset$C=O). These can act as charge traps. On sufficient exposure to corona with polyethylene, concentrations as great as the electret charge densities which have been measured can be reached. Thus, some of the stored charge may be due to the presence of these groups. A comparison with electrets formed by low energy electron bombardment in vacuum would help to verify this possibility.

The TSC technique offers a key to the comprehension of the fundamental mechanisms for charge storage and release, such as electron and ionic traps to form a space charge which exists near the surface of these electrets. It is these phenomena (along with hole formation) that govern electrical conduction in dielectrics. For example, since trap hopping is often invoked as a conduction mechanism, a knowledge of trap parameters is desirable. Thus, a study of charge storage yields valuable information about electrical conduction.

ACKNOWLEDGMENT

The authors wish to thank B. A. Gribbon of the Bell-Northern Research Laboratory for his valuable assistance with the experiments and Dr. G. Riess of the Ecole Superieure de Chemie de Mulhouse for supplying the polystyrene samples.

The subject matter was supported in part by financial assistance from the National Research Council of Canada through its Industrial Research Assistance Program.

REFERENCES

1. R. Creswell and M. Perlman, J. Appl. Phys. $\underline{41}$, 2365 (1970).
2. M. Perlman and R. Creswell, J. Appl. Phys. $\underline{42}$, 531 (1971).
3. M. Perlman and R. Creswell, "Thermal Current Study of Charge Traps in Insulators," Conference on Electrical Insulation and Dielectric Phenomena, NAS and NRC, Pennsylvania, October, 1970.
4. F. Frain and P. Murphy, Audio Eng. Soc. $\underline{18}$, 511 (1970).
5. G. Sessler and J. West, J. Polymer Sci. B, $\underline{7}$, 367 (1969); J. Electrochem. Soc. $\underline{115}$, 836 (1968).
6. C. Reedyk, "Electret Transducer Applied to the Telephone," 1970 Conference on Communications, San Francisco, and in press, IEEE Trans. on Electroacoustics.
7. C. Reedyk and M. Perlman, J. Electrochem. Soc. $\underline{115}$, 49 (1968).
8. T. Cowell and J. Woods, Brit. J. Appl. Phys. $\underline{18}$, 1045 (1967).
9. C. Bucci, et al., Phys. Rev. $\underline{148}$, 2, 816 (1966).
10. L. Badian, et al., Compt. Rend. $\underline{261}$, 2181 (1965).
11. G. Sessler, Bell Telephone Laboratories, private communication.
12. R. Partridge, J. Polymer Sci. A, $\underline{3}$, 2817 (1965).
13. T. Takamatsu and E. Fukada, Polymer J. $\underline{1}$, 1010 (1970).
14. D. Carlsson and D. Wiles, Can. J. Chem. $\underline{48}$, 2397 (1970).

EFFECT OF MORPHOLOGY ON THE DIELECTRIC PROPERTIES OF POLYSTYRENE-POLYETHYLENE OXIDE BLOCK COPOLYMERS

John M. Pochan[*] and Richard G. Crystal

Research Laboratories, Xerox Corporation, Xerox Square

Rochester, New York

INTRODUCTION

To date, very little work has been done on the dielectric properties of block copolymer systems, particularly the study of dielectric properties as a function of polymer morphology. Ishida, Shimada, Matsuura and Takayanagi (1) studied polystyrene-polymethacrylate block copolymers and noted that loss mechanisms associated with the individual components were readily observable, i.e. α_a of polystyrene (associated with T_g) and the α_a and β mechanisms of polymethacrylate. No mention was made, however, of sample morphology, and accordingly the effects of phase separation morphology on dielectric response.

In this study, morphological effects on the dielectric properties of polystyrene (PS) - poly(ethylene oxide) (PEO) block copolymers will be discussed. These block copolymers are unusual in a number of ways. They are composed of two dissimilar and incompatible polymers; one (PEO) is highly polar, semicrystalline and soluble primarily in polar solvents, while the other (PS) is non-polar, amorphous, and soluble primarily in non-polar solvents. Important thermal transitions associated with the individual homopolymers are shown in Table I. The morphologies can be controlled by casting from preferential and mutual solvents for the block components. Depending on composition and solvent history, morphologies can be obtained which vary from polystyrene as a matrix material surrounding poly(ethylene oxide) domains to interpenetrating layers of the two components to a poly-(ethylene oxide) matrix surrounding polystyrene

[*] To whom correspondence should be addressed.

domains. Crystallization of PEO can strongly influence the phase
separated structure, the appearance of which is dependent on solvent,
polymer composition, thermal history and molecular weight. Since
relatively few dielectric studies through the melting point of
crystalline polymers have been reported (2,3,4), the effects of
block composition and morphology on this transition phenomenon in
the PEO segment are of interest.

TABLE I

Thermal Transitions of PS and PEO

Polymer	T(°C)	Phenomenon
PS	100	T_g-Glass Transition
PEO	-20	T_g-Glass Transition
PEO	+65	T_m-Crystalline Melting Point

The synthesis and thermal (4), morphological (5) and rheo-
logical (6) studies on the polymers used in this study have been
reported. Erhardt, et al. (6), noted the effects of morphology on
the dynamic mechanical properties of several PS-PEO AB and ABA
block copolymers. As expected from mechanical measurements, the
mechanical PS T_g was detectable only when PS formed the matrix
phase, whereas calorimetry readily revealed the thermal PS T_g in
all cases.

EXPERIMENTAL

Two A-B type PS-PEO block copolymers were primarily used in
this study. Table II summarizes identification notation used in
this paper as well as molecular characterization and solvent history. Note that both samples contained a 3×10^4 M_n segment of PEO
coupled to segments of PS (3×10^4 and 7×10^4 M_n). We would expect
that equal PEO segment lengths would eliminate the M_w variable in
PEO crystallization. As morphology is markedly affected by composition and casting solvent, solvent histories are incorporated in
the sample notation for convenience. The solvent systems used
were chloroform, a mutual solvent for both polymers, and ethylbenzene, a PS preferential solvent. All attempts to prepare
necessarily flat, uniform dielectric samples cast from PEO preferential solvents failed due to crystallization effects - hot
pressing of samples could not be used because of its drastic effect
on phase separated morphologies (11).

TABLE II

Sample Notation and Characterization

Sample Notation	Weight % PS	Mn^*_{PS}	Mn^*_{PEO}	Mn^*	Casting Solvent
50-C	50	3	3	6	Chloroform
50-E	50	3	3	6	Ethylbenzene
70-C	70	7	3	10	Chloroform
70-E	70	7	3	10	Ethylbenzene

Sample preparation was as follows: weighed amounts of polymer were totally dissolved in the solvent of interest [in the case of a non-solvent for one component, very dilute solutions (< 1% weight/volume) would totally solubilize the soluble segment, leaving a somewhat turbid solution]. The solutions were then cast in teflon molds and evaporated very slowly so that a reproducible approximation of equilibrium conditions could be obtained. In the less volatile ethylbenzene system, this required drying periods of up to one week. As PEO is a moisture sensitive material, care was taken to store samples in a dry environment. All samples were dried in a vacuum oven at ambient temperature prior to measurement. Most samples were used 'as is' after drying; however, 50-E samples had to be machined to a uniform thickness. Sample thicknesses were in the range of 300 to 600 microns with standard deviations of less than five microns in any one sample. Mw's were measured by GPC before and after experimental runs to ascertain the degree of thermal decomposition. In all cases, this method showed little polymer decomposition at temperatures below 130°C. We were, therefore, careful to make dielectric measurements below this value.

The percent PEO crystallinity in samples was measured calorimetrically using a Perkin Elmer DSC-1B differential scanning calorimeter. All dielectric spectra were taken with a General Radio 1615-A transformer ratio arm capacitance bridge, coupled with a General Radio 1310-A oscillator and a type 1232-A tuned amplifier and null detector. Frequency capability was 30 to 10^5 Hz. A Balsbaugh Model MC-100 three terminal guarded electrode cell, modified for spring load capabilities, was used in conjunction with the aforementioned system. Temperature was controlled with a Delta Design Model 5750 environmental oven with temperature control accurate to one-half degree fahrenheit. As GPC data indicated little thermal or oxidative decomposition of the polymers below 130°C, the oven was not purged with inert gas.

Bulk morphologies of specimens used in this study were examined by transmission electron microscopy of one step replicas of samples fractured at liquid nitrogen temperatures. The replicas were prepared by vacuum shadowing with Pt-C followed by evaporation of a heavy layer of carbon. Replicas were removed by dissolution of the polymer in chloroform and examined in a Jeolco T7 instrument.

RESULTS AND DISCUSSION

Perhaps the most striking features of the dielectric properties of the PS-PEO blocks are first, the markedly different dielectric spectra obtained for samples cast from the mutual and the PS preferential solvents and second, the structural memory of morphology retained in the dielectric properties at temperatures as high as 30°C above the PS T_g.

Figures 1 and 2 show isothermal plots of ε' and ε'' vs log ν for samples 50-C. Only half the experimental isotherms have been plotted for the sake of clarity. Salient features observed in this figure are: the relatively high values of ε' and ε'', the magnitude with which the loss associated with PEO melting move into the experimental frequency range at temperatures below the PEO Tm, the observed ionic d.c. conductance of the sample at temperatures near and above the T_g of PS, and the "memory" effect in the dielectric properties at temperatures above the PS T_g. Similar plots for sample 70-C are almost identical to Figures 1 and 2, however, ε' for 70-C is approximately 20 and ε''_{max} about 6. These slightly lower values show the effect of the increased PS concentration on the dielectric properties. More subtle differences between 70-C and 50-C will be noted later.

The effect of ionic conduction is readily seen in Figure 2. At temperatures below PS T_g only slight increases in ε'' at low frequency are noted; however, at temperatures > PS T_g a large low frequency contribution to the loss is observed. This loss effect was of such magnitude that bridge balance was unobtainable in the high temperature - low frequency region. The curve for T = 130°C in Figure 2 is characteristic of this low frequency loss. In previous published studies, high temperature (T > 40°C) data in PEO was unobtainable due to ionic conductance (7) and experimentalists were content to study the α_a relaxation in PEO. As shown in Figures 1 and 2, studying PEO in the block copolymer configuration has permitted an extension of the viable temperature range by approximately 90°C. This range extension is a function of polymer morphology - samples 50-C, 70-C and 70-E were amenable to measurements to 130°C, while sample 50-E was not because of d.c. conduction at T > 70°C. The differences between these samples and the morphological reasons for the above noted effect will be discussed in this paper.

Although not included in the figures, the dielectric spectra of
sample 70-E differed drastically from the above. ε'' in the same
temperature-frequency region was more than an order of magnitude
lower in value (.34); peaking at a temperature of 62.4°C, ν = 300 Hz.
ε' at the same experimental values was ~3.2. Another interesting
feature of the 70-E spectra was the disappearance of the loss peak
at temperatures greater than the PEO melt temperature. This
phenomena will also be discussed in more detail later.

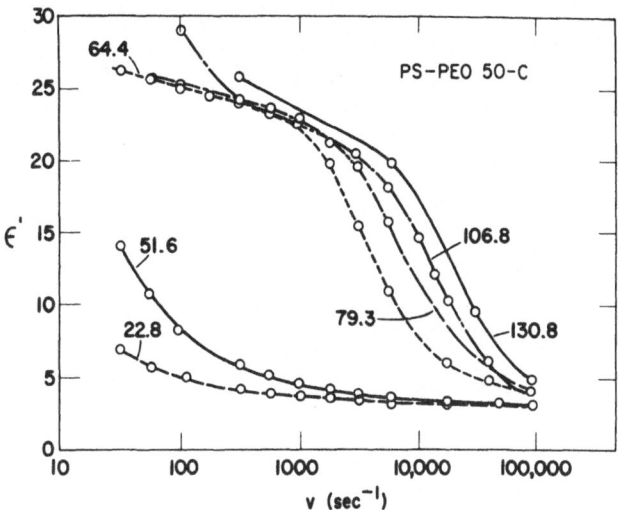

Figure 1. ε' vs log frequency at indicated temperatures for sample
50-C.

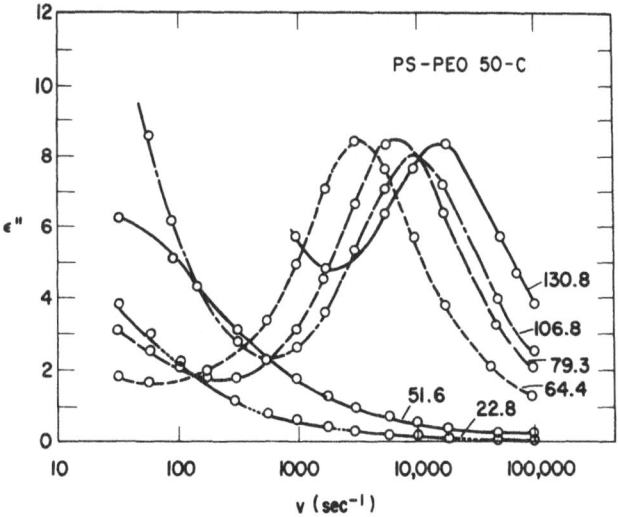

Figure 2. ε'' vs log frequency at temperatures indicated for sample
50-C.

Other important features of Figure 2 are the narrowness of the loss peak and its temperature insensitivity. At temperatures above T_m PEO, the half-width is approximately 1.2 to 1.3 decades of frequency for samples 50-C and 70-C, while below the PEO melting point the half width is temperature dependent; increasing with decreasing temperatures. The narrow line width is indicative of a single relaxation process as described by McCrum, et al. (8) That the process is not due to a single relaxation phenomenon can be seen in Figure 3, a Cole-Cole diagram for sample 50-C at a temperature of 70.7°C. The diagram is characteristic of many polymeric systems showing a slightly skewed appearance at high frequency and is not amenable to interpretation in terms of a single relaxation time. Cole and Davidson (9) analyzed data such as this in terms of an additional relaxation parameter γ.

$$\frac{\varepsilon^*(\omega) - \varepsilon_u}{\varepsilon_r - \varepsilon_u} = \frac{1}{(1 + \omega\tau_1)^\gamma} \tag{1}$$

where ε_u and ε_r are the high frequency unrelaxed and low frequency relaxed dielectric constant respectively, $\varepsilon^*(\omega)$ is the complex dielectric constant, ω is the angular frequency, τ_1 is a characteristic relaxation time, and γ is a variational parameter.

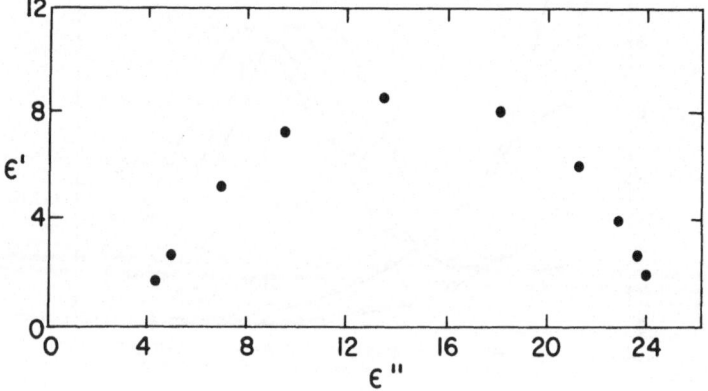

Figure 3.   Cole-Cole diagram for sample 50-C at 70.7°C.

A value of γ = .7 can be derived by extrapolating a first approximation to ε_r and ε_u. Thus a narrow distribution of relaxation times rather than a single one is controlling the process. Similar curves are obtained for samples 70-C and 50-C at temperatures above the PEO Tm; however, due to d.c. conductance no complete curves are available for the other samples. Figures 2 and 3 indicate that the dielectric increment and relaxation distribution functions (10) are relatively temperature insensitive at temperatures greater than the PEO melt temperature. The dielectric increment for this loss process is 15.2 and 22.5 for samples 70-C and 50-C respectively, showing $\Delta\varepsilon'$ almost proportional to the PEO composition.

Figure 2 also demonstrates the memory effect mentioned previously. At temperatures greater than the PS T_g the loss spectra retains its continuity, save for the d.c. conductance at low frequency, and shows little temperature effect. The spectra is reversible with only slight changes in ε' and ε''; however, since care was taken to attain high temperature spectra as rapidly as possible, changes in structural morphology were minimized. If more time had been taken in the higher temperature region, it is almost certain that large changes would have been noted; in particular, increased d.c. conduction as the polymer morphology changed. There is no evidence to suggest any other effect.

The morphologies of specific samples used in this study have been characterized by transmission electron microscopy. A brief description of each will be given. In addition, the reader is referred to two references (5,11) describing morphological control of these materials.

Sample 70-C consists of spherulites of PEO in which individual lamellae growing radially from the nucleation source are coated with PS. There appears to be a continuity of both PEO and PS throughout the sample matrix. See Figure 4.

Sample 50-C consists of lamellar layers of alternating amorphous PS and crystalline PEO. See Figure 5. The layers of PEO are approximately 120Å thick, indicating a single chain folded lamellar crystalline structure (5). The entire matrix of sample 50-C could be considered as a composite of many sandwiched capacitors oriented isotropically throughout the matrix. Again there is a continuity of PEO throughout the matrix. The alternating layer structure would ensure a minimum interphase surface to volume ratio geometry.

Sample 70-E consists of rods of PEO suspended in an amorphous PS matrix. The rods appear to be parallel to the cast sample surface with only slight deviations from this orientation. Of particular interest is the apparent slightly disordered hexagonal packing of the rods within the matrix. See Figure 6.

Figure 4.   Fracture replica electron micrograph from sample 70-C.
            Note lamellar spherulitic texture in which the crystal-
            line PEO lamellae are coated with amorphous PS.

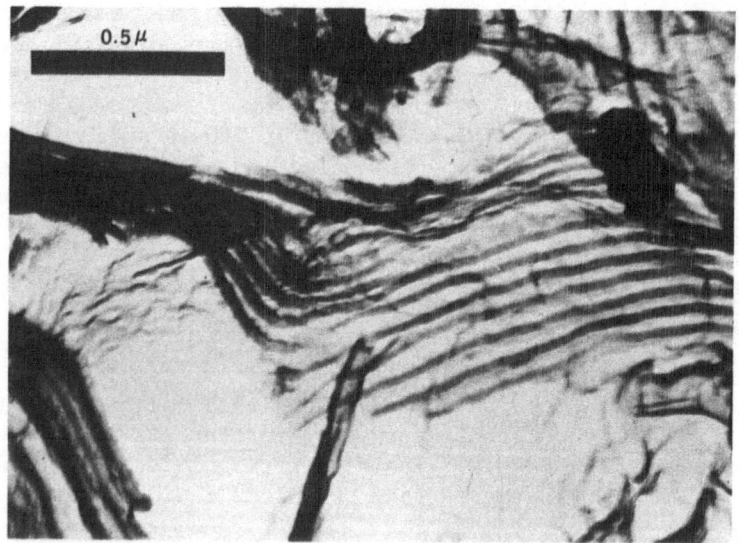

Figure 5.   Fracture replica electron micrograph from sample 50-C.
            Note the alternating layers of PS and PEO - a phase
            separated structure resulting in a minimum surface area
            between phases.

Figure 6. Fracture replica electron micrograph from sample 70-E.
Note the rod-like morphology of PEO suspended in a PS
matrix. The tops of the rods are quite visible and
appear to pack in a somewhat disordered hexagonal array.

While Sample 50-E has not been characterized, a morphology
consisting of unconnected PEO crystal fragments in a PS matrix would
be expected, based on previous studies. Our inability to study this
sample at temperatures greater than 70°C (due to high d.c. conduction)
would indicate that phase separation morphology was incomplete permitting large areas of highly conductive PEO to exist throughout
the matrix.

The sample morphologies are quite different and do show composition and solvent effects. A basic difference eminating from solvent history is the discontinuity of PEO domains upon casting from
ethylbenzene. This is particularly illustrated in calorimetric
measurements (4) in which PEO phases isolated from nuclei by the PS
matrix do not crystallize. The samples cast from ethylbenzene show
rods or crystallites of PEO which appear to be discrete and non-
contacting. It is this fundamental difference in structure that
must account for the large differences in dielectric spectra. In
spectra obtained from both samples exhibiting PEO phase continuity,
large ε' and ε'' measurements as well as narrow relaxation distributions are obtained, especially at high frequency and temperatures
greater than PEO Tm. In the case of the non-continuous PEO samples,
dielectric properties do not show the above effect. We can, therefore, attribute the drastic dielectric differences between solvent
systems to PEO matrix continuity.

The effect of PEO crystallinity can readily be used to explain observed variances in spectra. DSC crystallinity measurements show the solvent samples to have comparable crystallinity to that found by O'Malley, et al. (4), the degree of crystallinity being dependent primarily on polymer composition rather than solvent used in casting.

We would anticipate that degree of PEO crystal perfection also affects dielectric properties (12). This can be seen in Figure 7, a plot of ε' vs T at 1 KHz for the studied samples. A comparison of samples 50-C and 70-C shows quite different dielectric changes associated with the PEO melting. Sample 50-C exhibits a very sharp change in ε' over a narrow temperature range while sample 70-C shows a much broader temperature effect associated with the same phenomenon. Note that a rod-like phase separated structure or spherulitic phase separated structure would result in larger surface to volume ratios than layered structures. This greater interphase surface area with rods would result in more PS-PEO interfacial interaction and thus greater perturbation on the crystallizability of PEO (normalized, of course, for relative PEO content). Because of the higher surface to volume in 70-C as compared to sample 50-C, we would expect greater defects in the PEO crystal structure and lower degrees of crystallinity based on available PEO in the block polymer. These defect crystals would understandably melt over a wider temperature range than their more perfect counterparts. Sinnott (13) has shown this effect in polyethylene where he notes that more perfect crystals of polyethylene have higher and sharper melting points than less perfect crystalline specimens. Thus, crystal defect content and PEO amorphous content should cause the melting differences between samples 50-C and 70-C.

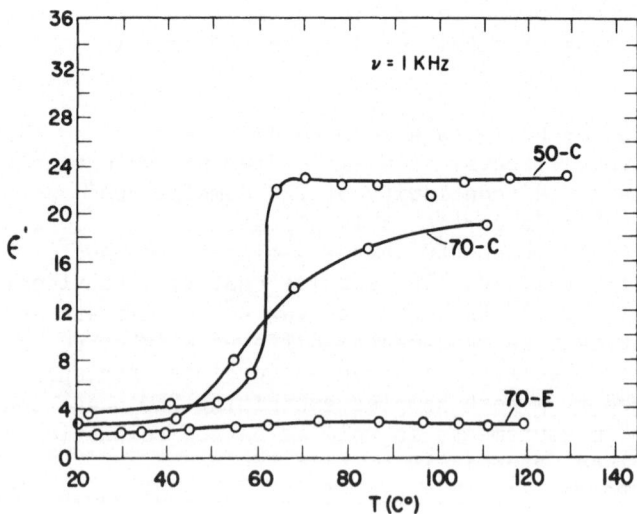

Figure 7. ε' vs temperature at 1 KHz for samples 50-C, 70-C, and 70-E.

The effect of phase separation morphology can also be seen in Figure 7, by comparing samples 70-C and 70-E. Sample 70-E exhibits only slight change in ε' (~2.5 → 4.0) upon PEO melting as compared to the much larger change in 70-C (~3.5 → 20). The effect of composition can also be seen in comparing 50-C with 70-C. Only the α_c process of PEO has been considered thus far. In sample 70-E the β relaxation of PEO was also observed at ~10^4 Hz and -16.5°C. Although sample 70-C was not examined at low temperature for this process, sample 50-C showed no β PEO loss process. Our inability to observe this process in sample 50-C could be a result of the higher PEO crystallinity in the samples. The data would be analogous to that of Ishida, et al. (14), who observed no β process in mats of PEO single crystals.

The data in Figure 7 for 70-E is dependent on sample preparation. Data shown is for a sample prepared as described above. We have also prepared the same system in a heated vacuum oven, effectively annealing the PEO crystallites. ε' values below and above PEO Tm are identical; however, the change in ε' with temperature is more abrupt in the annealed case. This again illustrates the effect of a more perfect crystallinity on the dielectric properties of block copolymers and semicrystalline polymers in general.

The effect of temperature on ε'' can be seen in Figure 8. As noted previously, samples 50-C and 70-C exhibit d.c. conduction effects at high temperature. Sample 50-C particularly shows this effect above the PS T_g where ε'' increases rapidly with temperature. The degree of crystalline perfection is again noted in differences in position of ε''_{max} for the samples. The relative temperature widths of 70-C and 70-E show that both samples contain crystal structure of approximately the same levels of imperfection, a concept confirmed by calorimetric measurements (4). The difference in ε''_{max} between these samples again shows the effect of morphology.

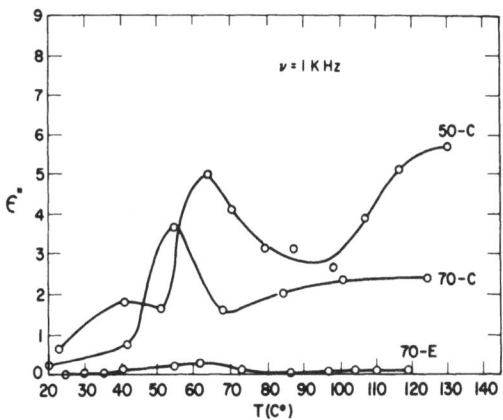

Figure 8. ε'' vs temperature at 1 KHz for samples 50-C, 70-C and 70-E.

The relaxation of PS has been difficult to observe experimentally in such materials because of the overriding effects of the PEO (both large loss effects and d.c. conduction). Plots as that illustrated in Figure 8 are one of the few ways that the PS mechanism can be detected. The broad low intensity peak beginning at ~90°C may be due to the PS T_g. It is difficult to tell (because of our degradation temperature restrictions) whether this loss is due to d.c. conduction caused by a partial phase inversion above the PS T_g or to the actual loss mechanism associated with the PS T_g. If it is the former, then the effect would be similar to that seen in 50-C.

Figures 9 and 10 show plots of the log max ν vs 1/T for samples 50-C, 70-C, and 70-E. In Figure 9 the data of Connor, et al. (7) for the β relaxation and the data of Ishida (19) et al. for the γ relaxation in PEO are also shown. This data has been plotted to show that if PS does not affect the PEO α_c process to any great extent, then neither the β nor γ transition will interfere with our observations. Figure 10 is an expanded plot of our data in the temperature range of interest. Table III contains a listing of the activation energies derived from our measurements along with the proposed mechanisms. All thermal data can be explained in terms of an Arrhenius activation equation. Those processes associated with PEO melting have been labeled α_c; however, the high temperature regions of samples 70-C and 50-C have not been defined although they are certainly due to the amorphous PEO state of the block. The low activation energy-high temperature phenomena are apparently only realized with morphologies generated by casting from the mutual solvent, chloroform, in which PEO domains are interconnected. The data for sample 70-E again shows the effect described previously - the disappearance of the loss peak associated with crystal melting. The effect of polymer composition on the high temperature process is also illustrated in Figure 10. For a given temperature, the relaxation associated with sample 70-C is observed at higher frequency than that of 50-C. This effect can be explained in terms of PE-PEO interfacial interactions. In the phase separation of two incompatible species in a block copolymer, an interdomain boundary containing both components can be anticipated. We would expect that relative to the volume of PEO, this area would be more prevalent in specimens with the higher interfacial surface to volume ratios such as in 70-C. We would also expect that higher PS content samples would show more of an effect from the PS matrix. The observed results would then indicate that interactions with the PS result in less PEO-PEO segment interactions allowing the dipolar PEO segments to react more easily to the applied electromagnetic field at a given temperature.

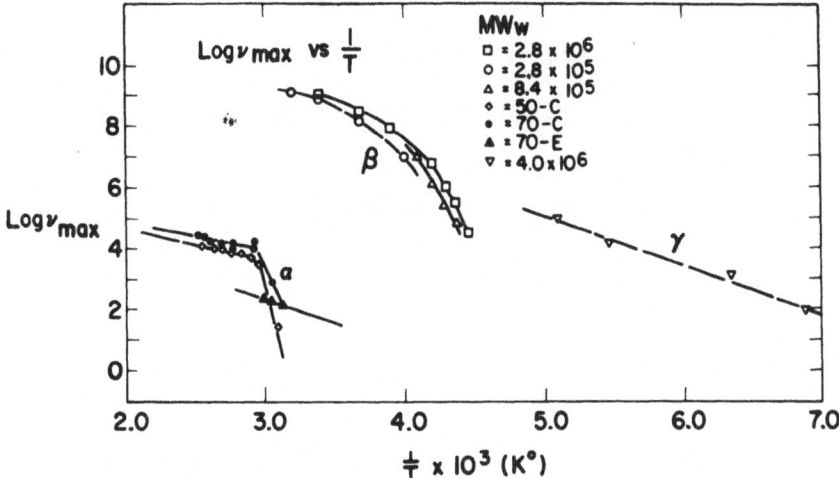

Figure 9. Log ν max vs 1/T for the α, β, and γ relaxation process
of PEO. β transition – data of Connor, et al. (7)
γ transition – data of Ishida, et al. (17)

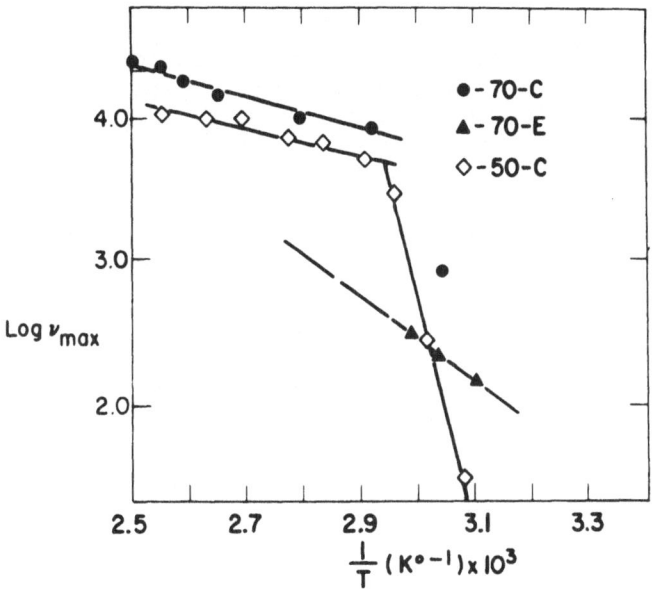

Figure 10. Log ν max vs 1/T for samples 50-C, 70-C and 70-E.

TABLE III

Activation Energies

Sample	Mechanism	E_a (Kcal/mole)
50-C	α_c	80
50-C		4
70-C		4
70-E	α_c	12

The energies associated with the crystalline melt phenomena
(α_c) vary between samples. The value of 80 Kcal/mole for sample
50-C compares favorably with values obtained dielectrically for
other crystalline polymers: 80 Kcal/mole for polychlorotrifluoroethylene (3), 150 Kcal/mole for oxidized polypropylene (2), and
45 Kcal/mole for polytetrafluoroethylene. Obviously these values
depend on the degree of crystal perfection and are relative. In
the case of block copolymer systems, this degree of perfection
is a function of morphology as reflected in data in Figure 7.
Sufficient data was not collected with sample 70-C to allow calculation of E_a; however, we would expect from the data and interpre-
tive reasoning thus far that the activation energy for 70-C would
be intermediate between that of 70-E and 50-C. This would lead us
to conclude that the activation energy of α_c for pure PEO may be
higher than that observed in any of our samples.

CONCLUSIONS

The dielectric properties of the PS-PEO block copolymer system
are strongly influenced by phase separation morphology. Variations
have been shown to be a function of sample preparation and composition. The degree of PEO crystalline perfection, as controlled by
morphology and composition, greatly influences the thermal characteristics of the α_c relaxation in PEO. We believe that this method
(use of block copolymers) has led to attainment of heretofore
unobtainable high temperature dielectric data and that the method
could be applied to the measurement of high temperature dielectric
properties of other semi-conductive polymers.

A relaxation mechanism associated with the PEO melt has been
identified and has been determined to be a function of sample preparation and composition. This process, at T>Tm PEO, shows characteristics of very narrow relaxation time distributions and is only

observable in samples containing a continuity of the PEO phase. The relaxation phenomena associated with the PS phase of the block has been shown to be effectively masked by the PEO dielectric properties.

ACKNOWLEDGMENTS

The authors would like to thank Dr. J. J. O'Malley for providing samples used in this study, Dr. G. Sitaramaiah for molecular characterization, and Dr. J. M. O'Reilly for his helpful suggestions and encouragement during the study.

REFERENCES

1. Y. Ishida, K. Shimada, K. Matsuura and M. Takayanagi, Kolloid Z. 200, No. 1, 51 (1965).
2. M. Kramers and K. E. Helf, Kolloid Z. 180, 114 (1962).
3. A. H. Scott, D. Scheiber, A. Curtis, J. Lauritzen and J. Hoffman, J. Res. Nat. Bur. Stds. 66A, 269 (1962).
4. J. J. O'Malley, P. F. Erhardt and R. G. Crystal, BLOCK COPOLYMERS, S. Aggarwal ed., Plenum Press, New York, 1970, p. 163.
5. R. G. Crystal, P. F. Erhardt and J. J. O'Malley, ibid., p. 179.
6. P. F. Erhardt, J. J. O'Malley and R. G. Crystal, ibid., p. 195.
7. T. M. Connor, B. E. Read and G. Williams, J. Appl. Chem. 74, 14 (1964).
8. T. M. Connor, B. E. Read and G. Williams, ibid., p. 110.
9. D. W. Davidson and R. H. Cole, J. Chem. Phys. 18, 1417 (1950).
10. T. M. Connor, B. E. Read and G. Williams, J. Appl. Chem. 114, 14 (1964).
11. R. G. Crystal in COLLOIDAL AND MORPHOLOGICAL PROPERTIES OF BLOCK AND GRAFT COPOLYMERS, G. Molau ed., Plenum Press, New York, 1971, p. 279.
12. Y. Ishida, J. Polymer Sci. A-2, 1, 1835 (1969).
13. K. M. Sinnott, J. Appl. Phys. 37, 3385 (1966).
14. Y. Ishida, M. Matsuo and M. Takayanagi, J. Polymer Sci, B, 3, 321 (1965).
15. F. Krum and F. H. Miller, Kolloid Z. 164, 8 (1959).

DIELECTRIC DISPERSION IN DILUTE SOLUTIONS OF SEVERAL PARA-SUBSTITUTED POLYSTYRENES*

B. Baysal, B. A. Lowry, H. Yu and W. H. Stockmayer

Department of Chemistry, Dartmouth College

Hanover, New Hampshire

ABSTRACT

Dielectric loss measurements are reported for polymers of several styrene derivatives in dilute benzene or toluene solution in the range 3-50 MHz and from about -10°C to +30°C. For polymers of 4-fluorostyrene and 4-chlorostyrene, as well as for a styrene/4-vinylpyridine 2/1 copolymer, the frequency of maximum dielectric loss at 300°K is about 30 to 40 MHz and the activation energy is about 5 kcal/mole. These results are in fair accord with acoustic and magnetic relaxation data on polystyrene solutions. Fragmentary results on polymers of p-methoxystyrene and p-fluoro-α-methylstyrene are described. Dipole moments in benzene are also reported.

Dielectric dispersion measurements on dilute solutions of polar chain molecules with dipoles rigidly and perpendicularly attached to the chain backbone offer a principal means of studying short range or "local" modes of irreversible chain motion (1). Examples of polymer chains of this type are polyoxyalkanes, polyvinyl halides, polymers of p-substituted styrenes, and many others. Such local chain motions are of obvious interest in connection with high-frequency or low-temperature mechanical properties of polymers and their solutions.

Although the literature on dielectric behavior of bulk polymers is voluminous, there is need for more data on polymer solutions. The effects of concentration, temperature and solvent in a variety of systems should be systematically explored. In this paper a few

*

Supported by the National Science Foundation.

329

results in very dilute solutions are reported for several p-substituted styrene polymers, thus extending earlier studies of p-chlorostyrene (2,3,4) and of styrene/p-chlorostyrene copolymers (3). Dipole moments are also reported for the same materials.

EXPERIMENTAL SECTION

Polymers were prepared by standard methods from commercially available monomers, freshly distilled under reduced pressure before use. Some details are given in Table I. No fractionations were performed, since the properties of interest here are independent (1) of molecular weight.

The polymerization of p-fluoro-α-methylstyrene appears not to have been previously described in the literature. A relatively high monomer freezing point (-26°C) and a low ceiling temperature conspire to limit conditions for polymerization. Our sample was obtained in 95% yield after 24 hours from a 20% solution in toluene at -78°C with 0.05M $BF_3 \cdot O(C_2H_5)_2$ as catalyst. These conditions were chosen in imitation of those used (5) to effect homogeneous cationic polymerization of α-methylstyrene, and may be presumed to have produced a strongly isotactic polymer (5). The polymer is soluble in benzene and toluene. Analysis: calcd. for $(C_9H_9F)_x$, F 13.95, C 79.39, H 6.66; found F 13.97, C 79.56, H 6.88.

A styrene/4-vinylpyridine 2/1 copolymer was prepared and analyzed in the manner of Fuoss and Cathers (6). Polymerization with benzoyl peroxide initiation was conducted at 85°C, the solution containing 20% toluene along with the monomers. The copolymer composition based on nitrogen analysis (4.68%) was 34.9 mole percent vinylpyridine, which agrees well with a figure (31.3 mole percent) calculated from the monomer reactivity ratios (7).

Solution viscosities were measured with Ubbelohde dilution viscometers. Light-scattering intensities and refractive-index increments were obtained with standard Brice-Phoenix equipment.

A General Radio Company Type 1620-A capacitance measuring assembly was used to measure dielectric constants at low frequencies (5 and 10 kHz), the solutions being confined in a Balsbaugh Laboratories (Duxbury, Massachusetts) Model 350-G three-terminal cell which requires about 55 ml of solution. The temperature was controlled to $\pm$ 0.05°C during these measurements, and all solutions were prepared by weighing both components. Solvents were middle fractions from two successive distillations.

TABLE I

Polymer Preparation

Monomer	Polymerization Conditions	[η] dl/g	M_w (Light Scattering)
p-Chlorostyrene	Bz_2O_2, 50°C	0.29 (T)	1.5×10^5 (T)
p-Fluorostyrene	Bz_2O_2, 50°C	(A) 0.20 (T) (B) 0.28 (T)	6.9×10^4 (T) 1.20×10^5 (T)
p-Methoxystyrene	Bz_2O_2, 105°C	0.59 (B)	3.5×10^5 (B)
p-Fluoro-α-methylstyrene	see text	0.124 (B)	5×10^4 (B)
Styrene/4-vinyl-pyridine	see text	1.13 (B)	8.6×10^5 (B)

Note: Viscosity and light scattering data are at 25°C in toluene (T) or benzene (B).

For dielectric loss measurements at higher frequencies (3 to 50 MHz) we used an a.c. bridge designed, constructed and loaned to us by W. B. Westphal and R. Charles of the M.I.T. Laboratory for Insulation Research. Two geometrically matched sample cells are connected in parallel with matched variable capacitors in the capacitance arms of the bridge; and another variable capacitor is in parallel with one of the resistance arms. The dielectric constant ε' and loss factor ε'' for a solution can be evaluated after the bridge has first been balanced with pure solvent in each sample cell and then balanced again after replacing solvent by solution in one arm. The method is essentially limited to low-loss solvents and solutions. Inductance effects set a practical upper limit to bridge balance at about 50 MHz. A General Radio Type 1330-A Oscillator served as the signal generator, and a Hallicrafters Model SX-62A receiver with R-48A speaker as the null detector. Temperatures of the sample cells were controlled by circulation of light oil from a thermostatted bath and measured with a thermocouple. The entire bridge assembly was kept in a dry box during all measurements to minimize the effects of moisture.

RESULTS AND DISCUSSION

Polarization

Low-frequency dielectric increments were measured at several concentrations in the range 1 to 3 percent, over a range of temperatures from about 10° to 50°C. Mean square dipole moments per repeat unit of the polymer chain were calculated from an appropriate form of the Guggenheim-Smith (8,9) equation:

$$<\mu^2>/x = \frac{27kTM_o}{4\pi N_A} \left(\frac{d\varepsilon/dc}{(\varepsilon_1 + 2)^2} - \frac{2n_1(dn/dc)}{(n_1^2 + 2)^2} \right) \tag{1}$$

in which M_o is the molar weight of a repeat unit, N_A is Avogadro's number, kT has its usual meaning, ε is static dielectric constant, n is refractive index for visible light (546 nm for the increments) and the subscript 1 refers to values for pure solvent. For the styrene/vinylpyridine copolymer the quantity M_o is taken as the weight of copolymer per mole of vinylpyridine units. Differences in vibrational ("atomic") polarization between solute and solvent have been omitted from Equation 1, but the effect is not large, at least for the solutions in benzene.

Results of the low-frequency measurements are given in Table II. Values of the repeat-unit moments m were selected from those found in McClellan's tables (10) for the following model compounds: p-chloro(ethylbenzene) 2.00D, p-fluorotoluene 1.82D, p-methoxytoluene 1.21D, and 4-ethylpyridine 2.65D. Accurate values of dipole moment for temperatures other than 25°C are not available, since we measured dn/dc only at that temperature; consequently, the temperature coefficients given in the last column of Table II are merely rough values.

A number of previous measurements on poly(p-chlorostyrene) (11-13) have given dipole-moment ratios $<\mu^2>/xm^2$ ranging from 0.53 to 0.8. The average of all the recorded ratios is 0.67 ± 0.10, in agreement with our value; but the meaning of such agreement is hard to gauge. For example, Burshtein and Stepanova (13) give a ratio of 0.71 in p-xylene at 25°C, but their figure comes from much lower values of $<\mu^2>/x$ and of m than ours. The same authors find a temperature coefficient $dln<\mu^2>/dT$ in toluene of about -4×10^{-3} deg^{-1}, in fair agreement with our result.

Very recently Mark (14) has calculated some conformational properties of poly(p-chlorostyrene) on the basis of the rotational-isomeric-state model, using the same conformational energies as for unsubstituted polystyrene chains (15). For slightly syndiotactic (i.e., free-radical initiated) polymers, he finds dipole-moment ratios lying between 0.65 and 0.8, and his temperature coefficients fall in the range -2 to -5×10^{-3}. Agreement with our data seems satisfactory. However, we must then register surprise at our lower experimental figures for poly(p-fluorostyrene) prepared under similar conditions. At present we have no explanation for the difference.

As to the results for the other polymers, the low dipole moment of the p-fluoro-α-methylstyrene polymer is doubtless connected to its highly stereoregular structure (5) while the high dipole-moment ratio for poly(p-methoxystyrene) may reflect the great flexibility of the methoxyl group uninfluenced by its neighbors.

Dispersion

Measurements in the 3-50 MHz range were made at concentrations of about 1 and 2 percent and over a temperature range from about -10 to near 30°C. The dielectric loss factor ε" was analyzed by graphical application of the empirical Fuoss-Kirkwood (16) relation:

$$\varepsilon'' = \varepsilon''_{max} sech[\alpha ln(f/f_m)] \qquad (2)$$

TABLE II

Dipole Moments at 25°C

Polymer and Solvent	$d\varepsilon/dc^a$	dn/dc^a	$<\mu^2>/x^b$	$m^{b,c}$	$<\mu^2>/xm^2$	-10^3 $d\ln<\mu^2>/dT$
p-Chlorostyrene (benzene)	2.60	0.103	2.55	2.00	0.67	~6
p-Chlorostyrene (CCl$_4$)	2.20	0.140	2.00	—	—	~4
p-Fluorostyrene (benzene)	1.87	0.050	1.68	1.82	0.51	~2
p-Methoxystyrene (benzene)	1.68	0.078	1.55	1.21	1.06?	~2
p-Fluoro-methylstyrene (benzene)	1.04	0.067	0.91	1.82	0.27	~2
" (CCl$_4$)	1.06	0.119	0.77	—	—	~1
Styrene/4-vinylpyridine (benzene)	2.30	0.084	5.12	2.65	0.73	~0

a Units: ml/g
b Dipole moments in Debyes
c See text

in which f is the frequency of observation, f_m the frequency at which ε'' passes through its maximum value, and α a parameter whose departure from unity measures the breadth of the loss peak. According to Equation 2, a plot of $\cosh^{-1}(\varepsilon''_{max}/\varepsilon'')$ against the logarithm of f should yield a straight line from which f_m and α can be determined. In practice, the loss tangent, $\tan\delta \equiv \varepsilon''/\varepsilon'$ can be used in place of ε'' in the above process, since the change in ε' through the loss region is small for such dilute solutions. Also, choice of a precise value of ε''_{max} or $\tan\delta_{max}$ is not crucial to the value of f_m obtained. An example of the fit obtained is shown in Figure 1, which displays the loss factors for a 2.11% solution of poly(p-fluorostyrene) at three temperatures. The drawn curves obey Equation 2 with constants obtained as above described.

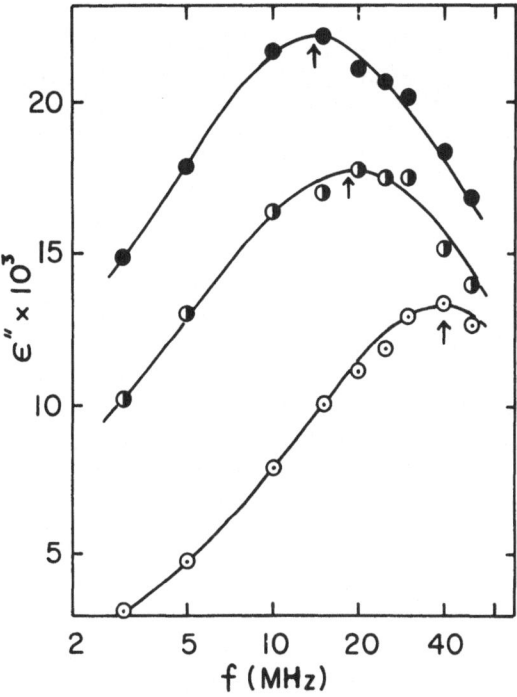

Figure 1. Dielectric loss factors ε'' at several temperatures of a 2.11 weight percent solution of poly(p-fluorostyrene), Sample A, in toluene. Open circles, -1.5°C (ordinate displaced upward by 3 units); filled circle, -9.5°C (ordinate displaced upward by 6 units). The drawn curves are fitted to the points according to Equation 2. Frequencies of maximum loss factor on these curves are shown by arrows.

Values of ε''_{max} calculated from the low-frequency dielectric increments by means of the relation $\varepsilon''_{max} = \alpha\Delta\varepsilon/2$ in general are within a few percent of the directly observed values. The breadth parameter α does not vary widely among the various solutions, there being no noticeable trends with concentration, temperature, or polymer sample. The average figure for all the solutions is $\alpha = 0.78 \pm 0.04$.

Values of $\log f_m$ for polymers of p-chlorostyrene, p-fluorostyrene and styrene/4-vinylpyridine in toluene or benzene are shown in Table III at several temperatures. The uncertainty in f_m is estimated at about 15 percent. It is seen at once these three polymers have very similar dielectric relaxation rates in toluene at a given temperature and also that they have similar temperature coefficients. In detail, it is seen that the p-fluoro polymer is slightly faster than the other two. Least-squares fits yield apparent activation energies for f_m of 4.5 ($\pm$ 0.5) kcal/mole for poly(p-fluorostyrene) and about 4.8 kcal/mole for the other two polymers.

The results for poly(p-chlorostyrene) are in good agreement with earlier dilute-solution work (2,3,4). The data of Mikhailov, Lobanov and Platonov (3) in toluene extend over a much wider range of temperature than ours, and so their given activation energy of 5.6 $\pm$ 0.2 kcal/mole for the interval 230–330°K is to be preferred. Combination of all the results gives

$$\log_{10}(2\pi f_m) = \log_{10}(\tau^{-1}) = 12.4 - (1220/T) \tag{3}$$

for poly(p-chlorostyrene) in toluene over the aforementioned range of temperature. The Russian workers (3) also studied styrene/p-chlorostyrene copolymers of several compositions and found a small but definite decrease in activation energy (from 5.6 to 5.2 kcal/mole) as the polar component of the copolymer was decreased from 100% to 10 mole percent. It may be concluded that to a first approximation the elementary relaxation process in a given solvent is essentially the same in all these styrene chains. Presumably the energy barriers are overwhelmingly steric in nature and not appreciably modified by p-substitution.

Earlier preliminary results for poly(p-chlorostyrenes) in the very viscous solvent o-terphenyl (1,17) indicated that there was considerable solvent participation in the relaxation process. The present bridge was inconvenient for highly viscous systems, but we did measure poly(p-fluorostyrene), Sample A, in two solvent mixtures at 25°C: for toluene/p-cymene (1/1) we find $\log f_m = 7.38$ and for

TABLE III

Maximum Dielectric Loss Frequencies

Polymer	Solvent	Temp. (°C)	$\log_{10} f_m$ (Hz)
pClSty	Toluene	-6.0	7.05 ($\pm$0.05)
		-0.5	7.20
		14.5	7.32
		25.0	7.50
pFSty[a]	Toluene	-9.5	7.11 A
		-6.0	7.24 B
		-1.5	7.28 A
		-0.5	7.26 B
		14.5	7.50 B
		27.4	7.58 A
Sty/4VP	Toluene	-8.6	7.00
		10.5	7.40
		27.0	7.55
Sty/4VP	Benzene	6.5	7.35
		10.9	7.42
		26.0	7.59

[a]Sample designations A and B as in Table I

toluene/<u>trans</u>-decalin (2/1) log f_m = 7.36. These relaxation times
are about 50% longer than in pure toluene, very nearly in the same
ratio as the solvent viscosities. This is scarcely an impressive
range of viscosity, but the effect is well outside experimental
error. It is tempting to appeal to the well-known Kramers theory
(18-20) of passage of a Brownian system over a potential barrier,
and indeed adaptations to the polymer problem have recently been
made by Helfand (21) and by Paul and Mazo (22). The relaxation
time predicted by this model is proportional to the friction coefficient of the diffusing entity, and thus (in a Stokes-law setting)
to the viscosity of the medium.

The activation energies found in the present work as well as
for other polymeric local dielectric relaxation processes in dilute
solution seem surprisingly low [for example, about 2.5 kcal/mole
for polyoxyethylene in benzene (2,23) or water (23) and about 3.6
kcal/mole for polyvinylchloride in tetrahydrofuran (24)]; but similar magnitudes are also given by other techniques, such as acousti-
cal (25) or nuclear magnetic (26,27) relaxation. If the Kramers
theory is to be taken literally, the energy barriers for local conformational changes must be lower than the experimental activation
energies by a contribution of perhaps 2 kcal/mole from the friction
coefficient. The remainder at first sight looks barely (if at all)
big enough to represent the barrier to internal rotation about a
single skeletal bond. At this point we encounter a conceptual difficulty. If the rate-determining step involved rotation around
only one skeletal bond, then the two attached chains would have to
move appreciably relative to each other through the solvent, and
this would give rise to a frictional resistance depending on chain
length (22), in contradiction to the experimental fact (1) that the
observed relaxation times are independent of molecular weight. If
the barriers were much higher, we could wriggle out of the difficulty by appealing to transition-state theory: the rate of passage
over the barrier in a given bond would not be diffusion-controlled
and therefore would be indifferent to the mobility of the attached
chain portions (28). Since the observed small barrier heights
appear to exclude the transition-state limit (29) (as do the viscosity effects mentioned earlier) we are driven back to search for
low barriers in the configuration space of two simultaneously rotating skeletal bonds (1,21,25). Thus far, there appear to be very
few conformational energy calculations on which to base any specu-
lations of this kind. An encouraging example is found in the work
of Gorin and Monnerie (30), who calculate a barrier of only about
3 kcal/mole between the two stable stereo-isomers g^-t and tg^+ of
<u>meso</u>-2,4-diphenylpentane. Further semi-empirical calculations for
non-equilibrium chain conformations are much to be desired.

Helfand's calculation (21) strongly suggests that processes of
the type just envisaged need not be much slower in long chains than
in oligomers. The data for poly(p-chlorostyrene) are consistent

with this view, in that the relaxation rate given in Equation 3
can be fitted to the simple Kramers expression with reasonable
values of the several inaccessible parameters needed.

Other methods of studying local chain relaxation processes
have given results in fair accord with the dielectric data here
reported. Bauer, Hässler and Immendörfer (25) have measured ultra-
sonic relaxation in moderately dilute polystyrene solutions in
benzene and carbon tetrachloride. They interpret their results in
terms of a three-state model and find an activation energy of 6.6
kcal/mole and a frequency factor of 1.4×10^{12} sec^{-1} for the rate
of transition from the two-fold upper level to the lower level of
their model, plus an equilibrium energy difference of 0.9 kcal/
mole. In the absence of definite geometrical identification of
these states, an estimate of the dielectric relaxation rate by means
of the well-developed formalisms of the site model (31,32) cannot
be completed, but in any case it is seen that the acoustic data
give a similar, though somewhat slower, rate than the dielectric
process.

Some years ago Odajima (33) and McCall and Bovey (34) observed
NMR relaxation times in polystyrene solutions. The latter authors
give a figure of 0.20 sec. at 30 MHz for the spin-lattice relaxation time T_1 of the meta-para protons at 25°C in tetrachloroethy-
lene, plus a temperature coefficient corresponding to an activation
energy of only 3 kcal/mole. The latter figure seems too low for
reconciliation with the dielectric and acoustic results; but the
value of T_1 is of an order of magnitude consistent with the dielectric data. This statement is based on a naive calculation (2)
from the Bloembergen-Purcell-Pound expression (35) for isotropic
rotational motion of the axes between interacting magnetic nuclei.
Real polymer systems present much more complicated problems (26,36)
and a value of T_1 at a single frequency is not sufficient for a
complete analysis.

Very recently Bullock, Butterworth and Cameron (27) have studied
ESR relaxation for spin-labelled polystyrene in toluene. From the
line widths they deduce orientational correlation times in very good
agreement with the NMR result of McCall and Bovey (34). Further,
they observe an activation energy of 4.3 kcal/mole, in pleasant
accord with our figure obtained from dielectric relaxation.

Finally, we describe the less definite results we obtained with
the two other polymers we studied. Poly(p-fluoro-α-methyl-styrene)
in both toluene and cymene gave a loss peak which narrowed percep-
tibly with increasing temperature but with a relatively low temperature coefficient for f_m, which at room temperature is in the neigh-
borhood of 30 MHz in toluene. The points were more widely scattered
than for the polymers of Table III, and we prefer not to report them
in detail.

Poly(p-methoxystyrene), with its labile methoxy group, would be predicted to show two distinct dispersion regions, one at about the same frequency as for the other p-substituted polystyrenes and the other, corresponding to rotation of OCH_3 about the ring-to-oxygen axis, at a much higher frequency (37) (above 10^{10} Hz). Our observations are consistent with this prediction, but are insufficient to verify it in detail: there is a loss peak, accounting for less than half of the total dipole polarization, at or near the upper frequency limit (50 MHz) of our apparatus. North and Phillips (4) studied a cation-initiated polymer of much lower chain length than that of our sample. In contrast to our observation, they do not find any appreciable relaxation in the 50 MHz region; but they do observe the shoulder (up to 3 x 10^9 Hz) of the predicted strong high-frequency peak.

ACKNOWLEDGMENT

We thank W. B. Westphal for loan of apparatus, and for his general interest and encouragement.

REFERENCES

1. See, for example, W. H. Stockmayer, Pure and Applied Chem. 15, 539 (1967).
2. W. H. Stockmayer, H. Yu and J. E. Davis, Polymer Preprints (ACS), 4 (2), 132 (1963).
3. G. P. Mikhailov, A. M. Lobanov and M. P. Platonov, Polymer Science U.S.S.R. 9, 2565 (1967).
4. A. M. North and P. J. Phillips, Trans. Faraday Soc. 64, 3235 (1968).
5. Y. Osumi, T. Higashimura and S. Okamure, J. Polymer Sci. Part A-1, 4, 923 (1966).
6. R. M. Fuoss and G. I. Cathers, J. Polymer Sci. 4, 97 (1949).
7. POLYMER HANDBOOK, ed. J. Brandrup and E. H. Immergut, Interscience, New York, 1966.
8. E. A. Guggenheim and J. E. Prue, PHYSICO-CHEMICAL CALCULATIONS, Interscience, New York, 1955, p. 106.
9. J. W. Smith, ELECTRIC DIPOLE MOMENTS, Butterworth, London, 1955, p. 60.
10. A. L. McClellan, TABLES OF EXPERIMENTAL DIPOLE MOMENTS, W. H. Freeman and Company, San Francisco, 1963.
11. P. Debye and F. Bueche, J. Chem. Phys. 19, 589 (1951).
12. H. A. Pohl and H. H. Zabusky, J. Phys. Chem. 66, 1390 (1962).
13. L. L. Burshtein and T. P. Stepanova, Polymer Science U.S.S.R. 11, 2885 (1969).
14. J. E. Mark, private communication (August 1971).
15. A. D. Williams and P. J. Flory, J. Am. Chem. Soc. 91, 3111 (1969).

16. R. M. Fuoss and J. G. Kirkwood, J. Am. Chem. Soc. 63, 385 (1941).
17. J. E. Davis, Ph.D. Thesis, M.I.T., 1960.
18. H. A. Kramers, Physica 7, 284 (1940).
19. S. Chandrasekhar, Revs. Mod. Phys. 15, 1 (1943).
20. H. C. Brinkman, Physica 22, 29 (1956).
21. E. Helfand, J. Chem. Phys. 54, 4651 (1971).
22. E. Paul and R. M. Mazo, Macromolecules 4, 424 (1971).
23. M. Davies, G. Williams and G. D. Loveluck, Zeits. f. Elektrochem. 64, 575 (1960).
24. L. deBrouckere and M. Mandel, Adv. Chem. Phys. 1, 77 (1958).
25. H.-J. Bauer, H. Hässler and M. Immendörfer, Disc. Faraday Soc. 49, 238 (1970).
26. J. E. Anderson, K.-J. Liu and R. Ullman, Disc. Faraday Soc. 49, 257 (1970).
27. A. T. Bullock, J. H. Butterworth and G. G. Cameron, European Polymer J. 7, 445 (1971).
28. H. Morawetz, Acc. Chem. Res. 3, 354 (1970).
29. It is interesting that the Kramers theory (16) leads to the transition-state result for sufficiently high curvature (rather than height) near the top of the barrier.
30. S. Gorin and L. Monnerie, J. Chim. Phys. Physicochim. Biol. 67, 869 (1970).
31. J. D. Hoffman and H. G. Pfeiffer, J. Chem. Phys. 22, 132 (1954).
32. G. Williams and M. Cook, Trans. Faraday Soc. 67, 990 (1971).
33. A. Odajima, J. Phys. Soc. Japan 14, 777 (1959).
34. D. W. McCall and F. A. Bovey, J. Polymer Sci. 45, 530 (1960).
35. N. Bloembergen, E. M. Purcell and R. V. Pound, Phys. Rev. 73, 679 (1948).
36. D. E. Woessner, B. S. Snowden, Jr. and G. H. Meyer, J. Chem. Phys. 50, 719 (1969).
37. K. Bergmann, D. M. Roberti and C. P. Smyth, J. Phys. Chem. 64, 665 (1960).

THE INFLUENCE OF IMPURITIES ON THE DIELECTRIC LOSSES

OF POLY (2,6-DIMETHYL-1,4-PHENYLENE ETHER)

C. W. Reed

General Electric Company, Research and Development

Center, Schenectady, New York

SYNOPSIS

Impurities can modify the dielectric loss of non-polar polymers in three ways - (1) by giving rise to ionic losses, (2) by modifying the intrinsic low loss characteristics of the polymers, and (3) by modifying the losses that result upon oxidation. We have attempted to shed some light on these areas by studying the polymer poly (2,6-dimethyl-1,4-phenylene ether) at frequencies from 0.05 Hz to 10 kHz at temperatures from room temperature to 230°C. Experimental samples of the polymer prepared differently have been examined and others have been "cleaned-up" by solvent extraction, acid washes, passage through activated alumina, etc. The low frequency losses, which increase rapidly at temperatures above 100°C, are most affected. We are able to attribute these directly to ionic conduction, and remnant catalysts have been identified as the primary source. At the glass transition temperature of the polymer, an activation energy for conduction exhibits a sharp change, which is considered due to a change in the energetics of the diffusion of ions in the glass and rubber phases. An order-of-magnitude calculation suggests as a model a very weakly dissociating system in the low dielectric constant polymer matrix. Though all are relatively low in magnitude, the intrinsic low-loss characteristics of these experimental samples vary considerably. We have been able to identify the role of impurities in such variations, but the origin has not yet been precisely established. Oxidation studies have provided evidence for more than one dissociating species in the polymer. The procedures for separating the different sources of loss and of identifying them are explained.

INTRODUCTION

With the growing interest in low-loss polymers, a knowledge of
the mechanisms responsible for the residual dielectric losses is
of increasing importance. In this connection, the role played by
impurities, in some instances giving rise to ionic losses (1,2,3),
in others modifying the intrinsic low loss characteristics, and in
others modifying the losses that result from oxidation, is poorly
understood. We have attempted to shed some light on these areas
in an experimental study of the polymer (2,6-dimethyl-1,4-phenylene
ether) (PPO* resin as manufactured by General Electric Company), at
frequencies from 0.05 Hz to 10 kHz at temperatures from room temperature to 230°C. Preliminary findings have been described previously
(4).

The strategy in this work has been to study polymer samples
modified by different methods of preparation and also by further
"cleaning-up," and then to relate dielectric loss+ behavior to
differences in the sample history. We have tried to characterize
the polymer as fully as possible and to minimize changes of the
polymer itself (loss of low or high molecular weight components),
although this has not always been possible. Intrinsic viscosity
was used as a measure of the extent of changes in polymer structure,
although gel permeation chromatography, with its ability to determine molecular weight distributions, would be the preferred method.
Our samples were compression molded to give a completely amorphous
polymer. The results are thus free from any influence of crystallinity.

A principal area of interest in this work has been in the
properties of ionic impurities in low-loss polymers. Basic information on the processes of ionic conduction in polymers will be
valuable in a description of some of the more applied problems, in
which there has hitherto been very little progress in understanding.
Such problems include charge storage (as in thermoplastic recording),
polarization, electrical breakdown, and the behavior of dielectric
loss at high electric fields. However, our knowledge of the processes of ionic conduction in low-loss polymers is fragmentary, at
best, and the field has received negligible attention compared to

---

*Registered trademark of the General Electric Company.

+Loss tangent or tan $\delta = \varepsilon''/\varepsilon'$ where ε'' is the dielectric loss
factor or simply the dielectric loss, and ε' is the dielectric constant. Dielectric losses, a material property, will be described
hitherto simply as losses, or in terms of the components of such
losses due to ionic conduction, secondary relaxation, α relaxation,
etc. Loss tangent will be used to describe bridge measurements of
such losses.

that of dielectric relaxation properties. This deficiency is essentially true for all thermoplastic polymers, both nonpolar, which typically have low dielectric loss and are exemplified by the present polymer, and polar.

The present work will show that low frequency a.c. measurements show a linear behavior which can be directly related to ionic conduction by use of a simple model, and afford information which has not been possible by d.c. measurement. With d.c. measurements electrode effects complicate the results and may even be dominant. It is often impossible to distinguish between slow non-ionic polarization, which may involve conformational changes of sections of polymer molecules themselves, and polarization (increase of the apparent dielectric constant) due to space-charge buildup. Also, once space-charge buildup occurs in a polymer, the internal fields are no longer those calculated from the external circuit, with consequent errors in measured quantities. To this author's knowledge, no d.c. conduction experiments have been performed on polymers where the internal field distribution was probed in a manner similar to that used for d.c. measurements on insulating liquids (5).

An essential requirement for a description of ionic conduction in polymers is to obtain details of the dissociation process. With thermoplastic polymers of low dielectric constant, both above and below the glass temperature, there have been no studies comparable to the classical work of Kraus and Fuoss in the 1930's on the conduction of weak electrolytes in solvents with low dielectric constant (6). It is also true that other independent measurements on polymers, which would help in a description of conduction, are lacking. Thus, measurements of charge carrier mobilities in thermoplastic polymers are almost nonexistent. As we shall see in an order-of-magnitude calculation, reliable values of charge carrier mobility are essential for an accurate description of the dissociation process. In the future, attempts will be made to describe the ion-pair contribution to conduction in the polymer. Hopefully at that point such data can be combined with electron injection data to permit an accurate calculation of the local electric field in such polymers, and this may be of major importance in an attempt to identify the mechanism of electrical breakdown in polymers.

Other areas examined cover the role of impurities in (1) producing weak, secondary dipolar relaxations and (2) modifying the losses that result from oxidation of the polymer. An important reason for looking at these areas is to identify the possible spurious influence of small amounts of polar impurity on the intrinsic secondary relaxations of the polymer.

SEPARATION OF DIELECTRIC LOSSES INTO CONTRIBUTIONS
FROM IONIC CONDUCTION, DIPOLAR RELAXATION, AND OTHER SOURCES

A primary task in this investigation has been to resolve the measured losses into contributions from different mechanisms and to identify these mechanisms. To do this the polymer sample can be treated as electrically equivalent to a capacitance (C_x) in parallel with a resistance (R_x), when the loss tangent and dielectric constant (ε') are given by

$$\tan \delta = \frac{1}{\omega R_x C_x} \tag{1}$$

and

$$\varepsilon' = \frac{C_x}{C_o} \tag{2}$$

where ω (= $2\pi f$) is the angular frequency and C_o is the capacitance if the space between the electrodes is filled with vacuum.

In general R_x is composed of two terms, R_i, a term due to ionic conduction, and R_d, a term due to non-ionic conduction. Since R_i and R_d are in parallel, $(R_x)^{-1} = (R_i)^{-1} + (R_d)^{-1}$. By definition, R_i will be a constant, independent of frequency at a given temperature. On the other hand, in general R_d will be a function of frequency, since it is primarily dipolar in nature. Hence, loss tangent is given by

$$\tan \delta = \frac{1}{\omega R_i C_x} + \frac{1}{\omega R_d C_x} \tag{3}$$

To obtain the contribution from ionic conduction, we must identify a region where $\tan \delta$ varies inversely with frequency. When this obtains, the frequency-dependent resistance R_d will be completely absent. Thus, on a log-log plot of $\tan \delta$ against frequency, a slope of -1 will be a test plot for ionic conduction. If the data fits the test plot then, knowing C_x, we can obtain R_i:

$$R_i = \frac{1}{2\pi f \, C_x \, \tan \delta} \tag{4}$$

Hence, knowing the sample dimensions, the resistivity can be obtained. This should be described as an a.c. resistivity, due entirely to ionic conduction, and not as a d.c. resistivity. The two are not necessarily the same, since the d.c. value varies with the time of voltage application, as a result of electrode polarization. As will be seen later, within the range of conditions covered by the present experiments, electrode polarization does not influence the a.c. resistivity.

The non-conduction losses have been loosely described as dipolar. Three sources can be identified. Above the glass temperature (T_g), non-conduction losses are due to the α relaxation which involves large conformational changes of the polymer. Below T_g, they are due to both weak dipolar relaxations arising from small amounts of polar impurities and the very uniform, low magnitude losses which span a very wide frequency range and are noted predominantly in non-polar polymers. For most polymers the first two sources are well-characterized but not so the third. Two theories have been proposed for the origin of these very uniform losses in other polymers, but neither is well confirmed. The first theory (7) considers as their origin the orientation of small dipole moments characteristic of the structure of non-polar polymers. For the present polymer this model would need to include contributions from dipole moments due to oxygen atoms in the polymer backbone, in addition to dipole moments due to hydrocarbon components. The second theory (8,9), which was applied to a variety of polymers including polyethylene and some oxide polymers, invokes a photon-phonon interaction that involves a coupling of the electromagnetic field to low frequency acoustic modes by means of anharmonic polar units. Since, as indicated below, it was not possible to distinguish between these two theories as the source of such losses in the present polymer, they will be referred to as background losses.

To obtain the losses due to ionic conduction, the contributions due to the α relaxation, above T_g, and due to both weak dipolar relaxation and background losses, below T_g, must be subtracted from the total measured losses. The following procedures were used:

1. Background Losses. At each measured frequency (0.1 Hz to 10 kHz) the total losses were plotted as a function of temperature (in the absence of weak dipolar losses, the total losses at each frequency increased continuously). Then, for the low frequencies (0.05 - 5 Hz), the lower temperature data, up to a temperature at which the losses showed a marked increase attributed (tentatively at first) to the onset of ionic conduction, were extrapolated to higher temperatures. Finally, the extrapolated higher temperature values were subtracted from the total losses, and the resulting losses were examined for their fit to the ionic conduction test plot. In an attempt to identify the nature of the background losses,

two types of temperature plots were tried: a plot of log tan δ
against 1/T (°K), where a linear behavior would indicate a tempera-
ture-independent distribution of relaxation times, and a plot of
log tan δ against log T, where linearity would indicate photon-
phonon interactions (7,8). For the temperature range covered,
however, both plots showed equally good linearity. Hence it was
impossible to tell which was more valid, or, therefore, to distinguish between these two possible sources of background loss.
Arbitrarily, 1/T plots were used for the present study.

     2. Weak Dipolar Losses. An example of the magnitude of weak
dipolar relaxations is shown later in Figure 7. The procedure for
obtaining the contribution of such losses was a special case of
the procedure for obtaining background losses. The data at each
measured frequency was plotted against 1/T, but with weak dipolar
relaxation present, the total losses at each frequency did not
increase continuously. By identifying the weak dipolar relaxations,
it was possible to include their contributions in the extrapolation of the low frequency data. The sum of background and weak
dipolar losses was then subtracted from the total losses.

     3. α Relaxation Losses. The same 1/T plots were also used
for obtaining the α relaxation losses. The low-frequency behavior
of such losses was inferred by comparison with the higher frequency
relaxation behavior, where ionic conduction is completely absent.
The procedure was attempted initially in order to correct the considerable deviation of the measured losses from the conduction test
plot above T_g. In the presence of such large deviations, it becomes difficult to say unambiguously that the data is uninfluenced
by interfacial effects. While the technique of inferring data is
not normally recommended, if it had resolved the conduction part of
the losses fairly clearly, it could be considered a useful device.
It was considered that the procedure might possibly work since, on
the basis of dielectric constant data described later, the α
relaxation contribution to these measurements is restricted to the
high frequency side of its peak. Thus the procedure attempts to
infer or deduce the α relaxation behavior near its peak from the
behavior closer to its tail. However, as will be described in the
results, considerable errors were present in the inferred low-
frequency α relaxation behavior, and the procedure was not successful.

     Fortunately, however, the need to correct for the α relaxation
contribution decreased with decreasing frequency, and it was thus
possible to resolve ionic conduction losses from the measured data
without subtracting the α relaxation losses.

     An example of these procedures for determining α relaxation
and background losses is described under the results.

EXPERIMENTAL

Poly (2,6-dimethyl-1,4-phenylene ether) was made by the oxidative coupling of 2,6 xylenol with a copper salt and amine as the catalyst system (Figure 1). The molecular weights ranged from 20,000 to 25,000. All samples were reprecipitated from toluene solution using methanol, and were then filtered and dried in a vacuum desiccator at 80°C for 48 hours. The purpose of this procedure was to eliminate dirt and to give a powder suitable for compression molding sheets 15-20 mil thick. The powder was compression molded under nitrogen at 280°C, at 5000 psi pressure for 1/2 minute, after preheating for 1 minute with no pressure.

All the samples of compression molded polymer were completely amorphous, determined by x-ray and calorimetric methods. Evaporated circular gold electrodes (4 cm diameter) were used. The dielectric cell for housing the specimens was made from stainless steel and was suitably guarded. Temperatures were measured with a thermocouple located in the dielectric cell.

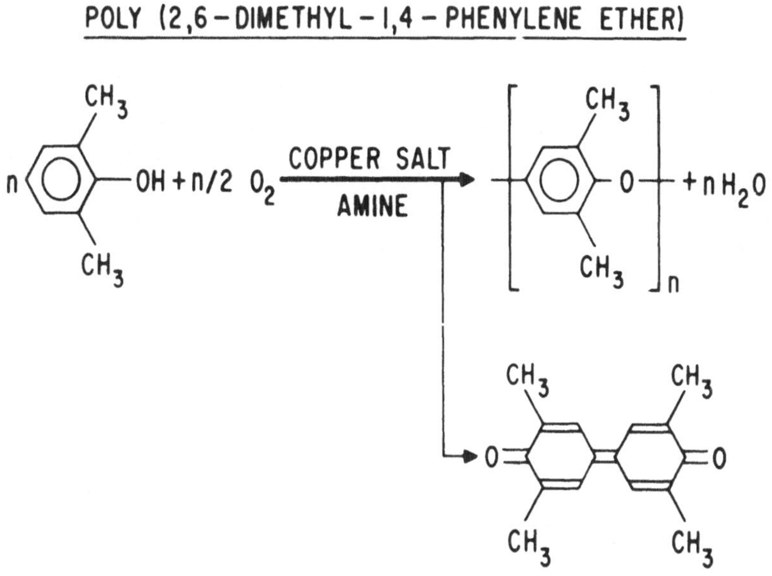

POLY (2,6-DIMETHYL-1,4-PHENYLENE ETHER)

Figure 1. Stoichiometry of polymer synthesis. (Diphenoquinones are a by-product of the reaction.)

Three bridges were used: (1) a low frequency bridge (0.05 Hz to 30 Hz) that uses d.c. operational amplifiers (10); (2) a Schering bridge (60 Hz); (3) a Model 1615 A General Radio bridge (120 Hz to 10 kHz). Each of the bridges was modified for measuring dielectric losses as low as tan δ = .00005, with an approximate accuracy of $\pm$ .00001. Below tan δ = .00005, the accuracy decreased rapidly.

RESULTS AND DISCUSSION

Losses Due to Ionic Conduction

Figure 2 shows a typical set of data obtained on a sample of the polymer. Loss tangent (tan δ) was measured as a function of frequency (.05 Hz to 10 kHz) at temperatures from room temperature to 230°C. Commencing at low frequency, as the temperature was raised, tan δ shows the characteristic increase with decreasing frequency often attributed to the onset of ionic conduction. Further increase of temperature causes an increase of the losses as shown. Let us consider the evidence for the role of ionic conduction, by first subtracting out the background and α relaxation losses, and then applying the test plot for ionic conduction described earlier.

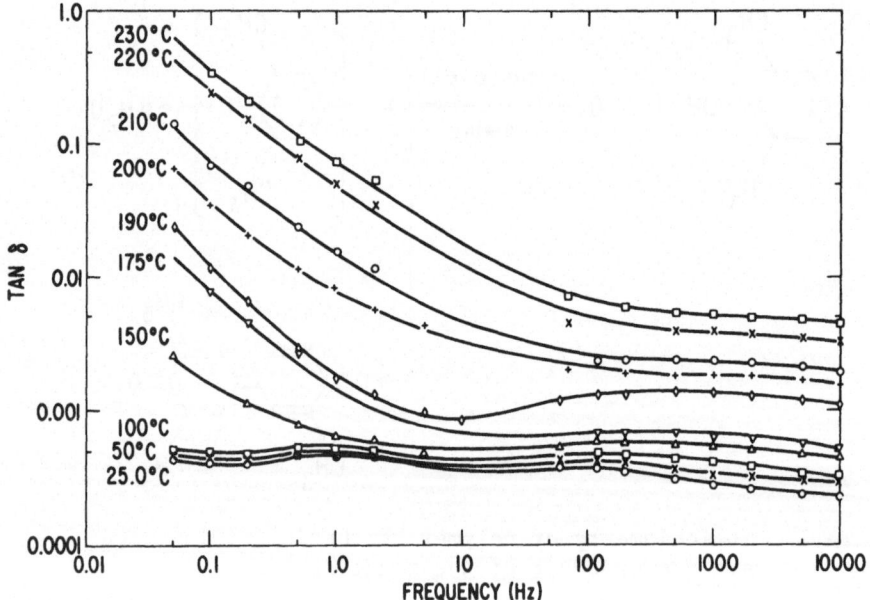

Figure 2. Loss tangent of the polymer as a function of frequency.

The procedure for obtaining the contributions from background and α relaxation losses is illustrated in Figure 3. For clarity, the procedure is shown for one frequency (0.1 Hz) only. The losses shown in Figure 3 are the measured losses taken from Figure 2. The data at 1 and 10 kHz, in a region where ionic conduction has no influence, shows two distinct regions: that at the lower tempera-ture is attributed to background losses, and that at the high tem-perature, we suggest, is related to the α relaxation. The plots at 1 and 10 kHz were used, as described earlier, to extrapolate the background losses and to infer the α relaxation losses at 0.1 Hz.

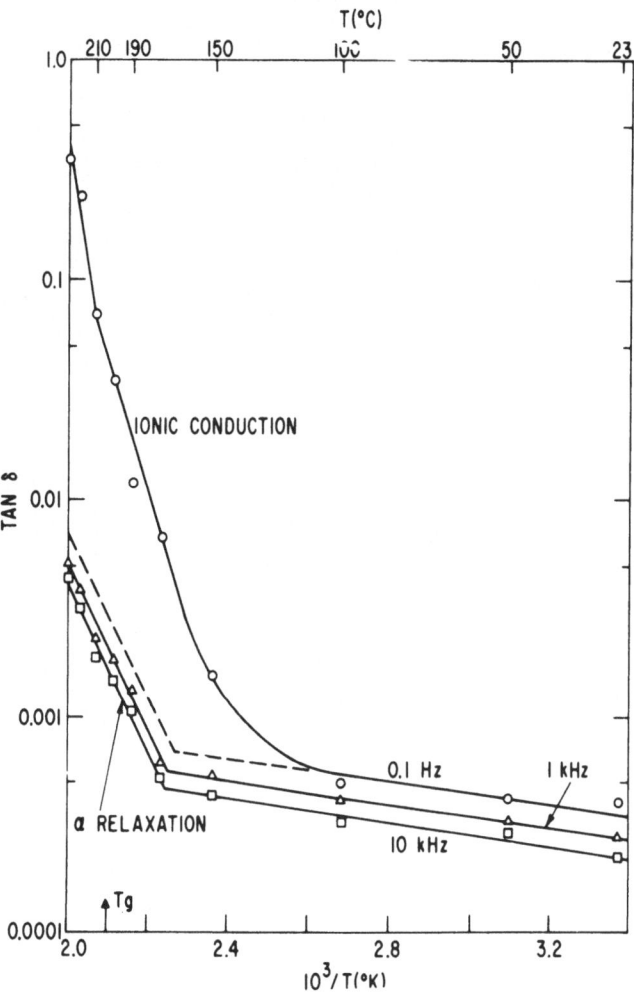

Figure 3.   Typical data for obtaining the contributions due to background losses and α relaxation losses at low fre-quencies where conduction losses are present. Hatched lines show the extrapolated background losses and inferred α relaxation losses at 0.1 Hz.

In Figure 4 the data of Figure 2 has been replotted with the
background and α relaxation losses subtracted from the measured
losses; the lines are drawn with slopes of -1, as required by
ionic conduction. Up to 210°C, the model holds well for all frequencies. This suggests that the extrapolated low temperature portion (100-175°C) of Figure 3 adequately describes the background
losses below T_g. By comparing Figures 2, 3, and 4, we note that
the corrections are most effective at 150°C. At 175 and 190°C,
the corrected higher frequency data in Figure 4 shows a good fit
to the test plot.

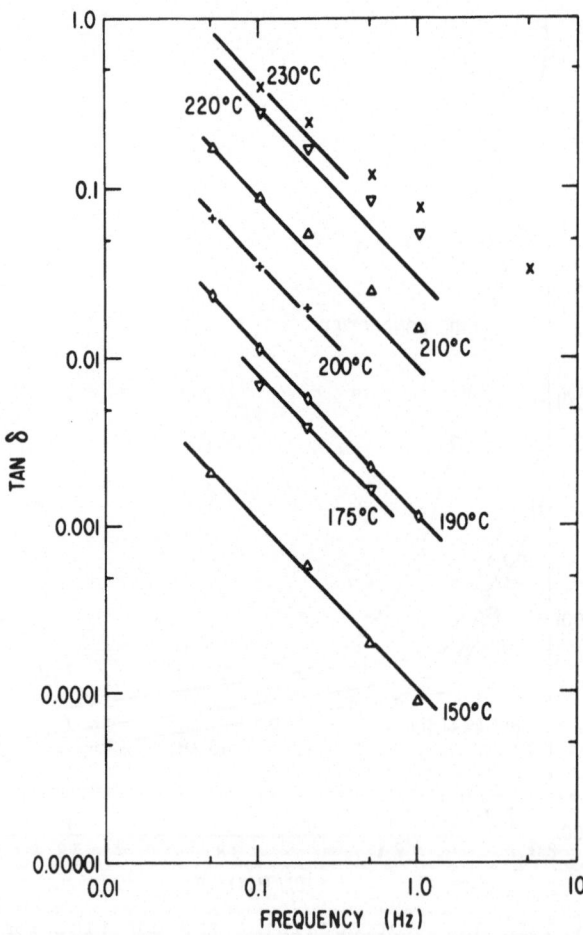

Figure 4.   Plot of tan δ against frequency for the polymer,
            corrected for the background and α relaxation losses.

Above 210°C, the corrected losses are greater than predicted at the high frequencies, but tend asymptotically toward the theoretical slope at low frequencies. On the basis of an examination of the dielectric constant behavior (see below), this discrepancy at higher frequencies is attributable to the inadequacy of the inferred α relaxation losses at the higher temperatures, rather than the presence of interfacial effects (often called Maxwell-Wagner effects) (10). However, the relative importance of the α relaxation losses decreases with decreasing frequency, and, as noted, at the lowest frequencies the data approach the correct theoretical slope. The fit is interpreted as positive evidence for ionic conduction up to 230°C. As expected, conduction increases with increasing temperature.

Dielectric Constant

The dielectric constant behavior corresponding to the loss data of Figure 2 permits additional analysis of the conduction process. Dielectric constant was calculated directly from the measured equivalent parallel capacitance and the sample dimensions. The data are shown in Figure 5; they are typical of the dielectric constant data obtained where ionic conduction was measured. At room temperature and at 100°C, ε' is essentially independent of frequency; at 150, 175, and 190°C the same was true (the data are omitted to emphasize the higher temperature data) with ε' decreasing with increasing temperature consistent with the decrease of the polymer density. The frequency-independent ε' confirms that we are observing a conduction process. However, at 210°C (i.e., slightly above the glass temperature, ~205°C) we note a steady, though small, increase of ε' with decreasing frequency, and at 230°C we see a slightly stronger increase of ε' with decreasing frequency.

Since there was no evidence of an inflection as usually found with a dipolar dispersion, in a preliminary interpretation the behavior at 210°C and 230°C was ascribed to a polarization due to ions. Moreover, the behavior was independent of electrode material (gold, silver paint, or stainless steel) and of the time of test at any one frequency, and consequently it was attributed to interfacial polarization, where ions are trapped at interfaces within the polymer rather than at the electrodes. However, further examination showed that the magnitude of the effect was independent of the magnitude of the applied voltage (up to a maximum applied voltage of 100 volts), which would appear to exclude any interfacial effect. Hence a more probable explanation is that we have the high frequency side of a dipolar relaxation covering an extremely wide frequency range, and the probable source is the α relaxation of the polymer. Due to the essentially non-polar nature of the polymer, the change in dielectric constant due to the α relaxation will be small. The magnitude of the changes in dielectric constant observed at 230°C

is consistent with such a weak relaxation. Somewhat surprising,
however, is the very broad nature of the α relaxation indicated by
the present results. Typically, amorphous polymers exhibit fairly
narrow α relaxations, whereas crystalline polymers show much broader
relaxations.

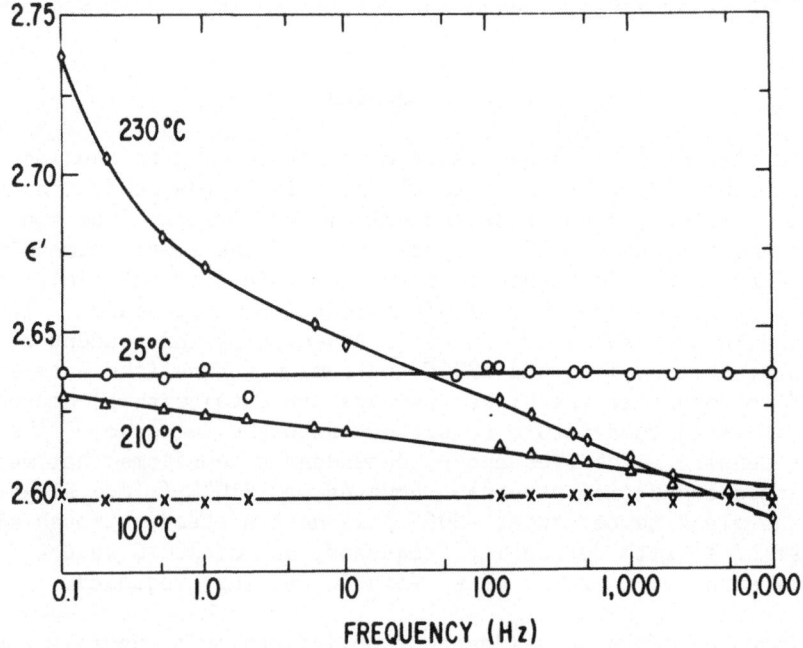

Figure 5. Plot of dielectric constant against frequency for the
polymer.

        In light of this interpretation of the dielectric constant
behavior above 200°C, interfacial effects can also be eliminated
as a potential source of the deviations for the ionic conduction
test plot at higher frequencies for temperatures above 200°C in
Figure 4. As suggested previously, the probable source of the
deviations is the inadequacy of the inferred α relaxation losses
above 200°C.

The frequency- and time-independent ε' below 200°C confirms that the ionic conduction is an equilibrium process and excludes the possibility of electrochemical decomposition. It is thus pictured that, in response to the alternating electric field, the ions are moving to and fro about some equilibrium position within the polymer lattice.

The stronger polarization occurring at the lower frequencies at 230°C was found to be time-dependent, and the values shown were measured after the bridge had been left running at 0.1 Hz with 100 volts applied to the sample for 30 minutes. The data therefore probably reflects some electrode polarization. In other measurements this problem was circumvented by making the low frequency measurements in as short a time as possible (a few minutes) and at lower applied voltages, whenever the temperatures were above T_g.

a.c. Resistivities

From the measured tan δ and capacitance of the sample, Equation 4 allows the calculation of an a.c. resistivity at each temperature. To minimize the deviations for the test plot occurring above 200°C, values of tan δ at 0.1 Hz were used. Capacitance values at 0.1 Hz were also used. The test plot assumes that C_x is constant. However, the variations in C_x were small and they do not cause much change in the calculated a.c. resistivities, which are largely determined by tan δ values. Hence there should be no serious objection to using the value of capacitance at 0.1 Hz. The values of a.c. resistivity are plotted in the usual Arrhenius manner in Figure 6 using the data from Figure 4. The activation energies obtained show two separate regions of conduction, one above 200°C and the other below 200°C.

A Differential Scanning Calorimeter measurement of T_g for the same sample of the polymer gave a value of 205-210°C. Presumably the change in activation energy occurring at 200°C reflects a modification in ionic conduction at T_g. Let us assume (1) that the activation energy is the sum of two terms, one for the dissociation of ion pairs and the other for the mobility of the ions, and (2) that the thermally activated dissociation is unchanged through T_g. Then the change in activation energy can be considered due to a change in the energetics for the ionic mobility, as the polymer changes from a glass to a rubber. Price and Dannhauser (12) consider that ions in polystyrene do not see the macroscopic viscosity of the polymer, rather a considerably lower microscopic viscosity. In this polymer, the motion of ions below T_g is more restricted in the glassy structure, where the free volume available is less, but the activation energy for such motion is less than in the rubber phase.

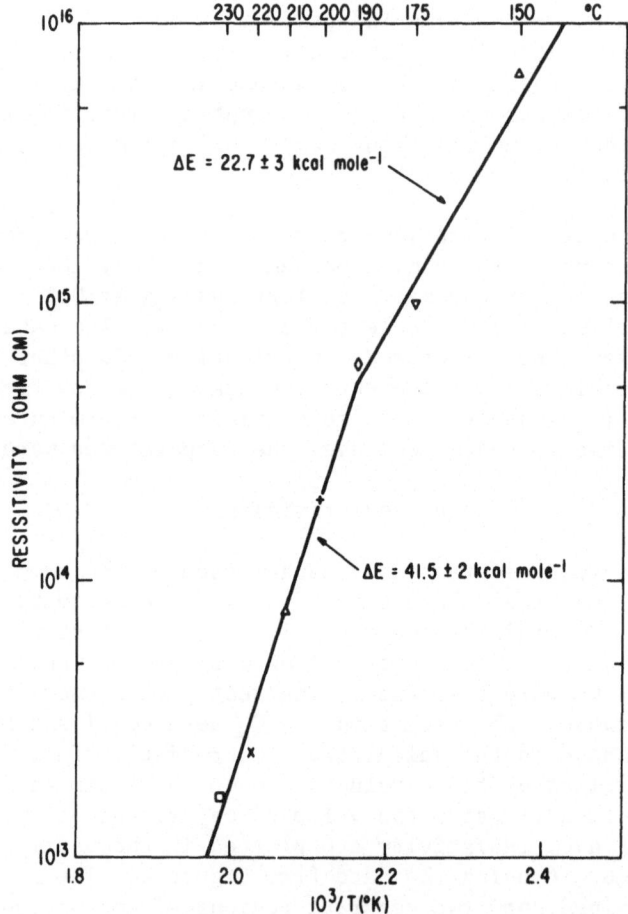

Figure 6. Resistivity as a function of reciprocal absolute
           temperature.

About a dozen samples of the polymer of various modification
showed a similar discontinuity in conduction near 200°C. The range
of activation energies measured was:

Below 200°C: ΔE = 17-28 kcal mole^{-1};
Above 200°C: ΔE = 41.5-45 kcal mole^{-1}.

These values of activation energy are consistent with the
activation energies found dielectrically for side group motion or
limited chain motion, below T_g, and for the motion of large chain
segments (α relaxation), above T_g. It is therefore considered
that, below T_g, the movement of ions responds to very localized
motions in the polymer, whereas, above T_g, such motion responds to
larger conformational changes. This view supports that of Price
and Dannhauser.

Source of Ionic Losses

The present work has attempted to remove impurities responsible for ionic losses, to identify the nature of the species responsible, and to obtain some understanding of the process of ionic
conduction in polymers. In this task it was regarded of primary
importance to be able to characterize the polymer so that any
changes which occurred could be followed. Attempts were made to
remove ionic species without otherwise influencing the polymer;
however, as will be indicated, this was not always possible. From
a practical standpoint, one would also like to be able to remove
ionic material in the simplest manner possible.

It has been possible to lower the losses due to ionic conduction by several chemical procedures and also by careful control of
the starting materials for synthesizing the polymer. The results
support the view that the low frequency losses are due to ionic
conduction, and they suggest that the losses are extrinsic in
nature, since they can be greatly reduced by the removal of residual catalyst. The predominant decrease in loss occurs when the
nitrogen content of the polymer is reduced. This suggests that the
ionic species may be amine salts formed from the amine used for
synthesis.

Table I summarizes the modifications used. The influence
upon conduction is show by the value of tan δ at 0.1 Hz and 150°C.
This is at a temperature below T_g, in the region where the losses
are directly related to conduction. Resistivity is proportional
to the reciprocal of the loss values, since all samples were of
approximately equal dielectric constant. The main features of our
results are noted below.

We were unable to remove copper to any measurable extent, which
indicates that copper is not involved in any of the variations in
tan δ shown. The purpose of refluxing with water and methanol was
to remove inorganics, or amines, or amine salts, all of which might
be physically bound within the polymer and difficult to remove
otherwise. Heating in a vacuum desiccator at 150°C also removed
amines but, as noted, resulted in an increase in intrinsic viscosity measured from solutions in chloroform at 25°C.

Intrinsic viscosity measurements also suggest that a column of
activated alumina (treatment 2 of Table I) removes lower molecular
weight molecules in addition to nitrogen-containing molecules.

TABLE I

Modifications of Polymer and Resulting Influence Upon Ionic Conduction

	Treatment	N Content (ppm)	Cu Content (ppm)	OH/ Chain	Tan δ at 0.1 Hz & 150°C	Comments
1.	None	800 ± 15	20 ± 5	1/3	.004	
2.	HCl wash, followed by passage through activated Al₂O₃	200 ± 15	30 ± 5	---	.0015	Intrinsic viscosity increased .49 → .52
3.	Reflux with water and methanol	450 ± 15	---	---	.002	
4.	Heat in vacuum desiccator at 150°C for 24 hours	630 ± 15	24 ± 5	---	.003	Intrinsic viscosity increased .49 → .54
5.	Pure monomer used for polymerization	30 ± 5	40 ± 5	1	.001	

    The polymer showing lowest loss due to ionic conduction was
that made from an exceptionally pure form of the 2,6 xylenol
(treatment 5). The polymer was not treated in any way, and we note
that it had a much lower nitrogen content than the other polymer
modifications. There is no direct correlation between the reduc-
tions in loss produced by reducing the nitrogen content of
samples using treatments 2, 3 and 4, and the loss of the sample
with very low initial nitrogen. This may result from the fact
that nitrogen may be present in the polymer not only as free amine
and amine salts (unreacted during polymerization), but also as
amine and amine salts incorporated into the polymer backbone. Thus,
to speculate, the contribution of different nitrogen-containing
species to the losses may be quite different, so that the 30 ppm
nitrogen content of treatment 5 could be entirely responsible for
tan δ = 0.001 (0.1 Hz and 150°C), yet the 200 ppm nitrogen content
of treatment 2 entirely responsible for tan δ = 0.0015.

    Future plans call for the deliberate addition of known con-
centrations of amine salts into the polymer, when the resulting
influence upon conduction and dielectric loss will be studied. Of
particular interest will be the influence of amines of different
size and shape on the activation energies for conduction above
and below T_g.

Relation Between Ionic Losses and Impurity Concentration

    From the relation tan δ = $1/\omega R_i C_x$ an estimate can be obtained
of the concentration of ions needed to give the losses measured
experimentally. It is of interest to compare such concentration
with the nitrogen content of the polymer. We start by taking the
standard equations for the resistance and capacitance of a dielectric:

$$R = \rho \frac{1}{A} \tag{5}$$

where R = resistance (ohms), ρ = resistivity (ohm cm), 1 = length
(cm), and A = area (cm^2), and

$$C = \frac{\varepsilon A}{4\pi(9 \times 10^{11})1} \tag{6}$$

where C = capacity (farads), ε = dielectric constant, A = area
(cm^2), and 1 = length (cm).

Combining Equations 5 and 6,

$$\tan \delta = \frac{4\pi(9 \times 10^{11})}{\omega\varepsilon\rho} \tag{7}$$

Assuming $\varepsilon = 2.55$, for $\tan \delta = 0.004$ at a frequency of 0.1 Hz we calculate $\rho = 1.76 \times 10^{15}$ ohm cm.

The concentration of ions is related to the resistivity by Equations 8 and 9 (assuming one carrier only):

$$J = \sigma E = \frac{E}{\rho} \quad \text{(Ohm's Law)} \tag{8}$$

where J = current density (amps/cm^2), E = field strength (volt/cm), ρ = resistivity (ohm cm), and σ = conductivity [(ohm cm)$^{-1}$], and

$$J = c(e)v = c(e)\mu E \tag{9}$$

where c = concentration (ions/cc), e = electronic charge, v = ion drift velocity (cm/sec), and μ = ion mobility (cm^2/volt sec).

Equations 8 and 9 combine to give a relationship between ion concentration, resistivity, and ion mobility:

$$c = \frac{1}{\rho(e)\mu} \tag{10}$$

Thus, a value for the mobility is needed in order to calculate ion concentration.

In the absence of any reliable values of charge carrier mobilities for the present polymer, an order of magnitude value for mobility can be obtained from the Nernst-Einstein relation:

$$D/\mu = kT/e \tag{11}$$

were D = diffusion constant (cm^2/sec), μ = mobility (cm^2/volt sec), k = Boltzmann constant (= 0.86×10^{-4} ev/°K), and T = absolute temperature (°K). The Nernst-Einstein relation has been found to hold for weak and strong electrolytes in solution, as would be expected since the forces restraining the movement of an ion under the influence of an electric field are the same as those which oppose thermal diffusion of the ion. Diffusion constant values are available for a number of simple gases in polymers, but the gases are all neutral molecules. It will be assumed that simple ions behave similarly and also that the polymer contains simple ions.

A typical range of diffusion constants for simple gas molecules in polymers is from 10^{-7} to 10^{-10} cm2/sec (12) giving mobilities from 2.7×10^{-6} to 2.7×10^{-9} cm2/volt sec (at 150°C). Hence Equation 10 gives a concentration range of 1.3×10^9 to 1.3×10^{12} ions/cc.

Let us now assume that each nitrogen atom is located on a salt molecule which dissociates to give one ion. By analysis there are approximately 800 ppm of nitrogen in the unmodified polymer, and this corresponds to approximately 1 nitrogen per polymer chain. Using Avogadro's Number (N = 6×10^{23} molecules/mole), and a molecular weight and density of 25,000 and 1 respectively, we calculate a concentration of polymer molecules of 2.4×10^{19} molecules/cc.

Hence, the concentration of dissociating salt molecules = 2.4×10^{19}/cc.

This figure is greatly in excess of the number of ions calculated from tan δ = 0.004 at 0.1 Hz (1.3×10^9 - 1.3×10^{12} ions/cc). It is therefore concluded that we have a very weakly dissociating system - undoubtedly due to the low dielectric constant - with a dissociation constant in the range 5×10^{-11} to 5×10^{-8}. This is somewhat lower than the values obtained for typical weak quaternary ammonium salts in low dielectric constant liquids (dissociation constant $\sim 10^{-6}$) and may suggest that the values of mobility are too high.

The value of this order of magnitude calculation is to support the model of a very weakly dissociating system in the low dielectric constant polymer matrix.

Dipolar Losses

Two weak dipolar losses attributable to impurities are shown in Figure 7. Sample A clearly does not show the two weak parts numbered 1 and 2, shown by sample B.

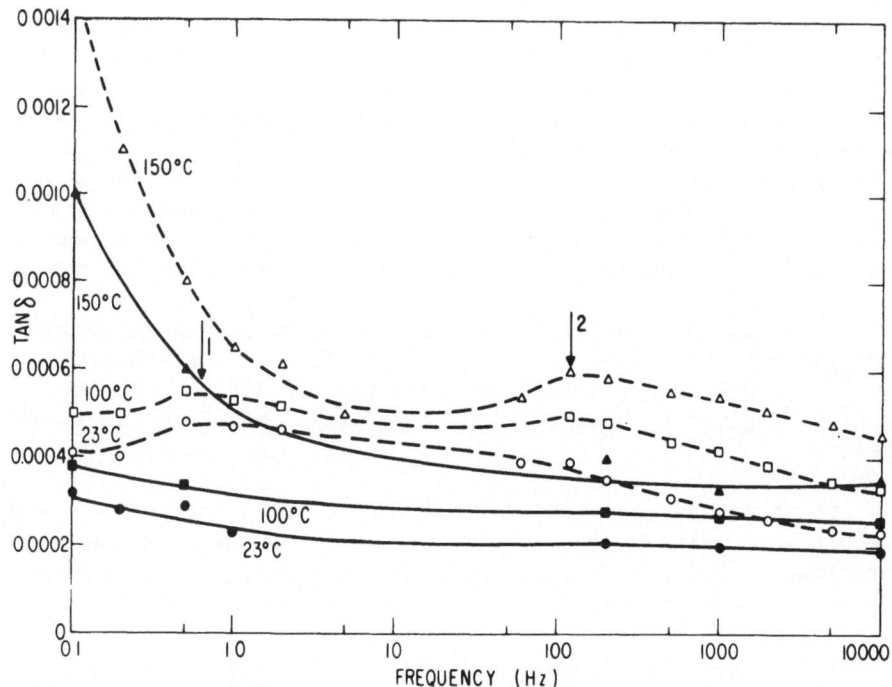

Figure 7. Loss tangent as a function of frequency showing weak
dipolar losses. Sample A ————; sample B – – – –.

The following possibilities have been eliminated as the origin
of these dipolar losses: (1) crystallinity (both samples were
amorphous); (2) molecular weight (sample A ~20,000, sample B
~25,000); (3) hydroxyl content (by IR measurement: sample A ~1 per
chain, sample B ~1/3 per chain); (4) diphenoquinones (these are
possible byproducts of polymerization (13) as indicated in
Figure 1; with sample B, diphenoquinones were extracted with
activated alumina and also added, each with no effect); and (5)
methyl side group rotation and limited motion of the polymer backbone (again the lack of crystallinity and the similar molecular
weights would make such motions nearly identical in both A and B).

Although we are unable to make a positive identification of
the source of the differences in background loss between samples A

and B, we can draw some general inferences about their origin.
Certainly the characteristic appearance of Figure 7 suggests that
the losses exhibited by sample A are due to one or more weak
dipolar dispersions superimposed upon the very uniform background
loss exhibited by sample B. Further, since the temperatures are
below the glass temperature of the polymer, we can say that these
weak dispersions cannot be due to large-scale rearrangements of
the polymer chain, but that they must be due to secondary relaxations resulting from motions within the polymer in the glass phase.
As indicated above, we have excluded as the origin of such secondary
relaxations both (1) the rotation of methyl side groups and (2)
the limited motions of the chain backbone [due to either crankshaft
or local mode mechanisms (14)]. This essentially confirms that the
dispersions shown by sample B cannot be intrinsic in nature, and we
conclude that they must be due to impurities. In addition to
isolated polar impurities within the polymer matrix, we must also
consider impurities incorporated within the polymer backbone, as
the potential origin of such dispersions. Indeed, as described
below, this may be the preferred description for the present polymer.

The only property we have been able to relate to the difference
in loss between sample A and sample B is the nitrogen content. Previously nitrogen has been linked to the region of ionic conduction,
but it can contribute also to dipolar losses because of its associated
dipole moment. Sample B had a nitrogen content of 800 ppm, whereas
sample A was a polymer made from pure monomer (treatment 5 of
Table I) and had a nitrogen content of 30 ppm. However, samples
with high nitrogen content continued to exhibit non-uniform losses
of approximately the same magnitude after the nitrogen content had
been reduced by clean-up to one-fourth the original value. An
explanation for this apparent inconsistency may be that only nitrogen which is incorporated (as amine) into the polymer backbone is
contributing to the non-uniform losses. Support for this explanation
is evident when the influence of impurities on dielectric constant
is examined later. Presumably such nitrogen is more difficult to
remove from the polymer than nitrogen present as free amine, and
thus the losses may remain unaffected by clean-up. If this explanation obtains we may expect to find the free amines contributing to
the losses at some other frequency. Assuming that the dipole associated with the free amine is less restricted in its motion, we will
need to look for such a dispersion at higher frequencies or lower
temperatures.

We have been able to find only one reference (15) where nitrogen
has been associated with a secondary relaxation in a polymer. This
was for nylon 66 where the motion of $-NH_2$ chain-end groups was
suggested responsible. However, later work (16) offer an alternative
explanation for the same relaxation. At this point we note that

dielectric loss measurements on non-polar polymers can be a sensitive complementary technique for studying polymer structure, in addition to providing information on the presence and role of impurities.

Oxidation

Impurities may also modify the dielectric losses of a polymer by influencing its oxidation. In order to examine such behavior, deliberate oxidation of polymer samples was effected by heating in oxygen at various temperatures.

Samples held in oxygen at, and below 190°C for a period of several days showed negligible change in dielectric loss. However, above 200°C the influence upon dielectric loss was quite evident and increased markedly with increasing temperature up to 240°C (above which no measurements were made). Figure 8 shows the loss spectrum after a sample of polymer with high nitrogen content (800 ppm) had been treated in oxygen at 230°C for 3 hours. Comparing with Figure 2, we note: (1) the low frequency losses, attributed to amine salts, have disappeared completely (presumably due to oxidative decomposition of amines) and (2) a very broad relaxation has appeared (a wide spectrum of relaxations may be indicated).

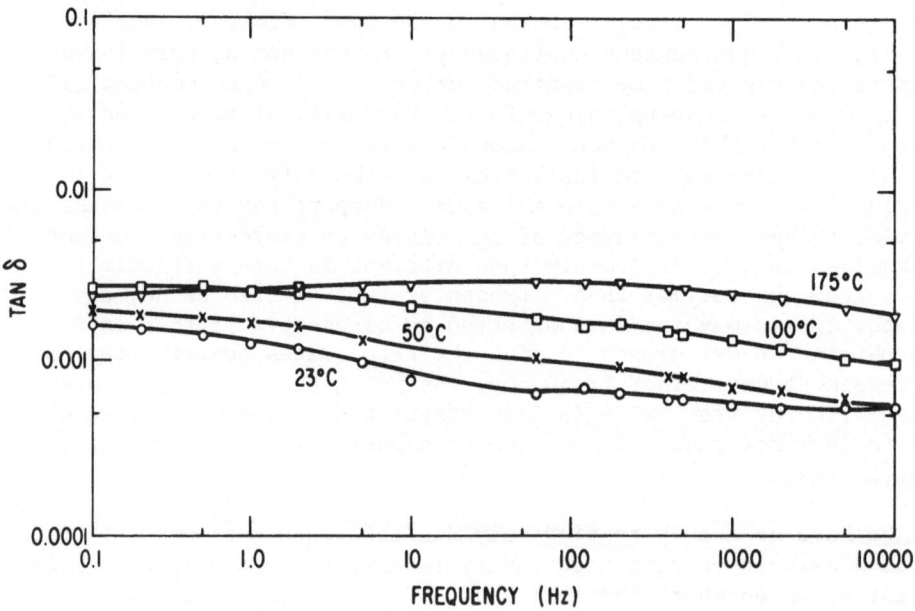

Figure 8.   Loss tangent as a function of frequency for sample of polymer previously heated in oxygen at 230°C.

When heated in oxygen at 230°C, a sample of polymer with low nitrogen content (~30 ppm) showed a similar (though slightly smaller in magnitude) behavior to that of Figure 8. (The very similar behavior of low and high nitrogen samples suggests that the oxidation not only oxidizes remnant amines, but also causes an oxidation of the polymer.) However, a completely different behavior was observed with the low-nitrogen sample at a slightly lower temperature of 210°C. To observe the behavior more closely, dielectric measurements were made in situ, with a sample without evaporated gold electrodes. Initially the losses due to ionic conduction decreased in magnitude, suggesting oxidative decomposition, but then they increased continuously with time until the experiment was terminated after 110 minutes. Meanwhile the higher frequency losses, which were similar to those of Figure 8, increased continuously. Figure 9 illustrates the behavior of tan δ for 0.1 Hz and 1 kHz for this sample. (Zero time was taken as the time at which the 0.1 Hz losses first started to decrease; at this time the temperature was 206°C, but in ~6 minutes it had reached 210°C). The results imply that at 210°C we get both oxidative decomposition of the existing (amine) source of ions and the oxidative formation of a new species capable of dissociating to form ions. However, as described above, these new species were not stable at 230°C. If the increasing portion of the 0.1 Hz losses is extrapolated back to zero time, we obtain tan δ = 0.0004 (subtracting out a background loss from the high frequency data). This suggests that whatever was being oxidized to give the increasing 0.1 Hz loss was already contributing (~10%) to the ionic losses prior to oxidation. This would mean that the oxidation is converting a weakly ionizing species into a somewhat more strongly ionizing species.

During normal measurements at 230°C in a nitrogen atmosphere, the changes in loss described above were completely suppressed, showing that oxygen is essential to these loss changes.

These results are of interest because they provide evidence that there is more than one type of dissociating species in the present polymer and that these species differ somewhat, though not greatly, in their oxidative stability. It does not seem unreasonable to speculate that these are all nitrogen-containing species, or that they are all amine salts of differing structure.

Influence of Impurities on Dielectric Constant

The results described by Figures 2-5 were for polymer samples containing high levels of residual catalyst. For samples with low impurity levels, the dielectric constants were lower. For example, for sample A of Figure 7, ε' was 2.55 at 23°C and 2.50 at 175°C. (As with high-impurity-level samples, the variation of ε' between 23 and 175°C was directly related to density change.) Purification

of the high-impurity-level polymer led to a decrease of the
dielectric constants (23–175°C) below T_g from the values of
Figure 5, yet, as noted previously, the weak secondary relaxations
shown in Figure 7 were unaffected. We conclude that, in the temperature range 23–175°C, differences in dielectric constant reflect
the contribution of impurities to low temperature (high frequency)
dipolar relaxations.

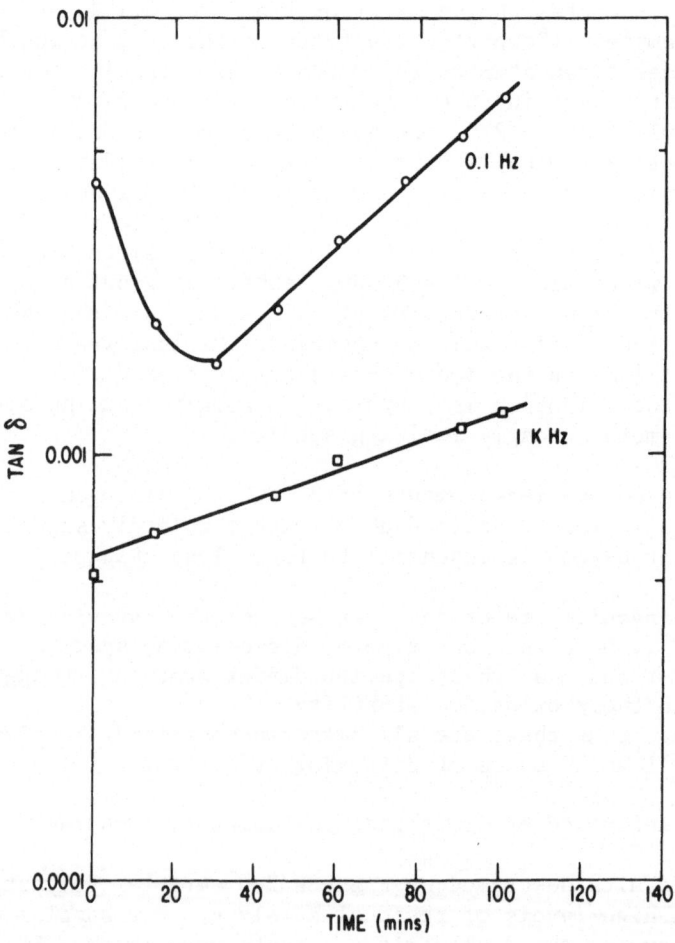

Figure 9.   Variation with time of loss tangent at 0.1 Hz and 1 kHz
            as sample of polymer heated in oxygen at 210°C.

It is noteworthy that the magnitude of the α relaxation losses
shown in Figures 2 and 4, and the contribution of such losses to
ε' (Figure 5), were also larger than for low-impurity-level samples.
As noted previously, this applies to data restricted to the high
frequency side of the α relaxation peak. However, it is significant that, in both high- and low-impurity-level samples, the α
relaxation losses indicated a very broad distribution of relaxations over the frequency range 0.1 Hz to 10 kHz. This suggests
that some of the polar character of the impurities is incorporated
into the polymer backbone (or, to speculate further, that some of
the remnant catalysts are polymerized into the polymer structure),
since we expect that discrete impurity molecules would give a
fairly narrow relaxation.

These observations of the dielectric constant and the associated
loss behavior of low- and high-impurity-level samples give further
weight to the suggestions made earlier that residual amines are present as both free amine and amine incorporated chemically in the
polymer backbone.

CONCLUSIONS

1. An a.c. resistance in parallel with an a.c. capacitance
is a valid model for the low frequency losses of poly (2,6-dimethyl-
1,4-phenylene ether).

2. At the glass transition temperature of the polymer, the
a.c. resistivity exhibits a sharp change in activation energy. We
suggest that this reflects a change in the energetics of the diffusion of ions in the glass and rubber phases.

3. The source of the low frequency losses in the polymer is
extrinsic in nature and is due to ionic conduction. The ionic conduction is an equilibrium process, free from electrochemical decomposition.

4. A probable source of the ionic conduction is due to the
dissociation of salts of remnant amines.

5. These amine salts are very weakly dissociated in the low
dielectric constant polymer.

6. The origin of differences in background losses in the
polymer can be attributed to impurities. These are tentatively
assigned as nitrogen-containing impurities, and they are possibly
the amines simultaneously responsible for ionic conduction.

7. Heating the polymer in air or oxygen at temperatures
above 200°C results in pronounced changes in dielectric loss in

the frequency range 0.1 Hz to 10 kHz. Consideration of these changes provides evidence for more than one dissociating species in the polymer.

8. A nitrogen atmosphere completely suppresses the changes in loss of the polymer at 230°C.

9. The amines are probably present both as discrete molecules and as amines polymerized into the polymer chains (i.e., as chemical irregularities in the chain).

10. In non-polar polymers, impurities contribute to dielectric losses via ionic conduction, dipolar relaxation (below T_g), and the α relaxation (above T_g). Such impurities contribute to the measured dielectric constant.

ACKNOWLEDGMENT

The author wishes to thank R. E. Burnett, D. M. White, and H. J. Klopfer for their considerable advice on the "clean-up" procedures and the handling of the polymer, and S. I. Reynolds for his instruction in the use of the low frequency bridge.

REFERENCES

1. W. Reddish, Pure and Appl. Chem. 5, 727 (1962).
2. K. A. Buckingham and W. Reddish, Proc. I.E.E. (6B) 114, 1810 (1967).
3. R. H. Partridge, J. Polymer Sci. 5, 205 (1967).
4. C. W. Reed, Nat. Acad. Sci.-Nat. Res. Council Publ. No. 1705, 136 (1969).
5. E. O. Forster, J. Chem. Phys. 40, 86 (1964).
6. C. A. Kraus and R. M. Fuoss, J. Am. Chem. Soc. 55, 21 (1933).
7. A. J. Curtis, J. Chem. Phys. 36, 3500 (1962).
8. E. M. Amrhein and F. H. Mueller, Trans. Faraday Soc. 64, 666 (1968).
9. E. M. Amrhein and F. H. Mueller, J. Am. Chem. Soc. 90, 3146 (1968).
10. S. I. Reynolds, General Electric Research and Development Center Report No. 67-C-301, August, 1967.
11. L. K. H. van Beek, "Dielectric Behavior of Heterogeneous Systems," Progress in Dielectrics, Vol. 7 (1967).
12. J. R. Price and W. Dannhauser, p. 33, 1967 Annual Report on Conference on Electrical Insulation and Dielectric Phenomena, N.A.S., Washington.
13. DIFFUSION IN POLYMERS, ed. J. Crank and G. S. Park, Academic Press, 1968, pp. 46-50.

14. G. F. Endres, A. S. Hay and J. W. Eustance, J. Org. Chem.
     $\underline{28}$, 1300 (1963).
15. ANELASTIC AND DIELECTRIC EFFECTS IN POLYMERIC SOLIDS,
     N. G. McCrum, B. E. Read and G. Williams, John Wiley, 1967,
     p. 180.
16. E. Rushton and G. Russell, Brit. Elec. Ind. Res. Assoc.
     Rept. L/T 355 (1956).
17. A. J. Curtis, J. Res. Nat. Bur. Std. $\underline{65A}$, 185 (1961).

INDEX

A

activation energy 18
activation volume 18,137
alkanes, α,ω-dibromo 100
anthrone 39
apparatus, density measuring 190

B

bead-spring model 46,74,99

C

charge carrier mobility 298,361
charge decay function 23
charge trap 296
Cole-Cole plots, 9,24,125,132,318
conduction mechanism 229,296,311,344
copolymers
    chlorostyrene-methylstyrene 8
    ethylene-acrylic acid 238
    ethylene-methacrylic acid 239
    fluoroethylene-propylene 297
    styrene-chlorostyrene 14
    styrene-ethylene oxide block 314
corona charging 296
correlation
    angular 12,47,73
    length 6,13,54,73
    spatial 1
    time dependent 12,32,111
    typal 1
crankshaft rotation 95,127,269

D

Debye θ temperature 141
dielectric apparatus
    description of 150
    operation of 185